Calculations and Simulations of Low-Dimensional Materials

Calculations and Simulations of Low-Dimensional Materials

Tailoring Properties for Applications

Ying Dai
Wei Wei
Yandong Ma
Chengwang Niu

WILEY-VCH

Authors

Prof. Ying Dai
Shandong University
School of Physics
Shandananlu 27
250100 Jinan
China

Prof. Wei Wei
Shandong University
School of Physics
Shandananlu 27
250100 Jinan
China

Prof. Yandong Ma
Shandong University
School of Physics
Shandananlu 27
250100 Jinan
China

Prof. Chengwang Niu
Shandong University
School of Physics
Shandananlu 27
250100 Jinan
China

Cover Image: © Yuichiro Chino/Getty Images

Library of Congress Card No.: applied for

British Library Cataloguing-in-Publication Data
A catalogue record for this book is available from the British Library.

Bibliographic information published by the Deutsche Nationalbibliothek
The Deutsche Nationalbibliothek lists this publication in the Deutsche Nationalbibliografie; detailed bibliographic data are available on the Internet at <http://dnb.d-nb.de>.

Print ISBN: 978-3-527-34909-8
ePDF ISBN: 978-3-527-83211-8
ePub ISBN: 978-3-527-83212-5
oBook ISBN: 978-3-527-83213-2

Typesetting Straive, Chennai, India
Printing and Binding CPI Group (UK) Ltd, Croydon, CR0 4YY

C113432_300722

Contents

Preface

In modern society, theoretical calculation and simulation are advancing the development of physics, chemistry, and materials science, etc., along with the great successes in computer technology. After the successful exfoliation of graphene from graphite by Sir Andre Geim and Konstantin Novoselov, then various two-dimensional materials are predicted, synthesized, transferred, and characterized, and we are witnessing the coming of the ear of two-dimensional materials. In comparison to the corresponding bulk counterparts, materials in low dimension exhibit unprecedented novel chemical and physical properties. Meanwhile, many novel phenomena as well as promising potential devices are proposed based on low-dimensional materials.

Among these intriguing physics in two-dimensional lattices, for example, the valley physics, excitonic effects, and Rashba effect attract special interest from the standpoints of both fundamental and applied level. Thus, there is an impressing progress in the research on these exciting topics. For the magnetic property, it was extensively studied in conventional bulk systems, as it had long been considered to survive in low-dimensional systems due to the thermal fluctuations according to the Mermin–Wagner theorem. In recent years, with the discovery of long-range magnetic order in two-dimensional lattice, magnetic materials in two dimensions spurred extensive attention. Currently, there are many works devoted to investigate the magnetic properties in two-dimensional materials. Also, the past decades have witnessed the evolution of topological phases and topological materials which reshape our understanding of physics and materials. First-principles calculations based on the density functional theory provide effective descriptions of topological phases and play important roles in predicting realistic topological materials, from insulators to semimetals, and from nonmagnetic systems to magnetic ones. In fields of photochemistry and photoelectrochemistry, the theoretical calculations and simulations are so powerful in designing, screening, and characterizing the photocatalysts and electrocatalysts of high stability, selectivity, and activity. Indeed, calculation and simulation are guiding, leading, and moving forward these fields.

The book is divided into eight chapters. In Chapter 1, density functional theory and its derivatives (e.g., the tight-binding method) are briefly introduced, and in Chapter 2, the recent development of valley physics and Rashba effect in two-dimensional materials from the first-principles are reviewed. In Chapter 3, calculations and simulation results for low-dimensional ferromagnetic materials are discussed. Chapter 4 provides a gentle introduction of topological phases and topological materials especially in two dimensions. The concepts of different topological states are discussed using first-principles band structures and topological invariants. In Chapter 5, ingredients in the many-body Green's function perturbation theory for calculating the excited-state properties of low-dimensional materials are summarized, and in Chapter 6, time-dependent density functional theory and nonadiabatic molecular dynamics are introduced for those who are studying photoexcited charge carrier dynamics. Process of how to calculate and simulate photocatalytic reaction is presented in Chapter 7, and electrochemical reactions from calculations and simulations are reviewed in Chapter 8.

This book would not have been possible without help from many people.

Jinan, China
Summer, 2022

Ying Dai
Wei Wei
Yandong Ma
Chengwang Niu

1

An Introduction to Density Functional Theory (DFT) and Derivatives

1.1 The Problem of a N-electron System

The density functional theory (DFT) is first resulted from the work by Hohenberg and Kohn [1], wherein the complicated individual electron orbitals are substituted by the electron density. Namely, the DFT is entirely expressed in terms of the functional of electron density, rather than the many-electron wave functions. In this case, DFT significantly reduces the calculations of the ground state properties of materials. That is why DFT is useful for calculating electronic structures, especially with many electrons. As the foundation of DFT, two theorems are proposed by Hohenberg and Kohn [1]. The first theorem presents that the ground state energy is a functional of electron density. The second theorem shows that the ground state energy can be achieved by minimizing system energy on the basis of electron density.

It should be noted that, although Hohenberg and Kohn point out there are relations between properties and electron density functional, they do not present the exact relationship. But fortunately, soon after the work of Hohenberg and Kohn, Kohn and Sham simplified the many-electron problems into a model of individual electrons in an effective potential [2]. Such a potential contains the external potential and exchange-correlation interactions. For exchange-correlation potential, it is a challenge to describe it rigorously.

The simplest approximation for treating the exchange-correlation interaction is the local density approximation (LDA) [3], wherein the exchange and correlation energies are obtained by the uniform electron gas model and fitting to the uniform electron gas, respectively. LDA can provide a realistic description of the atomic structure, elastic, and vibrational properties for a wide range of systems. Yet, because LDA treats the energy of the true density using the energy of a local constant density, it cannot describe the situations where the density features rapid changes such as in molecules [4, 5]. To address this problem, the generalized gradient approximation (GGA) is proposed [6–8], which depends on both the local density and the spatial variation of the density. And in principle, GGA is as simple to use as LDA. Currently, in the vast majority of DFT calculations for solids, these two approximations are adopted.

Calculations and Simulations of Low-Dimensional Materials: Tailoring Properties for Applications, First Edition. Ying Dai, Wei Wei, Yandong Ma, and Chengwang Niu.

By considering the Born–Oppenheimer and non-relativistic approximations, the effective Hamiltonian of a N-electron system in the position representation can be given by,

$$H(r_1, r_2, \dots r_N) = \hat{T} + \hat{V}_{\text{ext}} + \hat{V}_{\text{ee}} = -\frac{1}{2}\sum_i \nabla_i^2 + \sum_i \hat{v}_{\text{ne}}(r_i) + \frac{1}{2}\sum_i \sum_{j \neq i} \frac{1}{|r_i - r_j|} \tag{1.1}$$

The first term is kinetic energy operator. The second term is an external potential operator. In systems of interest to us, the external potential is simply the Coulomb interaction of electrons with atomic nuclei:

$$\hat{v}_{\text{ne}}(r_i) = -\sum_\alpha \frac{Z_\alpha}{|r_i - R_\alpha|} \tag{1.2}$$

where the r_i is the coordinate of electron i and the charge on the nucleus at R_α is Z_α. The third term of Eq. (1.1) is the electron-electron operator. The electronic state can be obtained by the Schrödinger equation:

$$H(r_1, r_2, r_N)\Psi(r_1, r_2, r_N) = E\Psi(r_1, r_2, r_N) \tag{1.3}$$

Here, $\Psi(r_1, r_2, r_N)$ is a wave function in terms of space-spin coordinates. Apparently, the wave function is antisymmetric under exchanging the coordinates. Under Dirac notation, the Eq. (1.1) can be expressed in representation-independent formalism:

$$H|\Psi\rangle = E|\Psi\rangle \tag{1.4}$$

In principle, the ground state energy E_0 of the N-electron system can be found based on the variational theorem, which is obtained by the minimization:

$$E_0 = \min_\Psi \langle\Psi|\hat{H}|\Psi\rangle \tag{1.5}$$

Here, the search is over all antisymmetric wave functions Ψ. In this regard, better approximations for Ψ can readily result in the ground state energy E_0 of the N-electron system, but the computational cost would be very high. Therefore, the direct solution is not feasible. To address this issue, DFT is developed, which is based on a reformulation of the variational theorem in terms of electron density.

We know that $|\Psi|^2 = \Psi^*\Psi$ represents the probability density of measuring the first electron at r_1, the second electron at r_2, … and the Nth electron at r_N. By integrating $|\Psi|^2$ over the first $N - 1$ electrons, the probability density of the Nth electron at r_N is determined. Then the probability electron density that defines any of the N electrons at the position r is given by:

$$\rho(r) = N \int \dots \int \Psi^*(r_1, r_2, r_N)\Psi(r_1, r_2, \dots r_N) d_{r_2} \dots d_{r_N} \tag{1.6}$$

And the electron density is normalized to the electron number:

$$\int \rho(r) dr = N \tag{1.7}$$

The energy of the system is expressed as:

$$E = \langle\Psi|\hat{H}|\Psi\rangle = \langle\Psi|\hat{T}|\Psi\rangle + \langle\Psi|\hat{V}_{\text{ext}}|\Psi\rangle + \langle\Psi|\hat{V}_{\text{ee}}|\Psi\rangle = T + V_{\text{ext}} + V_{\text{ee}} \tag{1.8}$$

Here,

$$T = \langle\Psi|\hat{T}|\Psi\rangle = \sum_i \int \Psi^*(r_1, r_2, r_N)\left(-\frac{1}{2}\nabla_i^2\right)\Psi(r_1, r_2, r_N)d_{r_1}d_{r_2}\dots d_{r_N} \quad (1.9)$$

$$V_{\text{ext}} = \langle\Psi|\hat{V}_{\text{ext}}|\Psi\rangle = \sum_i \int \Psi^*(r_1, r_2, r_N)\hat{v}_{\text{ne}}(r_i)\Psi(r_1, r_2, \dots r_N)d_{r_1}d_{r_2}\dots d_{r_N}$$

$$= \int v_{\text{ne}}(r)\rho(r)dr = V_{\text{ext}}[\rho] \quad (1.10)$$

1.2 The Thomas–Fermi Theory for Electron Density

Before discussing the Hohenberg–Kohn theorems, we first introduce the Thomas–Fermi theory. The Thomas–Fermi theory is important as it gives the relation between external potential and the density distribution for interacting electrons moving in an external potential:

$$\rho(r) = \gamma(\mu - v_{\text{eff}}(r))^{3/2} \quad (1.11)$$

$$v_{\text{eff}}(r) \equiv v_{\text{ne}}(r) + \int \frac{\rho(r')}{|r - r'|}dr' \quad (1.12)$$

Here,

$$\gamma = \frac{1}{3\pi^2}\left(\frac{2m}{\hbar^2}\right)^{3/2} \quad (1.13)$$

and μ is the r independent chemical potential. The second term is Eq. (1.12) is the classical electrostatic potential raised by the density $\rho(r)$. Based on the Thomas–Fermi theory, Hohenberg and Kohn build up the connection between electron density and the Schrödinger equation. And in the following, we will introduce the two Hohenberg–Kohn theorems, which lie at the heart of DFT.

1.3 The First Hohenberg–Kohn Theorem

By replacing the external potential $v_{\text{ne}}(r)$ with an arbitrary external local potential $v(r)$, the corresponding ground state wave function Ψ can be found by solving the Schrödinger equation. Based on the obtained wave function, the ground state density $\rho(r)$ can be computed. And obviously, two different local potentials would give two different wave functions and thus two different electron densities. This gives the map:

$$v(r) \to \rho(r) \quad (1.14)$$

Based on the Thomas–Fermi theory, Hohenberg and Kohn demonstrated that the preceding mapping can be inverted, namely, *the ground state electron density $\rho(r)$ of a bound system of interacting electrons in some external potential $v(r)$ determines the potential uniquely*:

$$\rho(r) \to v(r) \quad (1.15)$$

This is known as the first Hohenberg–Kohn theorem.

To demonstrate this theorem, we consider two different local potentials $v_1(r)$ and $v_2(r)$, which differ by more than the constant. These two potentials yield two different ground state wave functions Ψ and Ψ', respectively. And apparently, these two ground state wave functions are different. Assume $v_1(r)$ and $v_2(r)$ correspond to the same ground state wave function, then

$$\hat{H}|\Psi\rangle = E_0|\Psi\rangle \tag{1.16}$$

$$\hat{H}'|\Psi\rangle = E_0'|\Psi\rangle \tag{1.17}$$

By subtracting Eq. (1.17) from Eq. (1.16), we can obtain:

$$(\hat{V}_1 - \hat{V}_2)|\Psi\rangle = \left(E_0 - E_0'\right)|\Psi\rangle \tag{1.18}$$

which can be expressed in position representation,

$$\sum_i [\hat{v}_1(r_1) - \hat{v}_2(r_i)]\Psi(r_1, r_2, r_N) = \left(E_0 - E_0'\right)\Psi(r_1, r_2, r_N) \tag{1.19}$$

This suggests that

$$v_1(r) - v_2(r) = \text{const} \tag{1.20}$$

thus in contradiction with the assumption that $v_1(r)$ and $v_2(r)$ differ by more than a constant. Accordingly, two different local potentials that differ by more than the constant cannot share the same ground state wave function, which demonstrate the map:

$$v(r) \to \Psi \tag{1.21}$$

Then, we demonstrate the map:

$$\Psi \to \rho(r) \tag{1.22}$$

Let Ψ and Ψ' be the ground state wave functions corresponding to $v_1(r)$ and $v_2(r)$, respectively. Assuming that Ψ and Ψ' exhibit the same ground state electron density $\rho(r)$, then the variational theorem gives the ground state energy as:

$$\begin{aligned} E_0 &= \langle\Psi|\hat{H}|\Psi\rangle < \langle\Psi'|\hat{H}|\Psi'\rangle = \langle\Psi'|\hat{H}' + \hat{V}_1 - \hat{V}_2|\Psi'\rangle \\ &= E_0' + \int [v_1(r) - v_2(r)]\rho(r)dr \end{aligned} \tag{1.23}$$

$$\begin{aligned} E_0' &= \langle\Psi'|\hat{H}'|\Psi'\rangle < \langle\Psi|\hat{H}'|\Psi\rangle = \langle\Psi|\hat{H} - \hat{V}_1 + \hat{V}_2|\Psi'\rangle \\ &= E_0 + \int [v_2(r) - v_1(r)]\rho(r)dr \end{aligned} \tag{1.24}$$

By subtracting Eq. (1.23) from Eq. (1.24), we can obtain:

$$E_0' - E_0 < E_0 - E_0' \tag{1.25}$$

This makes no sense. This finally leads to the conclusion that there cannot exist two local potentials differing by more than an additive constant that has the same ground state density.

1.4 The Second Hohenberg–Kohn Theorem

According to the first Hohenberg–Kohn theorem, the ground state density $\rho(r)$ determines the local potential $v(r)$, and in turn determines the Hamiltonian. Therefore, for a given ground state density $\rho_0(r)$ that is generated by a local potential, it is possible to compute the corresponding ground state wave function Ψ_0. That is to say, Ψ_0 is also a unique functional of $\rho_0(r)$:

$$\Psi_0 = \Psi_0[\rho_0] \tag{1.26}$$

According to Eq. (1.26), the ground state energy E_0 is also a functional of $\rho_0(r)$:

$$E_0 = E_0[\rho_0] \tag{1.27}$$

Hohenberg and Kohn define the universal density functional:

$$F[\rho] = \langle \Psi[\rho] | \hat{T} + \hat{V}_{\text{ee}} | \Psi[\rho] \rangle \tag{1.28}$$

Here, $\Psi[\rho]$ is any ground state wave function corresponding to the ground state density $\rho(r)$. By combining Eq. (1.10), the total energy functional can be defined as:

$$E[\rho] = F[\rho] + \int v_{\text{ne}}(r)\rho(r)dr \tag{1.29}$$

From the Ritz principle, we have:

$$E_0 = \min_{\rho} E[\rho] = \min_{\rho} \left\{ F[\rho] + \int v_{\text{ne}}(r)\rho(r)dr \right\} \tag{1.30}$$

This is known as the second Hohenberg–Kohn theorem.

1.5 The Kohn–Sham Equations

For a system with noninteracting electrons, the effective Hamiltonian n can be given by:

$$\hat{H}_{\text{s}} = \hat{T} + \hat{V}_{\text{s}} \tag{1.31}$$

The corresponding Schrödinger equation is:

$$\left(-\frac{1}{2}\nabla^2 + \hat{v}_s(r)\right)\psi_i(r) = \varepsilon_i\psi_i \tag{1.32}$$

Then the density is given by:

$$\rho_{\text{s}}(r) = \sum_i |\psi_i(r)|^2 \tag{1.33}$$

Here, the single particle orbital $\psi_i(r)$ is constructed based on the effective potential $v_{\text{s}}(r)$.

The total energy can be expressed as:

$$E_{\text{s}}[\rho] = T_{\text{s}}[\rho] + V_{\text{s}}[\rho] \tag{1.34}$$

The first term is the kinetic energy of the noninteracting electrons, which is given by:

$$T_s[\rho] = \langle \Psi[\rho]|\hat{T}|\Psi[\rho]\rangle = \sum_i \left\langle \psi_i(r) \mid -\frac{1}{2}\nabla^2 \mid \psi_i(r)\right\rangle \tag{1.35}$$

The second term is the effective potential, which is given by:

$$V_s[\rho] = \langle \Psi[\rho]|\hat{V}_s|\Psi[\rho]\rangle = \int \hat{v}_s(r)\rho(r)dr \tag{1.36}$$

Accordingly, the total energy can be given by:

$$E_s[\rho] = \sum_i \varepsilon_i \tag{1.37}$$

By combining Eqs. (1.36) and (1.37), the kinetic energy can be expressed as:

$$T_s[\rho] = \sum_i \left\langle \psi_i(r)\right| - \frac{1}{2}\nabla^2 \left|\psi_i(r)\right\rangle = \sum_i \varepsilon_i - \int \hat{v}_s(r)\rho(r)dr \tag{1.38}$$

Using the method of Lagrange multipliers, we can obtain the following equation:

$$\mu = \frac{\delta E_s[\rho]}{\delta\rho(r)} = \frac{\delta T_s[\rho]}{\delta\rho(r)} + \frac{\delta V_s[\rho]}{\delta\rho(r)} = \frac{\delta T_s[\rho]}{\delta\rho(r)} + v_s(r) \tag{1.39}$$

Solution of Eq. (1.39) yields the density $\rho_s(r)$.

The classical electrostatic interaction energy is given by:

$$E_H[\rho] = \frac{1}{2}\int V_H(r)\rho(r)dr \tag{1.40}$$

And

$$V_H(r) = \int \frac{\rho(r')}{|r - r'|}dr' \tag{1.41}$$

Then the following equation is obtained:

$$\begin{aligned} E[\rho] = T[\rho] + V_{ee}[\rho] + V[\rho] &= T_s[\rho] + E_H[\rho] \\ &+ (T[\rho] - T_s[\rho] + V_{ee}[\rho] - E_H[\rho]) + V[\rho] \end{aligned} \tag{1.42}$$

The third term is exchange and correlation energy functional:

$$E_{xc} = T[\rho] - T_s[\rho] + V_{ee}[\rho] - E_H[\rho] \tag{1.43}$$

The exchange-correlation potential is defined as:

$$V_{xc} = \frac{\delta E_{xc}[\rho]}{\delta\rho(r)} \tag{1.44}$$

Using the method of Lagrange multipliers, we can obtain the following equation:

$$\mu = \frac{\delta E[\rho]}{\delta\rho(r)} = \frac{\delta T_s[\rho]}{\delta\rho(r)} + V_H(r) + V_{xc}(r) + v_{ne}(r) \tag{1.45}$$

And solution to Eq. (1.45) is $\rho(r)$.

Therefore, given the relation:

$$v_s(r) \equiv V_H(r) + V_{xc}(r) + v_{ne}(r) \tag{1.46}$$

We have:

$$\rho_s(r) \equiv \rho(r) \tag{1.47}$$

We then arrive at the Kohn–Sham equation:

$$\left(-\frac{1}{2}\nabla^2 + \hat{V}_H(r) + \hat{V}_{xc}(r) + \hat{v}_{ne}(r)\right)\psi_i(r) = \varepsilon_i\psi_i \tag{1.48}$$

Solving this equation gives the orbital and then the density of the original interacting system.

The exchange and correlation functional can be written as:

$$E_{xc}[\rho] = \int \varepsilon_{xc}(r'; \rho)\rho(r')dr' \tag{1.49}$$

Here, $\varepsilon_{xc}(r';\ \rho)$ is the exchange-correlation energy density. And the exchange-potential is defined as:

$$V_{xc}(r;\rho) = \frac{\delta E_{xc}[\rho]}{\delta\rho(r)} = \varepsilon_{xc}(r;\rho) + \int \frac{\delta\varepsilon_{xc}(r';\rho)}{\delta\rho(r)}\rho(r')dr \tag{1.50}$$

Then the total energy can be given by:

$$E[\rho] = \sum_i \varepsilon_i - E_H[\rho] + E_{xc}[\rho] - \int V_{xc}(r;\rho)\rho(r)dr \tag{1.51}$$

From preceding equation, we can see that, except for the exchange-correlation functional, all the aforementioned expressions are exact. In practice, we have to use approximations for exchange-correlation potential, as the exact form is unknown.

1.6 The Local Density Approximation (LDA)

LDA is one of the most widely used and simplest approximations for E_{xc}. In LDA, the exchange-correlation functional is approximated as:

$$E_{xc}^{LD}[\rho] = \int \rho(r)\varepsilon_{xc}^{unif}[\rho(r)dr] \tag{1.52}$$

Here, $\varepsilon_{xc}^{unif}[\rho(r)]$ is the exchange-correlation energy per electron in homogeneous electron gas at density ρ. LDA works well for homogeneous electron gas, and thus is valid for systems where electron density does not change rapidly.

The exchange-correlation energy density can be broken down into two parts:

$$\varepsilon_{xc}^{unif}[\rho(r)] = \varepsilon_{x}^{unif}[\rho(r)] + \varepsilon_{c}^{unif}[\rho(r)] \tag{1.53}$$

The first term is the exchange term, which is given by:

$$\varepsilon_{x}^{unif}[\rho(r)] = \text{Const.} \times \rho^{1/3}(r) \tag{1.54}$$

While for the second term of Eq. (1.53), it is the correlation density, which does not have an analytic formula. However, the correlation energies can be obtained numerically from quantum Monte Carlo (QMC) calculations by Ceperley and Alder [9].

And for spin-polarized systems, the spin-up and spin-down densities are taken as two independent densities in the exchange-correlation energy. And in this case, the Eq. (1.52) can be expressed as:

$$E_{xc}^{LSD}[\rho] = \int \rho(r)\varepsilon_{xc}^{unif}[\rho_{\downarrow}(r), \rho_{\downarrow}(r)]dr \tag{1.55}$$

For calculating the electronic structure, LDA approach is estimated to be successful. However, for some systems, it does not work. As a result, many efforts are devoted to improve it. One of them is to include the gradient of the density in the exchange correlation functional, as we will show next.

1.7 The Generalized Gradient Approximation (GGA)

To introduce the gradient of the density in the exchange correlation functional, the gradient expansion approximation (GEA) is first proposed. Starting from the uniform electron gas, a slowly varying external potential $v(r)$ is introduced. And then the exchange-correlation energy is expanded in terms of the gradients of the density:

$$E_{xc}^{GEA}[\rho] = E_{xc}^{LD}[\rho] + \int C_{xc}(\rho(r))\rho(r)^{4/3}\left(\frac{\nabla\rho(r)}{\rho(r)^{4/3}}\right)^2 dr \tag{1.56}$$

Here, $C_{xc}(\rho)$ is the sum of the exchange and correlation coefficients of the gradient expansion. As the reduced density gradient is small, GEA approach should be superior to LDA approach. However, because the reduced density gradient can be large in some region of space for real systems, GEA is shown to be worse than LDA.

To overcome the shortcomings of GEA, the GGA is developed. In GGA, the exchange-correlation functional is approximated as:

$$E_{xc}^{GGA}[\rho] = \int f(\rho(r), \nabla\rho(r))dr \tag{1.57}$$

Here, f is some function. Many GGA functionals have been proposed, including B88 [10], Lee-Yang-Parr (LYP) [11], PW91 [12], and Perdew–Burke–Ernzerhof (PBE) [13] exchange-correlation functionals.

1.8 The LDA+U Method

While LDA and GGA are estimated to be able to deal with the many systems and phenomena, they do not work well for the systems with rare-earth and late-transition metal elements. This is because the effective single-particle methods are applicable for highly delocalized band states but not for strongly localized states. For the d and f electrons of rare-earth and late-transition metal elements, they essentially retain their atomic character in solids. As a consequence, standard DFT functionals such as LDA, local spin density approximation (LSDA), and GGA itinerant d states and metallic ground state for many transition metal oxides, for which semiconducting behavior is demonstrated experimentally. For improving these issues, LDA+U method is developed [14–16]. Here, if not specified elsewhere, +U indicates a Hubbard, and LDA indicates the standard DFT functionals, i.e. LDA, LSDA, and

GGA. The idea of LDA+U method is on the basis that the strongly correlated electronic states (i.e. d and f sates) are treated by the Hubbard model, and the rest of the valence electrons are described by the standard DFT functionals. Therefore, the total energy within LDA+U method can be given by:

$$E_{\text{LDA+U}}[\rho] = E_{\text{LDA}}[\rho] + E_{\text{Hub}}\left[\left\{\rho^{I\sigma}_{mm'}\right\}\right] - E_{\text{dc}}[\{\rho^{I\sigma}\}] \tag{1.58}$$

Here, the first term is the standard DFT total energy functional being corrected, the second term represents the Hubbard Hamiltonian to model correlated states, and the third is the double-counting term. The LDA+U method can well describe the electronic properties of the Mott insulators and increase the band gaps in the Kohn–Sham spectrum.

1.9 The Heyd–Scuseria–Ernzerhof Density Functional

In standard DFT, the Fock exchange energy is computed based on a local energy density and its derivatives. However, the exact form for the Fock exchange energy is known as nonlocal from the Hartree–Fock theory. To improve the accuracy, the PBE exchange energy should be mixed with a fraction of the exact nonlocal Fock exchange energy, giving rise to the hybrid functionals, such as the Heyd–Scuseria–Ernzerhof (HSE) functional [17]. The HSE exchange-correlation energy is given by:

$$E_{\text{xc}}^{\text{HSE}} = \alpha E_{\text{x}}^{\text{HF,short}}[\mu] + (1-\alpha)E_{\text{x}}^{\text{PBE,short}}[\mu] + E_{\text{x}}^{\text{PBE,long}}[\mu] + E_{\text{c}}^{\text{PBE}}[\mu] \tag{1.59}$$

Here, E_{x}^{HF} and $E_{\text{x}}^{\text{PBE}}$, respectively, represent the exact Fock exchange energy and the PBE exchange energy. $E_{\text{c}}^{\text{PBE}}$ is the PBE correlation energy. μ is the range-separation parameter. α is the mixing parameter. From Eq. (1.59), it can be seen that the HSE functional is split into short- and long-range terms. In this case, it can improve the accuracy, while avoiding the computational cost.

1.9.1 Introduction to Tight-Binding Approximation

Consider in a single atom there are multiple atomic orbitals $\varphi_{\text{m}}(\mathbf{r})$ with m being the orbital indices. Here, $\varphi_{\text{m}}(\mathbf{r})$ must be eigenfunctions of the Hamiltonian of that single atom H_{atom}. When we place it in a crystal with plenty of atoms, the wave function of different atoms overlap each other to form a different wave function. Due to that $\varphi_{\text{m}}(\mathbf{r})$ is not a real eigenfunction for a Hamiltonian of crystal and we need to find out what the true eigenfunctions are. If the overlap of one atom on another is small enough, we can still assume that electrons are tightly bound to the corresponding atoms, which is exactly the reason why we call it as tight-binding approximation. The approximate Hamiltonian is $H(\mathbf{r}) = H_{\text{atom}}(\mathbf{r}) + \Delta U(\mathbf{r})$, whose Bloch wave function can be taken as a combination of all the isolated orbitals: [18, 19]

$$\psi_{\text{m}}(\mathbf{r}) = \sum_{\mathbf{R_n}} b_{\text{m}}(\mathbf{R_n})\varphi_{\text{m}}(\mathbf{r} - \mathbf{R_n}),$$

where $\mathbf{R_n}$ denotes all lattice points and $b_{\text{m}}(\mathbf{R_n})$ is just a coefficient number for orbital m. In the presence of translation symmetry, coefficient numbers can be replaced by a Bloch form, which gives:

$$\psi_{\mathrm{m}}(\mathbf{r}) \approx \frac{1}{\sqrt{N}} \sum_{\mathbf{R_n}} e^{i\mathbf{k}\cdot\mathbf{R_n}} \varphi_{\mathrm{m}}(\mathbf{r} - \mathbf{R_n}).$$

1.9.2 Matrix Elements of Tight-Binding Hamiltonian

To get Hamiltonian in the momentum space, we shall do a basis transformation such as: [20]

$$\psi_{\mathbf{k}}(\mathbf{r})|H|\psi_{\mathbf{k}}(\mathbf{r}) = \frac{1}{N} \sum_{i,j} e^{-i\mathbf{k}\cdot(\mathbf{R}_i - \mathbf{R}_j)} \langle \varphi(\mathbf{r} - \mathbf{R}_i)|H|\varphi(\mathbf{r} - \mathbf{R}_j)\rangle.$$

When $\mathbf{R}_i$ is equal to $\mathbf{R}_j$, we will find the onsite energy represented for the atomic energy shift due to the overlap of other atoms, which can be given as:

$$\beta_i = \langle \varphi(\mathbf{r} - \mathbf{R}_i)|H|\varphi(\mathbf{r} - \mathbf{R}_i)\rangle.$$

If $\mathbf{R}_i$ is not equal to $\mathbf{R}_j$, hopping energy between different lattice sites can be defined as:

$$t_{ij} = -\langle \varphi(\mathbf{r} - \mathbf{R}_i)|H|\varphi(\mathbf{r} - \mathbf{R}_j)\rangle.$$

1.9.3 Matrix Elements with the Help of Wannier Function

Usually, Bloch wave functions are not orthogonal, which may result in some problems. To resolve that we should define the orthogonal Wannier function as:

$$w(\mathbf{r} - \mathbf{R}_i) = \frac{1}{\sqrt{N}} \sum_{\mathbf{k}} e^{-i\mathbf{k}\cdot\mathbf{R}_i} \psi_{\mathbf{k}}(\mathbf{r}).$$

By using a bra-ket notation, the Hamiltonian in the real space takes the form of: [21]

$$H = \beta_i \sum_{\mathbf{R}_i} |\mathbf{R}_i\rangle\langle\mathbf{R}_i| - t_{ij} \sum_{i \neq j} |\mathbf{R}_i\rangle\langle\mathbf{R}_j|,$$

where β_i and t_{ij} denote the onsite energy and hopping energy, respectively. To get the energy of Hamiltonian, we shall do a basis transformation into the momentum space, which is implemented by a Fourier transform:

$$H = \sum_i \beta_i \sum_{\mathbf{k}} |\mathbf{k}\rangle\langle\mathbf{k}| - t_{ij} \sum_{i \neq j} \frac{1}{N} \sum_{\mathbf{k}} e^{-i\mathbf{k}\cdot(\mathbf{R}_i - \mathbf{R}_j)} |\mathbf{k}\rangle\langle\mathbf{k}|.$$

1.9.4 Example for a Graphene Model

By using graphene lattice [22] as an example, we give a detailed description about how to use tight-binding method. Here, we consider two atoms with s orbital in a single unit cell of unit lattice constant, whose lattice vector can be written as:

$$a_1 = \frac{1}{2}\left(3, \sqrt{3}\right), a_2 = \frac{1}{2}\left(3, -\sqrt{3}\right).$$

Each atom is connected with three nearest-neighbor atoms with a distance of:

$$\delta_1 = \frac{1}{2}\left(1, \sqrt{3}\right), \delta_2 = \frac{1}{2}\left(1, -\sqrt{3}\right), \delta_3 = (-1, 0).$$

In the momentum space, we consider the onsite energy for two s orbitals is ϵ_1. The nearest-neighbor hopping strength is t, which leaves a Hamiltonian matrix as follows:

$$H = \begin{vmatrix} \epsilon_1 & -t\sum_j e^{ik\delta_j} \\ -t\sum_j e^{-ik\delta_j} & \epsilon_1 \end{vmatrix}.$$

We can get the energy dispersion with an implementation of diagonalization, which results in:

$$E(k_x, k_y) = \epsilon_1 \pm t\sqrt{1 + 4\cos\left(\frac{\sqrt{3}k_y}{2}\right)\cos\left(\frac{3k_x}{2}\right) + 4\cos^2\left(\frac{\sqrt{3}k_y}{2}\right)}.$$

1.10 Introduction to $k \cdot p$ Perturbation Theory

A single electron in a periodic potential $V(\mathbf{r})$ obeys a Schrödinger equation, and such a form can be written as [23, 24]:

$$\left[\frac{\mathbf{p}^2}{2m} + V(\mathbf{r})\right]\psi_{(\mathbf{k},\mathbf{r})} = E\psi_{(\mathbf{k},\mathbf{r})},$$

where the eigenvalues and eigenfunctions can be written as $E_n(\mathbf{k})$ and $\psi_n(\mathbf{k}, \mathbf{r}) = e^{i\mathbf{k}\cdot\mathbf{r}}u_n(\mathbf{k}, \mathbf{r})$. The periodic part that is called as a Bloch function satisfies such an equation as:

$$H_0(\mathbf{k})u_n(\mathbf{k}, \mathbf{r}) = E_n(\mathbf{k})u_n(\mathbf{k}, \mathbf{r}),$$

where the Hamiltonian located at momentum point of $\mathbf{k}_0$ can be given as: $H_0(\mathbf{k}_0) = \frac{\mathbf{p}^2}{2m} + \frac{\hbar}{m}\mathbf{k}_0 \cdot \mathbf{p} + \frac{\hbar^2\mathbf{k}_0^2}{2m} + V(\mathbf{r})$. If the eigenvalues and eigenfunctions are assumed to be solved for point as $\mathbf{k}_0$, we can get the solutions of nearby points such as $\mathbf{k} = \mathbf{k}_0 + \delta\mathbf{k}$ through the following equation:

$$\left[H_0(\mathbf{k}_0) + \frac{\hbar}{m}\delta\mathbf{k}\cdot\mathbf{p}\right]u_n(\mathbf{k}, \mathbf{r}) = \left[E_n(\mathbf{k}_0) + \frac{\hbar^2}{2m}\left(\mathbf{k}_0^2 - \mathbf{k}^2\right)\right]u_n(\mathbf{k}, \mathbf{r}).$$

Then, the perturbation Hamiltonian gains a form of $H'(\mathbf{k}) = \frac{\hbar}{m}\delta\mathbf{k}\cdot\mathbf{p}$. With the help of perturbation theory, we can solve the energy of nearby points under two different situations, which is illustrated in the following chapter.

1.10.1 Solution for Non-degenerate Bands

If the bands are not degenerate for $\mathbf{k}_0$, the perturbated eigenvalues of the points $\mathbf{k} = \mathbf{k}_0 + \delta\mathbf{k}$ can be given by [25, 26]:

$$E_n(k) = E_n(k_0) + \frac{2}{2m}\left(k_0^2 - k^2\right) + \frac{\hbar}{m}\delta k\langle u_{nko}(r) \mid p \mid u_{nko}(r)\rangle$$
$$+ \frac{\hbar 2}{m^2}\sum_{n'\neq n}\frac{\langle u_{nk_0}(r) \mid p \mid u_{n'k_0}(r)\rangle\langle u_{n'k_0}(r) \mid p \mid u_{nk_0}(r)\rangle\delta k^2}{E_n(k_0) - E_{n'}(k_0)},$$

where the first term and second term of right-hand side are zero order approximation, and the third term and fourth term serve as the first and second order approximation. Here, we give an explicit example to describe how to get a Hamiltonian based on the following equation. We assume the eigenvalues and eigenfunctions of $\mathbf{k}_0 = 0$ are known by first-principles calculations or experiment. Consider a cubic lattice with a point group of O_h and two bands $|u_{1\mathbf{k}}(\mathbf{r})\rangle$ and $|u_{2\mathbf{k}}(\mathbf{r})\rangle$ that transform as Γ_1^+ and Γ_1^- representation, and the symmetry representation of H' is the vector representation as Γ_{15}^-.

For the first order approximation, the direct product $\Gamma_1^+ \otimes \Gamma_{15}^- \otimes \Gamma_1^+ = \Gamma_{15}^-$ changes the representation Γ_1^+ into Γ_{15}^-, resulting in a vanishing matrix. While for the second order approximation, the direct product $\Gamma_{15}^- \otimes \Gamma_1^+ = \Gamma_{15}^-$ limits the $|u_{n'\mathbf{k}_0}(\mathbf{r})\rangle$ to the Γ_{15}^- representation, which is just the band representation of antibonding p bands. There are three basis x, y, and z for Γ_{15}^- representation; only three terms can exist as a cross term of $\langle \Gamma_1^+ | p_x | x \rangle$, $\langle \Gamma_1^+ | p_y | y \rangle$, and $\langle \Gamma_1^+ | p_z | z \rangle$ according to the selection rules. Finally, we can get the total eigenvalues as:

$$E_1(\mathbf{k}) = E_1(0) + \frac{\hbar^2\mathbf{k}^2}{2m} + \frac{\hbar^2\mathbf{k}^2}{m^2} \sum_{n' \neq 1} \frac{|\langle u_{1\mathbf{k}_0}(\mathbf{r}) | p_x | x \rangle|^2}{E_1(0) - E_{n'}(0)}.$$

1.10.2 Solution for Degenerate Bands

If the bands are degenerate for $\mathbf{k}_0$, the eigenfunction must be a linear combination of degenerated bands. Here, we assume band i is degenerate with band j, and the first order perturbation equation can be written as: [27]

$$\begin{vmatrix} \langle i|H_0 + H'|i\rangle - \varepsilon & \langle i|H_0 + H'|j\rangle \\ \langle j|H_0 + H'|i\rangle & \langle j|H_0 + H'|j\rangle - \varepsilon \end{vmatrix} = 0.$$

The four terms on left-hand side can change the forms as $\langle i|H_0 + H'|i\rangle = E_i^0 - \varepsilon$, $\langle j|H_0 + H'|j\rangle = E_j^0 - \varepsilon$, $\langle i|H_0 + H'|j\rangle = (\hbar/m)\mathbf{k} \cdot \langle i|\mathbf{p}|j\rangle$, and $\langle j|H_0 + H'|i\rangle = (\hbar/m)\mathbf{k} \cdot \langle j|\mathbf{p}|i\rangle$. The solution yields:

$$\varepsilon(\vec{\kappa}) = \frac{E_i^0 + E_j^0}{2} \pm \frac{1}{2}\sqrt{\left(E_i^0 - E_j^0\right)^2 + \frac{4\hbar^2}{m^2}\mathbf{k} \cdot \mathbf{p}_{ij}^{\leftrightarrow} \cdot \mathbf{k}},$$

where $\mathbf{p}_{ij}^{\leftrightarrow}$ is a third order tensor. For a cubic lattice, such a tensor has a form of:

$$\mathbf{p}_{ij}^{\leftrightarrow} = \begin{pmatrix} p_{ij}^2 & 0 & 0 \\ 0 & p_{ij}^2 & 0 \\ 0 & 0 & p_{ij}^2 \end{pmatrix}.$$

1.10.3 Explicit Hamiltonian of $k \cdot p$ Perturbation Theory

Generally, any 4×4 Hamiltonian can be expanded with 16 Dirac matrices as:

$$H = \epsilon(\mathbf{k})I + \sum_i d_i(\mathbf{k})\Gamma_i + \sum_{ij} d_{ij}(\mathbf{k})\Gamma_{ij},$$

where I is the identify matrices, and the five Dirac Γ_i matrices can be defined as $\Gamma_1 = \sigma_1 \otimes \tau_1, \Gamma_2 = \sigma_2 \otimes \tau_1, \Gamma_3 = \sigma_3 \otimes \tau_1, \Gamma_4 = 1 \otimes \tau_2, \Gamma_5 = 1 \otimes \tau_4$, and $\Gamma_{ij} = [\Gamma_i, \Gamma_j]/2i$ with σ_i and τ_i are two sets of Pauli matrices.

First, we should write the symmetry matrix in the basis we choose. Taken Bi_2Se_3 as an example, [28, 29] we choose four states as:

$\left|P1^+_-, \frac{1}{2}\right\rangle, \left|P2^-_+, \frac{1}{2}\right\rangle, \left|P1^+_-, -\frac{1}{2}\right\rangle$, and $\left|P2^-_+, -\frac{1}{2}\right\rangle$. According to the transformation formula, four symmetry matrices D(R) can be constructed as:

1. Time-reversal symmetry: $T = i\sigma_y K \otimes 1$.
2. Threefold rotation symmetry along z axis: $R_3 = e^{i(\sigma_3 \otimes 1)\pi/3}$.
3. Twofold rotation symmetry along x axis: $R_2 = i\sigma_1 \otimes \tau_3$.
4. Inversion symmetry: $P = 1 \otimes \tau_3$.

Second, we apply the symmetry matrices into 16 Dirac matrices to get the representation of matrices, which is implemented by $D(R)\Gamma_i D(R)^{-1}$. In the presence of time-reversal symmetry, we only need to take into consideration one identity matrix and five Dirac Γ_i matrices.

Third, we combined the Dirac matrices with corresponding polynomials of the momentum k who share get the Hamiltonian. It can be written as:

$$H = \epsilon_{\mathbf{k}} + M(\mathbf{k})\Gamma_5 + B(k_z)\Gamma_4 k_z + A(k_\parallel)(\Gamma_1 k_y - \Gamma_2 k_x),$$

where $\epsilon_{\mathbf{k}} = C_0 + C_1 k_z^2 + C_2 k_\parallel^2$, $M(\mathbf{k}) = M_0 + M_1 k_z^2 + M_2 k_\parallel^2, B(k_z) = B_0 + B_2 k_z^2$, and $k_\parallel^2 = k_x^2 + k_y^2$.

References

1 Hohenberg, P. and Kohn, W. (1964). Inhomogeneous electron gas. *Phys. Rev.* 136: B864–B871.

2 Kohn, W. and Sham, L.J. (1965). Self-consistent equations including exchange and correlation effects. *Phys. Rev.* 140: A1133–A1138.

3 Jones, R.O. and Gunnarsson, O. (1989). The density functional formalism, its applications and prospects. *Rev. Mod. Phys.* 61: 689–746.

4 Grossman, J.C., Mitas, L., and Raghavachari, K. (1995). Structure and stability of molecular carbon: importance of electron correlation. *Phys. Rev. Lett.* 75: 3870–3873.

5 Zupan, A., Blaha, P., Schwarz, K., and Perdew, J.P. (1998). Pressure-induced phase transitions in solid Si, SiO2, and Fe: Performance of local-spin-density and generalized-gradient-approximation density functionals. *Phys. Rev. B* 58: 11266–11272.

6 Perdew, J.P., Burke, K., and Ernzerhof, M. (1996). Generalized gradient approximation made simple. *Phys. Rev. Lett.* 77: 3865–3868.

7 Perdew, J.P., Burke, K., and Wang, Y. (1996). Generalized gradient approximation for the exchange-correlation hole of a many-electron system. *Phys. Rev. B* 54: 16533–16539.

8 Becke, A.D. (1997). Density-functional thermochemistry. V. Systematic optimization of exchange-correlation functionals. *J. Chem. Phys.* 102: 8554–8560.

9 Ceperley, D.M. and Alder, B.J. (1980). Ground state of the electron gas by a stochastic method. *Phys. Rev. Lett.* 45: 566–569.

10 Becke, A.D. (1988). Density-functional exchange-energy approximation with correct asymptotic behavior. *Phys. Rev. A* 38: 3098–3100.

11 Lee, C., Yang, W., and Parr, R.G. (1988). Development of the Colle-Salvetti correlation-energy formula into a functional of the electron density. *Phys. Rev. B* 37: 785–789.

12 Colle, R. and Salvetti, O. (1975). Approximate calculation of the correlation energy for the closed shells. *Theor. Chim. Acta* 37: 329–334.

13 Miehlich, B., Savin, A., Stoll, H., and Preuss, H. (1989). Results obtained with the correlation energy density functionals of becke and Lee, Yang and Parr. *Chem. Phys. Lett.* 157: 200–206.

14 Anisimov, V.I. and Gunnarsson, O. (1991). Density-functional calculation of effective Coulomb interactions in metals. *Phys. Rev. B* 43: 7570–7574.

15 Anisimov, V.I., Zaanen, J., and Andersen, O.K. (1991). Band theory and Mott insulators: Hubbard U instead of Stoner I. *Phys. Rev. B* 44: 943–954.

16 Anisimov, V.I., Solovyev, I.V., Korotin, M.A. et al. (1993). Density-functional theory and NiO photoemission spectra. *Phys. Rev. B* 48: 16929–16934.

17 Heyd, J., Scuseria, G.E., and Ernzerhof, M. (2003). Hybrid functionals based on a screened Coulomb potential. *J. Chem. Phys.* 118: 8207–8215.

18 Papaconstantopoulos, D.-A. (1986). *Handbook of the Band Structure of Elemental Solids*. Springer.

19 Madelung, O. (2012). *Introduction to Solid-State Theory*, vol. 2. Springer Science & Business Media.

20 Wieder, B.-J., Bradlyn, B., Wang, Z.-J. et al. (2018). Wallpaper fermions and the nonsymmorphic Dirac insulator. *Science* 361: 246.

21 Liu, G.-B., Shan, W.-Y., Yao, Y. et al. (2013). Three-band tight-binding model for monolayers of group-VIB transition metal dichalcogenides. *Phys. Rev. B* 88: 085433.

22 Kane, C.-L. and Mele, E.-J. (2005). Quantum spin hall effect in graphene. *Phys. Rev. Lett.* 95: 226801.

23 Luttinger, J.-M. (1955). Quantum theory of cyclotron resonance in semiconductors: general theory. *Phys. Rev.* 102: 1030.

24 Pikus, G.-E. (1962). A new method of calculating the energy spectrum of carriers in semiconductors. I. Neglecting spin-orbit interaction. *Sov. Phys. JETP* 14: 898.

25 Bir, G.-L. and Pikus, G.-E. (1975). *Symmetry and Strain-Induced Effects in Semiconductors*. New York: Wiley.

26 Ivchenko, E.-L. and Pikus, G.-E. (1995). *Superlattices and Other Heterostructures*. Berlin: Springer.

27 Voon, L.-C.-L.-Y. and Willatzen, M. (2009). *The kp Method: Electronic Properties of Semiconductors*. (Springer Science & Business Media.

28 Liu, C.-X., Qi, X.-L., Zhang, H.-J. et al. (2010). Model Hamiltonian for topological insulators. *Phys. Rev. B* 82: 045122.

29 Zhang, W., Yu, R., Zhang, H.-J. et al. (2010). First-principles studies of the three-dimensional strong topological insulators Bi_2Te_3, Bi_2Se_3 and Sb_2Te_3. *New J. Phys.* 12: 065013.

2

New Physical Effects Based on Band Structure

2.1 Valley Physics

Valley refers to the local minimum in the conduction band or the local maximum in the valence band in the momentum space. Given the systems with two or more valleys, if the valleys can be polarized and detected, the valley can form a new degree of freedom of carriers in addition to charge and spin. Analogy to charge for electronics and spin for spintronics, the possibility of utilizing valley degree of freedom as information carrier gives rise to the concept of valleytronics [1–3]. The idea of using valley degree of freedom dates back to the studies in the late 1970s, which investigated the valley behavior and inter-valley coupling in silicon inversion layers [4–6]. Moreover, by tuning the intravalley exchange and correlation, the electrons would preferentially occupy the valleys, thereby generating the valley polarization. Subsequently after the works on silicon inversion layer, several other systems like the AlAs quantum well, silicon heterostructures, bismuth, and diamond are also shown to feature the valley feature [7–11]. Despite these extensive efforts, there lacks an intrinsic physical property that correlated with the valley occupancy. This is in sharp contrast with the case of spin, where the magnetic moment, the spin optical selection rule, and the spin-orbit coupling (SOC) can be employed as the intrinsic properties associating with the spin. Therefore, the utilization of the valley index is severely limited as compared with the spintronics, and the intrinsic property associated with the valley occupancy is under exploration.

The recent rise of 2D materials [12–14] provides an unprecedented opportunity to address the aforementioned issues. The first 2D system harboring the valley physics is proposed in graphene (Figure 2.1) [1]. In graphene with broken inversion symmetry, the band edges of the band structure lie at the $+K$ and $-K$ points of the 2D Brillouin zone, forming two degenerate by nonequivalent valleys. Importantly, the carriers in the $+K$ and $-K$ valleys are subjected to opposite Berry curvatures and orbital magnetic moment. These two quantities can be used to distinguish the valleys. The Berry curvature can behave like effective magnetic fields, which would result in an anomalous velocity perpendicular to an electric field. This is a Hall effect, which is referred to as the valley Hall effect [15]. While for the orbital magnetic moment, it results from self-rotating, so an energy shift can be expected under

Calculations and Simulations of Low-Dimensional Materials: Tailoring Properties for Applications, First Edition. Ying Dai, Wei Wei, Yandong Ma, and Chengwang Niu.

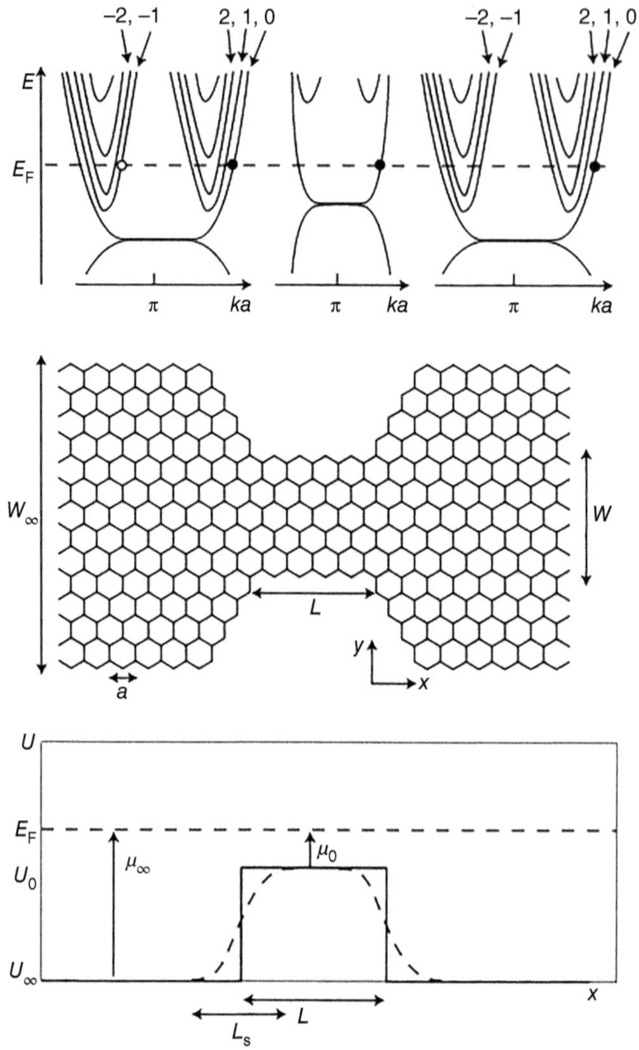

Figure 2.1 Schematic diagram of the valley filter based on graphene. Top panel: Dispersion relation in the wide and narrow regions. An electron in the first valley (modes $n = 0, 1, 2, \ldots$) is transmitted (filled circle), whereas an electron in the second valley (modes $n = -1, -2, \ldots$) is reflected (open circle). Middle panel: Honeycomb lattice of carbon atoms in a strip containing a constriction with zigzag edges. Bottom panel: Variation of the electrostatic potential along the strip for the two cases of an abrupt and smooth potential barrier (solid and dashed lines). The polarity of the valley filter switches when the potential height, U_0, in the constriction crosses the Fermi energy, E_F. Source: Rycerz et al. [1]/with permission of Springer Nature.

the magnetic field. It should be noted that, under the inversion symmetry, the $+K$ and $-K$ valleys can be transformed into each other, making it not possible for distinguishing the valleys. Therefore, breaking the inversion symmetry is a necessary condition for utilizing Berry curvature and orbital magnetic moment to distinguish the valley occupancy. Currently, several approaches are proposed to break the inversion symmetry of graphene, including the creation of edge modes [16] and defect lines [17]. However, these approaches are challenging in experiments, limiting their practical applications.

After discovering the valley behaviors in graphene, the monolayer group-VI transition metal dichalcogenides (TMDs), i.e. MoS_2, WS_2, $MoSe_2$, and WSe_2, are identified as the most promising platform to study the novel valley-contrasting physics [18–21]. In monolayer TMDs, the M atomic layer is sandwiched between two X atomic layers, which breaks the inversion symmetry naturally (Figure 2.2a). Moreover, there are two degenerate but nonequivalent valleys locating at the $+K$ and $-K$ points, generating two valleys. Therefore, under an in-plane electric field, the carriers in $+K$ and $-K$ valleys would be accumulated at the opposite edges of the sample, which is related to the opposite Berry curvatures at the two valleys (Figure 2.2c). More importantly, as the valleys are dominated by the d orbitals of the M atoms, the valleys experience a valley spin splitting due to the strong SOC strength within M-d orbitals [22]. And due to the in-plane character of the d orbitals for the valleys in the valence bands, they have a very large valley spin splitting. While for the valleys in the conduction bands, because of the out-of-plane character, the valley spin splitting

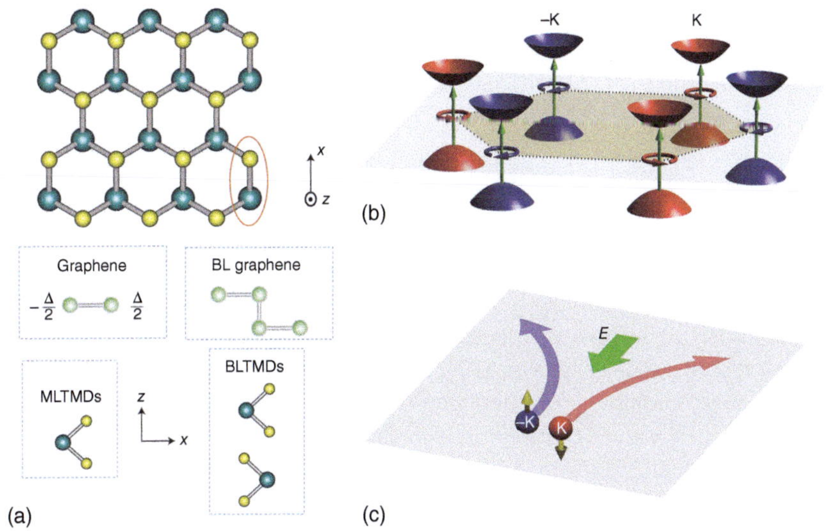

Figure 2.2 Crystal structure and valley physics in monolayer TMDs. (a) 2D hexagonal lattice. (b) Valley contrasting optical selection rules in a 2D hexagonal lattice with broken inversion symmetry. The interband transition in valley K ($-K$) couples to $\sigma+$ ($\sigma-$) circularly polarized light only (circular arrows). (c) Diagram of the valley Hall effect in monolayer TMDs. Arrows suggest the pseudo-vector quantities, i.e. Berry curvature or orbital magnetic moment, of the carriers. Source: Xu et al. [18]/with permission of Springer Nature.

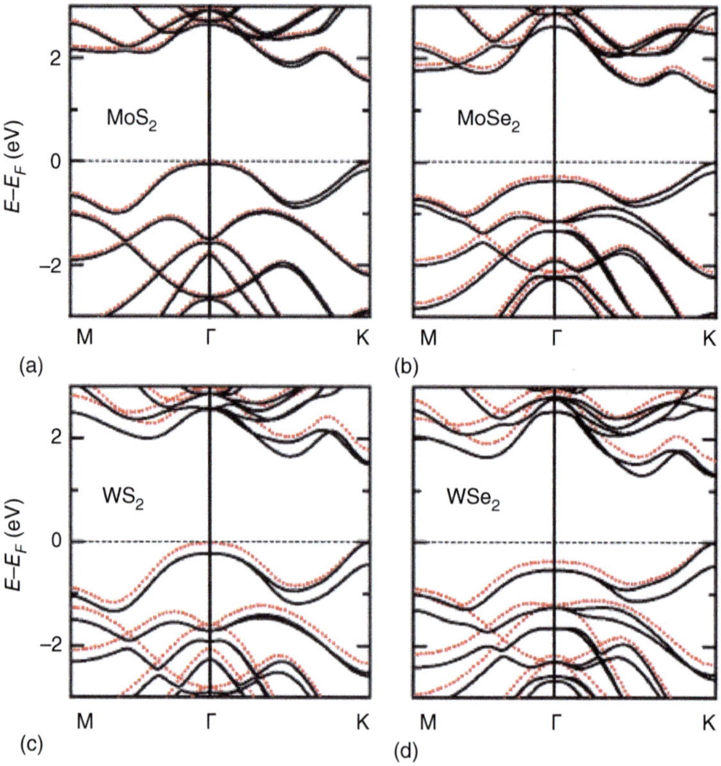

Figure 2.3 Electronic band structures calculated for the MX_2 monolayer systems with (solid line) and without (dotted line) inclusion of the spin-orbit interaction. Source: Zhu et al. [22]/with permission of American Physical Society.

is rather small (Figure 2.3). Under such splitting, we can manipulate the carrier with one spin in one valley. This indicates novel spin and valley physics that are absent in graphene. Considering these merits, monolayer TMDs are considered as one of the most promising 2D valleytronic materials.

Besides graphene and monolayer TMDs, many other 2D materials presenting the valley-contrasting physics are proposed. For example, using first-principles calculations, we reported that monolayer H-Tl_2O is a compelling 2D valleytronic material with spin-valley coupling [23]. Soon after the work on monolayer H-Tl_2O, we discovered another class of 2D valleytronic materials in monolayer MN_2X_2 (M = Mo, W; X = F, H), which are shown to be dynamically and thermally stable. Monolayer MN_2X_2 is an indirect gap semiconductor with the valence band maximum locating at the $+K$ and $-K$ points, forming two degenerate but nonequivalent valleys in the valence band [24].

Recently, Lu et al. proposed that hexagonal monolayer $MoSi_2N_4$ and $MoSi_2As_4$ exhibit a pair of valleys at the $+K$ and $-K$ points [25]. The circularly polarized photons can be adopted to distinguish the $+K$ and $-K$ valleys. Arising from the valley-contrasting Berry curvature, the intriguing valley Hall and spin Hall effects can be realized. In addition to the traditional two level valleys, the valleys in monolayer $MoSi_2As_4$ are shown to be multiple folded. This suggests an additional intrinsic degree of freedom for monolayer $MoSi_2N_4$ and $MoSi_2As_4$. Moreover, monolayer $MoSi_2N_4$ and $MoSi_2As_4$ have several advantages over monolayer TMDs. For example, while monolayer TMDs suffer from the limited size of growth [26] and low electron/hole mobility [27], which restricts them for further quantum-transport applications, monolayer $MoSi_2N_4$ and $MoSi_2As_4$ were successfully synthesized with large size up to 15 mm × 15 mm [28] and is predicted to be with a large electron/hole mobility of four to six times larger than that of monolayer MoS_2 [29]. Therefore, monolayer $MoSi_2N_4$ and $MoSi_2As_4$ are believed to be promising candidates for the potential of valleys to be applied in multiple information processing.

After discovering these promising valleytronic materials, the next step is to explore how to use these materials for applications in valleytronic devices. To make a valleytronic device, the crucial step is to lift the degeneracy between the $+K$ and $-K$ valleys, thereby producing the valley polarization [15]. In 2D valleytronic materials, the valley-contrasting orbital magnetic moment is accompanied by a valley-contrasting optical selection rule. In detail, the interband transitions in the vicinity of the $+K$ and $-K$ valleys correlate with the right- and left-handed circularly polarized, see Figure 2.2b [18]. To this end, the method of optical pumping is proposed to achieve the valley polarization in monolayer TMDs. Cui et al. found that the photoluminescence in monolayer MoS_2 has the same helicity as the circularly polarized component of the excitation laser [30]. This is a strong signature of the valley polarization induced by optical pumping. Below temperature 90 K, a high photoluminescence circular polarization is obtained, and it decays with temperature, see Figure 2.4. To confirm that the polarized photoluminescence is related to the valley instead of spin, the Hanle effect is investigated. They showed that the persistent photoluminescence polarization is observed in monolayer MoS_2 when traversing the magnetic field. This confirms that the polarized photoluminescence is attributed to the valley

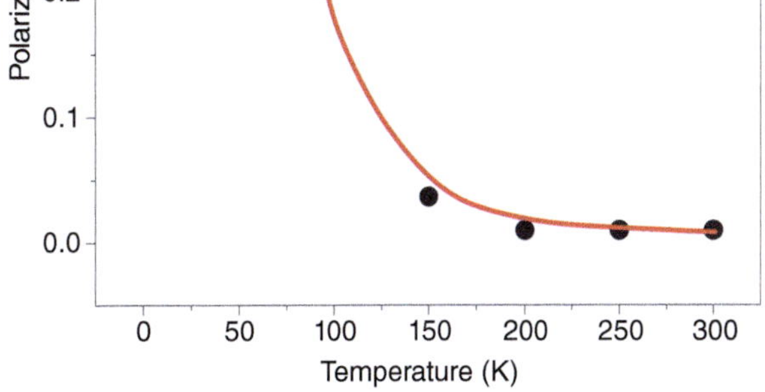

Figure 2.4 The variation of degree of circular polarization P with temperature. For fitting the relationship, assuming an inter-valley scattering proportional to the phonon population is assumed. Source: Zeng et al. [30]/with permission of Springer Nature.

polarization as a transverse magnetic field causes the spin to precess. Furthermore, for bilayer MoS_2 with inversion symmetry, the photoluminescence is found to be unpolarized by applying the same excitation condition, which again confirms the optically pumped valley polarization in monolayer MoS_2. Similar to this work, Mak et al. also identified the viability of optical control of valleys and the valleytronics in monolayer MoS_2 [20]. Besides these two experimental works, there are also many other experimental breakthroughs in optical pumped valley polarization. For more detail, please see the review of Ref. [18].

Although valley polarization is successfully achieved through optical pumping, it is subjected to the quite short lifetime of carriers. Considering this point, optical pumping is unable to robustly tune the valley, making it not applicable for developing practical information devices. An alternative approach for realizing the valley polarization is to break time-reversal symmetry. That is because the time-reversal symmetry requires the spins at the $+K$ and $-K$ valley to be energetically degenerate, but opposite, forming the valley-spin locking relationship. By breaking the time-reversal symmetry, the valley degeneracy would be lifted. In fact, lifting the valley degeneracy in monolayer TMDs has been achieved by applying an external magnetic field in some experimental works [31, 32]. However, this approach is shown to be rather modest. For instance, the valley polarization is estimated to be 0.1–0.2 $meV\,T^{-1}$ [31, 32]. This is suitable for the development of valleytronic devices where the large valley polarization is required, analogous to large spin polarization for spintronics. Recently, the research efforts for breaking the time-reversal symmetry in this field have been devoted to introducing foreigner atom doping and proximity-induced magnetic interaction, and many breakthroughs have been made. In addition, with the recent discovery of 2D magnetic materials, the intrinsic valley polarization is also identified, which attracts more and more attention. In the following, we will discuss the recent theoretical developments on the spontaneous valley polarization, valley polarization induced by foreign atom doping, and valley polarization in van der Waals heterostructures.

2.1.1 Spontaneous Valley Polarization

Spontaneous valley polarization was first proposed in 2013 [33]. Using the tight-binding model, Feng et al. proposed the theory of spin and valley physics of an antiferromagnetic honeycomb lattice. In the antiferromagnetic honeycomb lattice, there is an emergent electronic degree of freedom of carriers, which is characterized by the product of spin and valley indices ($s{\cdot}\tau$). This gives rise to the $s{\cdot}\tau$-dependent optical selection rule. When the spin-valley coupling is weak, namely the valley degeneracy is considered to be kept, the system would exhibit an $s{\cdot}\tau$-selective circular dichroism (CD). Illuminated by the left-polarized light, the spin-up electrons at the $+K$ valley and spin-down electron at the $-K$ valley would be excited to the conduction band. And by illustrating the right-polarized light, the spin-down electrons at the $+K$ valley and spin-up electron at the $-K$ valley would be excited to the conduction band. For the case with strong spin-valley coupling, the valley degeneracy is lifted, and thus the gaps at the $+K$ and $-K$ points are different.

In this regard, by tuning the frequency of the polarized light, one spin from one valley can be excited. Besides the optical selection rule, the Berry curvature is also $s{\cdot}\tau$-dependent. Charge carrier with opposite $s{\cdot}\tau$ indices would exhibit opposite transversal anomalous velocities. After establishing the theory of spin and valley physics in an antiferromagnetic honeycomb lattice, Feng et al. suggested one actual material of monolayer manganese chalcogenophosphates, $MnPX_3$, X = S, Se [33]. By calculating their band structure, they showed that the gaps at $+K$ and $-K$ differ by 43 meV, indicating the spontaneous valley polarization and the $s{\cdot}\tau$-dependent physics.

After discovering the spontaneous valley polarization in the antiferromagnetic honeycomb lattice, the spontaneous valley polarization in the ferromagnetic lattice is also proposed. Using the two-band k·p model, Duan et al. also proposed that the coexistence of SOC and intrinsic exchange interaction can result in that the valley polarization occurs spontaneously [34]. They further predicted one such real material of monolayer 2H-VSe_2.

Similar to the anomalous Hall effect, Duan et al. defined the valley Hall effect in ferromagnetic material as anomalous valley Hall effect [34]. Note that the charge Hall current is easier to be detected experimentally, and the anomalous valley Hall effect would provide a promising way to realize the valley-based information storage. On the basis of the anomalous valley Hall effect, Duan et al. proposed the possible electrically reading and magnetically writing memory devices [34]. This significantly advances the practical applications of valleytronics.

Based on the preceding two works [33, 34], the spontaneous valley polarization is well established. However, for monolayer $MnPX_3$, they show the $s{\cdot}\tau$-selective, rather than the τ-selective, CD, and Berry curvature. And valley is not an independent degree of freedom of carriers any more. While for monolayer 2H-VSe_2, this monolayer phase is not stable in experiments. Therefore, the candidate materials with spontaneous valley polarization are still under exploration. In the following, we will discuss the systems that are recently identified with spontaneous valley polarization.

Using first-principles calculations, Lu et al. investigated the valley physics in monolayer $VAgP_2Se_6$ [35]. They found that monolayer $VAgP_2Se_6$ can be easily exfoliated from its bulk counterpart. Monolayer $VAgP_2Se_6$ is intrinsically spin polarized. Monolayer $VAgP_2Se_6$ is a ferromagnetic semiconductor. The easy magnetization axis is along the out-of-plane, with a magnetocrystalline energy of 1.5 meV per unit cell, enabling the possible stable ferromagnetism. The spin-polarized band structure of monolayer $VAgP_2Se_6$ without SOC shows that the up-spin and down-spin channels are split significantly, and the bands near the Fermi level are dominated by the up-spin states and are characterized by the two massive Dirac cones with a band gap. This forms two degenerate valleys. Upon considering SOC, the gap at $+K$ valley decreases to 29 meV and increases to 44 meV at the $-K$ valley, thus lifting the valley degeneracy. Based on these results, Lu et al. designed a valley pseudospin field effect transistor. And different from the conventional transistors, such transistor could carry information of not only the electrons but also the valley pseudospins, which opens an approach to an avenue for realizing the low-power and high-performance valleytronic devices in the future.

Compared with monolayer TMDs, monolayer transition metal dinitrides (TMDNs) usually exhibit distinct properties because of the fewer valence electrons. For instance, monolayer MoN_2 is a ferromagnetic semiconductor, while monolayer MoS_2 is a semiconductor. Thus, monolayer TMDNs are considered as promising high-capacity electrode materials for alkali-ion batteries [36]. Wang et al. systematically studied the electronic properties of hydrogenated monolayer NbN_2 (H-NbN_2H_2), and found that they present spontaneous valley polarization. To investigate the magnetic ground state, they constructed a 2×2 supercell, and considered the ferromagnetic and the possible antiferromagnetic states. The simulated temperature-dependent magnetization of monolayer H-NbN_2H_2 and the Curie temperature are found to be 225 K. Based on the band structure with SOC of monolayer H-NbN_2H_2, there are two inequivalent valleys locating at the $+K$ and $-K$ points. Up including SOC but without magnetization, the valley spin splitting occurs. The magnitudes of spin splitting energies at the two valleys are same, but the signs are opposite. Therefore, two valleys are still degenerate at the nonmagnetic state, protecting by the time-reversal symmetry. When further including the exchange interaction, the valley polarization occurs spontaneously.

As Nb and V belong to the same family and monolayer VSe_2 is identified as a typical example with spontaneous valley polarization, monolayer NbX_2 (X = S, Se) is also expected to be with such interesting valley physics. We then investigated the valleytronic properties of monolayer NbX_2 [37]. The research on bulk NbX_2 dates back to 1980s [38]. More importantly, the synthesis of monolayer NbX_2 has been reported experimentally [39–42], indicating that any properties predicted in it show high experimental feasibility. The crystal structure of monolayer NbX_2 exhibited a hexagonal lattice with D_{3h} space group. Concerning Nb atom, its electronic configuration is $4d^4 5s^1$. When bonding with the six X atoms, four electrons are transferred to the X atom. In this case the low-lying orbital of d_{z2} is half-occupied, forming a magnetic moment of 1 μ_B. The magnetic moment is mainly distributed on the Nb atom. To investigate its magnetic ground state, we studied the energies of the ferromagnetic and antiferromagnetic states and found that the ferromagnetic state is the ground state. Considering the important role of out-of-plane magnetization in valley polarization, we studied the magnetocrystalline anisotropy energy of monolayer NbS_2 and $NbSe_2$, which is estimated to be 8.1 and 5.4 meV per unit cell, respectively.

As monolayer NbS_2 and $NbSe_2$ share similar electronic properties, we only discuss the band structures of monolayer $NbSe_2$. Without considering SOC and exchange interaction, one spin-degenerate band crosses the Fermi level. And this band has dominated Nb-d_{z2}, d_{x2-y2}, d_{xy} orbitals. By taking exchange interaction into consideration, the spin degeneracy is lifted. In this condition, monolayer $NbSe_2$ becomes semiconducting with an indirect band gap. Its top valence band locates at the $+K$ and $-K$ points, forming two degenerate, but inequivalent, valleys. Different from the case of monolayer $NbSe_2$, the top valence band maximum of monolayer NbS_2 does not locate at the $+K$ and $-K$ point, which however can be tuned to the $+K$ and $-K$ points under external strain. Upon further introducing SOC, the valley at the $-K$ point shifts above the $+K$ point, thus generating the valley polarization. And the valley polarization is 219 meV, which is large enough for further manipulating the valley feature, highly desirable for practical applications. And by reversing the magnetization orientation, the valley polarization can also be reversed, forming a valley polarization of −219 meV. The large valley polarization in monolayer $NbSe_2$ can be attributed to the strong SOC strength in the Nb atoms. Similar to monolayer $NbSe_2$, due to the intrinsic magnetic exchange interaction and strong SOC strength, a spontaneous valley polarization of 156 meV is also realized in monolayer NbS_2. Besides, the strain effect on the valleytronic properties of monolayer NbX_2 are also investigated. For more details, please refer to Ref. [37].

Recently, Janus monolayer TMDs were synthesized experimentally [43], which has attracted great interest. Quite naturally, the searching for spontaneous valley polarization in traditional monolayer TMDs can be extended to such Janus TMDs. One such example is the Janus monolayer VSSe [44]. Du et al. found that, similar to monolayer VSe_2, monolayer VSSe also exhibits a pair of valleys at the $+K$ and $-K$ points [44]. Upon including SOC, the spontaneous valley polarization occurs in monolayer VSSe. Besides the spontaneous valley polarization, monolayer VSSe exhibits a strain-driven 90° lattice rotation with an extremely high reversal strain (73%), suggesting an intrinsic ferroelasticity. Besides VSSe, monolayer VSeTe is also identified with spontaneous valley polarization [45].

Duan et al. investigated the valleytronic properties in another Janus structure of monolayer VClBr [46]. Analogues to Janus TMDs, monolayer VClBr exhibits two phases, i.e. H and T phases. Here, we only discuss the H phase, as it features the valley physics. Figure 2.5a presents the crystal structure of monolayer VClBr, which shows a sandwich layer structure of Cl-V-Br. The lattice constant is found to be 3.563 Å. The ground state is demonstrated to be ferromagnetic. Figure 2.5b shows the band structure of monolayer VClBr with considering magnetic exchange interaction but without SOC. The top valence band locates at the $+K$ and $-K$ points, producing two valleys. For the bottom of the conduction band, it locates near the Γ point. This forms an indirect gap of 0.61 eV. According to the orbital analysis, the valleys are mainly contributed by the V-3d_{xy} and V-3d_{x2-y2}, while for the lowest conduction at the $+K$ and $-K$ points, it is dominated by the $V-3d_{yz}$ and $V-d_{xz}$. This is different from the case of monolayer VSe_2 and VSSe. When taking SOC into consideration, the $+K$ valley shifts above the $-K$ valley, forming the spontaneous valley polarization, see Figure 2.5c. The valley polarization is found to be about 70 meV. And the valley polarization can be reversed upon reversing the magnetization orientation, see Figure 2.5d. Duan et al. also investigated the relationship between the magnetism and chiral optical band gap at the $+K$ and $-K$ points.

As mentioned earlier, Janus TMD structure is obtained by hydrogenating the nonmetal atoms. In fact, hydrogenating the metal atoms in 2D materials also receives more and more attention. As an example, we constructed such a system of $TiVI_6$ and demonstrated that it is a promising valleytronic material with spontaneous valley polarization [47]. The electronic configurations of Ti and V, respectively, are $3d^24s^2$ and $3d^34s^2$. By bonding with the I atoms, the electrons of Ti are completely denoted to I and V atoms, resulting in the oxidation states of +4 and +2 for Ti and V atoms, respectively. Under the octahedral crystal field, the d orbit splits into three groups, i.e. a(d_{z2}), e_1(d_{xy}, d_{x2-y2}), and e_2(d_{xz},d_{yz}). Obviously, the magnetic moment on each

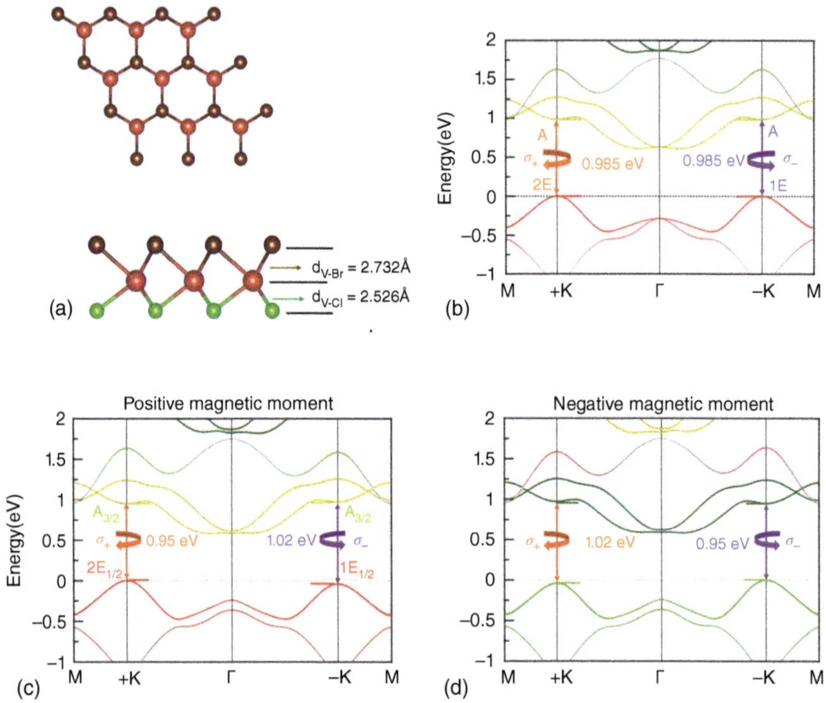

Figure 2.5 (a) Top and side views of the crystal structure of monolayer VClBr. (b) Band structures of monolayer VClBr with considering exchange interaction and without considering SOC. (c) Band structure of monolayer VClBr with considering both exchange interaction and SOC; (d) is the same as (c) but with the opposite magnetization direction. E_F is set to 0 eV. Source: Zhao et al. [46]/with permission of American Physical Society.

V atom is 3 μ_B, while it is 0 μ_B for the Ti atom. The ferromagnetic state is found to be more stable than the other three antiferromagnetic configurations. Magnetocrystalline anisotropy energy of monolayer $TiVI_6$ is also calculated, which is found to be 11 meV per unit cell. This suggests that the spins in monolayer $TiVI_6$ are favorably aligned along the out-of-plane direction. Moreover, because the easy axis is along the out-of-plane direction, the Ising model can be employed to estimate Curie temperature. Based on the Monte Carlo simulation, the Curie temperature is estimated to be 18 K, comparable with that of monolayer CrI_3 [48].

The spin-polarized band structure of monolayer $TiVI_6$ with considering SOC shows that there is an energy gap for both the up-spin and down-spin channels. Therefore, monolayer $TiVI_6$ is a ferromagnetic semiconductor. Although its conduction band minimum locates at the Γ point, its valence band maximum lies at the $+K$ and $-K$ points, leading to two valleys. Considering its broken inversion symmetry, the monolayer $TiVI_6$ is 2D valleytronic material. The band structure of monolayer $TiVI_6$ with including SOC indicates that the degeneracy of the $+K$ and $-K$ points is lifted. Therefore, the spontaneous valley polarization is obtained in monolayer $TiVI_6$. The valley polarization of monolayer $TiVI_6$ is found to be 22 meV. The orbital-resolved band structure with considering SOC shows that the band edges are dominated by the d orbitals of Ti and V atoms. The valley polarization is resulted from the SOC strength within the d orbitals. Besides, the strain effect on the electronic properties of monolayer $TiVI_6$ is also investigated. It is worth emphasizing that when increasing the compressive strain to −4%, the top valence bands at the Γ point shifts above the valleys, which would deform the valley behaviors. Under such condition, monolayer $TiVI_6$ is not suitable for practical valleytronic applications. Accordingly, strain can be used to manipulate the switching of the valley behavior.

Using first-principles calculations, we studied the spintronic and valleytronic properties of monolayer $LaBr_2$ [49]. The study on bulk $LaBr_2$ dates back to 1989 [50]. It exhibits a hexagonal layered structure with the space group of P63/mmc. Therefore, monolayer $LaBr_2$ can be easily obtained experimentally via exfoliation. Figure 2.6a shows the crystal structure of monolayer $LaBr_2$. Apparently, it presents a hexagonal lattice and contains a formula unit in each unit cell. The spin-polarized calculations show that monolayer $LaBr_2$ is spin polarized with a total magnetic moment of 1 μ_B per unit cell. And the magnetic moment is mainly distributed on the La atom. Such spin polarization can be attributed to the valence electronic configuration of La atom. By denoting two electrons to the Br atoms, one electron is left to occupy the lower d_{z2} orbital, leaving other higher orbitals empty. The ferromagnetic configuration is found to be most stable among these three configurations. Such magnetic coupling is sought into the Goodenough-Kanamori-Anderson rule. In monolayer $LaBr_2$, the La-Br-La angle approximates 90°. Based on the Goodenough-Kanamori-Anderson rule, the exchange interaction between the magnetic moment would be dominated by the ferromagnetic superexchange interaction.

The band structure of monolayer $LaBr_2$ without considering spin polarization and SOC is shown in Figure 2.6b. It can be seen that there is one spin-degenerated band crossing the Fermi level, leading to the metallic feature. This band is mainly contributed by the La-$d_{z2,x2-y2,xy}$ orbitals and slightly from Br-s/p orbitals. When considering both spin polarization and SOC, as shown in Figure 2.6c, the spin degeneracy of this band is lifted, which results in a semiconducting feature. There monolayer $LaBr_2$ is a ferromagnetic semiconductor. The magnetocrystalline anisotropy energy is also calculated for monolayer $LaBr_2$, which indicates that easy axis for monolayer $LaBr_2$ is normal to the 2D plane. From the band structure of monolayer $LaBr_2$ shown in Figure 2.6c, we can see that there is a pair of valleys locating at the $+K$ and $-K$ points in both the conduction and valence bands. As the lowest conduction band locates at the M points, only the valley at the valence band is discussed. Upon further taking SOC into consideration, the degeneracy between the $+K$ and $-K$ valleys are lifted, achieving the interesting valley polarization. When the magnetization is reversed, the valley polarization is also reversed. The valley polarization of monolayer $LaBr_2$ is found to be 33 meV, which is equivalent to the case of a valley degenerate material exposed to an external magnetic field of 192 T. Moreover, because monolayer $LaBr_2$ shows high experimental feasibility with an ultralow exfoliation energy, the spontaneous valley polarization is ready to be verified in experiment.

Similar to monolayer $LaBr_2$, monolayer Nb_3I_8 also has its layered bulk counterpart [51, 52] and was proposed to exhibit the spontaneous valley polarization [53]. Bulk

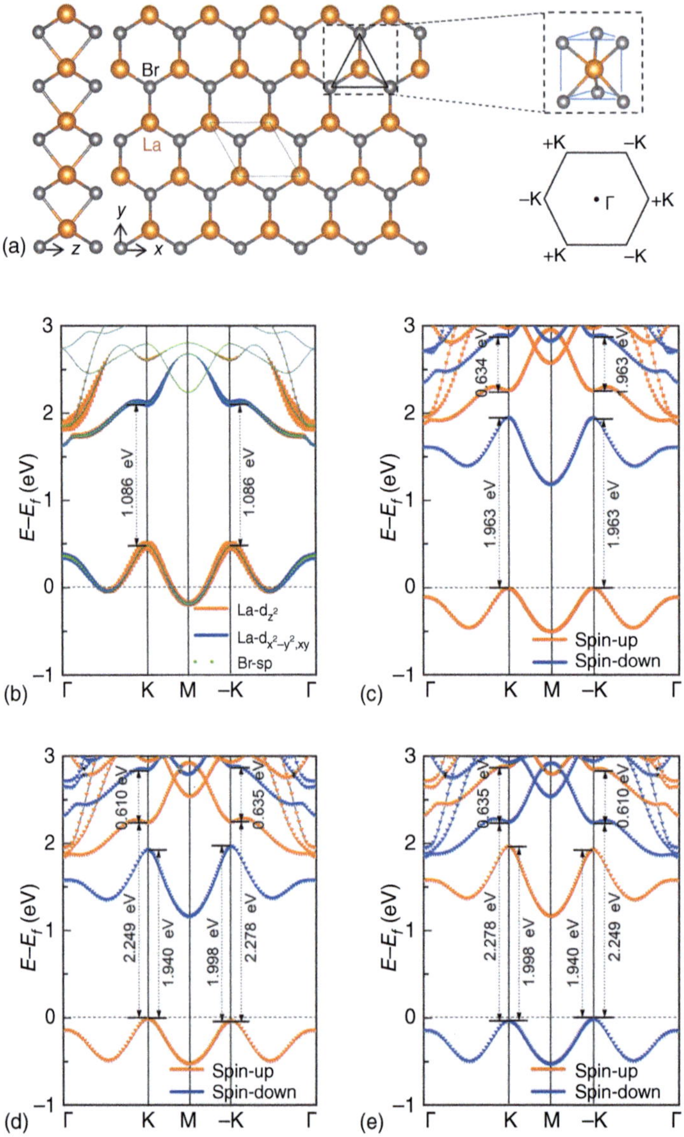

Figure 2.6 (a) Side and top views of the crystal structure of monolayer $LaBr_2$. Insets in (a) give the trigonal prismatic geometry and the 2D Brillouin zone. (b) Band structure of monolayer $LaBr_2$ without considering spin polarization and SOC. (c) Band structure of monolayer $LaBr_2$ with considering spin polarization but without considering SOC. (d) Band structure of monolayer $LaBr_2$ with considering both spin polarization and SOC; (e) is same as (d) but with downward magnetization. E_F is set to 0 eV. Source: Zhao et al. [49]/with permission of AIP Publishing.

Nb_3I_8 exhibits a space group of R-3m. When exfoliated into the monolayer form, the space group is transferred into P3m1. Each unit cell of Nb_3I_8 contains three Nb and eight I atoms. Each three Nb atoms is assembled into a Nb_3 trimer in monolayer Nb_3I_8. Accordingly, monolayer Nb_3I_8 features two different Nb-Nb distances: one is 3.02 Å for the intratrimer distance and one is 4.63 Å for the intertrimer distance. This gives rise to a distorted octahedral coordination environment for the Nb atoms. Moreover, in monolayer Nb_3I_8, the Nb_3 trimers form a Kagome lattice. Therefore, the inversion symmetry of monolayer Nb_3I_8 is broken. Normally, the Nb-d orbitals split into two groups, i.e. e_g and t_{2g}. By forming the Nb_3 trimer, the d orbitals split into more energy levels. Concerning the t_{2g} orbital, it splits into 1e, $1a_1$, 2e, $2a_1$, 3e, and $1a_2$ orbitals. For the Nb_3 trimer, the three Nb atoms share seven electrons, leading to a magnetic moment of 1 μ_B. Accordingly, monolayer Nb_3I_8 is spin polarized. After examining all the possible magnetic configurations, the ferromagnetic state is estimated to be the ground state. In addition, monolayer Nb_3I_8 is shown to favor an in-plane magnetization orientation. However, the magnetic anisotropy energy is only 0.5 meV, which can be easily tuned to the out-of-plane direction via external approaches. Therefore, in this work, Nb_3I_8 with out-of-plane magnetization is discussed.

The spin-polarized band structure of monolayer Nb_3I_8 without considering SOC shows that the up-spin and down-spin channels are separated significantly, leaving the up-spin bands near the Fermi level. Moreover, monolayer Nb_3I_8 exhibits a semiconducting feature with a direct band gap, with the band edges locating at the $+K$ and $-K$ points. Accordingly, monolayer Nb_3I_8 has a pair of valleys locating at the $+K$ and $-K$ points in both the conduction and valence bands. By further taking SOC into consideration, the valley degeneracy between the $+K$ and $-K$ valleys is broken, resulting in the spontaneous valley polarization. Such fascinating spontaneous valley polarization is related to the combined effect of strong SOC strength and the magnetic exchange interaction. In addition to the two pairs of valleys near the Fermi level, there is another pair of valleys locating in the second-lowest conduction band. The valley polarization is found to be −95 meV. Besides, the valley feature of monolayer Nb_3I_8 is demonstrated to be sensitive to the spin orientation. When spin lies along in-plane direction, the valley polarization is excluded in the presence of SOC. And the spin lies along the out-of-plane direction, the valley polarization in monolayer Nb_3I_8 is realized.

Using first-principles calculations, Yang et al. showed that monolayer VSi_2N_4 [54] is a promising valleytronic material with spontaneous valley polarization. In monolayer VSi_2N_4, the V, Si, and N atomic layers are sandwiched in the sequence of N-Si-N-V-N-Si-N. Central V atom coordinates with six neighboring N atoms, forming a trigonal prismatic geometry. The lattice constant of monolayer N-Si-N-VN-Si-N is found to be 2.88 Å. It exhibits a hexagonal lattice with the D_{3h} space group. Therefore, the inversion symmetry is also broken in nature. The calculated phonon spectra show that monolayer VSi_2N_4 is stable. Different from the synthesized $MoSi_2N_4$ and WSi_2N_4, monolayer VSi_2N_4 is spin polarized. The bond angle of the V-N-V is found to be 90.26°. On the basis of the Goodenough-Kanamori-Anderson rule, the exchange coupling between the V atoms is dominated by the ferromagnetic superexchange coupling. After examining the relative stability of the ferromagnetic and antiferromagnetic coupling using first-principles calculations, the ferromagnetic state is more favorable than the antiferromagnetic state by 29.70 meV. For examining its magnetization easy axis, the magnetocrystalline anisotropy energy of monolayer VSi_2N_4 is calculated, which is found to favor the in-plane magnetization. Therefore, in principle, monolayer VSi_2N_4 is a 2D XY magnet. In such system, the magnetic order can be stablished at the finite size limit, and the Curie temperature is 307 K.

Due to the local character of the V-d orbital, the PBE+U method is employed here. When $U_{eff} = 0$ eV, monolayer VSi_2N_4 exhibits a metallic feature. By increasing the U_{eff} to 3 eV, it transforms into a semiconductor with a direct gap. The valence band maximum and conduction band minimum both locate at the $+K$ and $-K$ points, indicating the valley features. Figure 2.7a displays the band structure of monolayer VSi_2N_4 with including SOC. It can be seen that the valley degeneracy is preserved. Upon tuning the orientation of magnetization from in-plane to out-of-plane, a valley polarization of 63.11 meV is achieved in monolayer VSi_2N_4, as shown in Figure 2.7b. It should be noted that the energy barrier for tuning the magnetization orientation is estimated to be 63.99 μeV. This is not so feasible in experiment. To this end, the strain effect is applied to monolayer VSi_2N_4. Under the strain of 4–8%, the magnetization orientation is tuned from in-plane to out-of-plane. Under such condition, the valley polarization can be achieved in monolayer VSi_2N_4.

The last candidate material for realizing spontaneous valley polarization that we introduce here is monolayer GdI_2. Using first-principles calculations, Feng et al. reported that monolayer GdI_2 is a 2D valleytronic material with spontaneous valley polarization [55]. Bulk GdI_2 is a ferromagnetic material with a large magnetic moment of 7.33 μ_B per Gd atom [56–58]. It exhibits a layered structure with van der Waals interaction dominating the interlayer coupling. To examine the experimental feasibility of the exfoliation of monolayer GdI_2, the exfoliation energy is calculated.

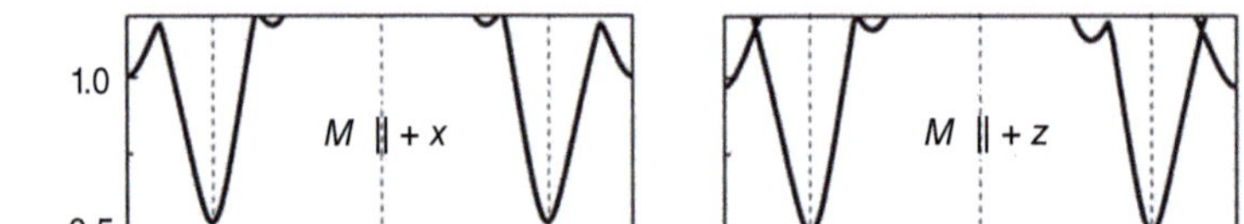

Figure 2.7 The band structure of monolayer VSi_2N_4 with considering both spin polarization and SOC when the magnetization is along (a) $+x$ and (b) $+z$. E_F is set to 0 eV. Source: Cui et al. [54]/with permission of American Physical Society.

The exfoliation energy is found to be 0.33 J m^{-2}, suggesting that it is experimentally achievable. The crystal structure of monolayer GdI_2 is similar to that of monolayer H phase MoS_2. The lattice constant is optimized to be 4.099 Å. The thermal and dynamical stability of monolayer GdI_2 is confirmed using the phonon spectra calculations and molecular dynamic simulations. In monolayer GdI_2, each Gd atom denotes two electrons to the I atoms, resulting in the electronic configuration of $4f^7 5d^1$. This leads to a magnetic moment of 8 μ_B per Gd atom. After investigating all the possible magnetic orders, the ferromagnetic ground state is established for monolayer GdI_2.

A significant spin splitting is observed in the band structure of monolayer GdI_2, forming an indirect band of 0.589 eV. Interestingly, the band edges of the conduction and valence bands come from different spin channels, which suggests that monolayer GdI_2 is a bipolar magnetic semiconductor. The valence band maximum lies at the K point, while the conduction band minimum locates at the M point. Therefore, there are two valleys locating at the $+K$ and $-K$ points in the valence band, suggesting monolayer GdI_2 a valleytronic material. When considering SOC, the valley degeneracy in the valence band is lifted, giving rise to a spontaneous valley polarization of 149 meV in monolayer GdI_2. This large valley polarization is equivalent to a magnetic field of 745–1490 T. And by reversing the magnetization orientation, the valley polarization is also reversed. Such sizeable valley polarization is attributed to the joint effects of the intrinsic magnetic exchange field and strong SOC strength.

Currently, several 2D valleytronic materials have been proposed, which, however, is still rare. And a large part of these candidate materials suffer from the in-plane magnetization. Therefore, candidate materials with spontaneous valley polarization, especially with large valley polarization and experimental feasibility, are still under exploration.

2.1.2 Valley Polarization by Foreign Atom Doping

To realize valley polarization, foreign atom doping is an effective approach as atom doping technologies have been well established in experiment. Based on this point, Schwingenschlögl et al. investigated the valley physics in Mn-doped monolayer MoS_2 on the basis of first-principles calculations [59]. Concerning monolayer MoS_2, it is considered as one of the most promising valleytronic material, but its inequivalent valleys are energetically degenerate. Therefore, external approaches are needed to lift this degeneracy. Mn atom is known as a magnetic transition metal atom, which is usually used to induce spin polarization. For the Mn_{Mo} configuration, the total magnetic moment is 1 μ_B, which can be attributed to the fact that Mn has one more electron than Mo atom. While for the Mn_S configuration, the total magnetic moment is 1 μ_B. The spin-polarized band structure of monolayer MoS_2 without considering SOC exhibits that its band edges locate at the $+K$ and $-K$ points. The spin-polarized band structure of Mn_{Mo} configuration without considering SOC shows that the bands are spin polarized. By taking SOC into consideration, in Mn_{Mo} configuration, the valley polarization is realized. When the magnetization orientation is inverted, the valley polarization is inverted. There are some impurity bands located in the band gap, which are occupied by the state from Mn.

Although valley polarization is realized in Mn-doped MoS_2, the valleys lie far away from the Fermi level. In this case, the valleys are not applicable for practical applications. Is it quite natural to consider if the valley polarization can be realized in Cr-doped MoS_2? If yes, would the valleys locate at around the Fermi level? Schwingenschlögl et al. then systematically investigated the valleytronic properties of Cr-doped MoS_2 [60]. Despite the isoelectronic nature of Cr and Mo atoms, in the Cr-doped MoS_2, the magnetic moment is found to be 2.1 μ_B per Cr atom. This is sought into the fact that Cr^{4+} has two valence electrons. And such spin polarization is related to the electronic correlations, as repeated calculations without considering the on-site Coulomb interaction give a non-spin-polarized state. In fact, the introducing of Cr atom in monolayer MoS_2 not only results in the magnetic moment on the Cr site but also impacts the atoms neighboring it significantly. And the total magnetic moment of the Cr-doped MoS_2 is 0 μ_B. When doping the V rather than the Cr atom, a total magnetic moment of 1 μ_B is obtained. That is because V has one more electron with respect to Mo.

The band structure of monolayer MoS_2 is shown in Figure 2.8a. It shows a direct band gap with the band edges locating at the $+K$ and $-K$ point. This results in two pairs of valleys in the conduction and valence bands. In the valence band, the SOC-induced valley spin splitting is significant, which is estimated to be 143 meV. While for the conduction band, the valley spin splitting is neglectable. The band structure of Cr-doped MoS_2 without considering SOC is displayed in Figure 2.8b. This allows to identify the states from different spin channels. There are three up-spin states locating above the Fermi level and one down-spin state locating below the Fermi level. Figure 2.8c shows the band structure of Cr-doped MoS_2 with considering SOC. It only slightly differs from that without considering SOC except for the $+K$ and $-K$ points in the valence band. The valley polarization is realized in Cr-doped MoS_2, as shown in Figure 2.8c. However, the valleys are separated by one flat band from the Fermi level. Obviously, this is not what we want. Then Schwingenschlögl et al. investigated the valleytronic properties of V-doped MoS_2 [60].

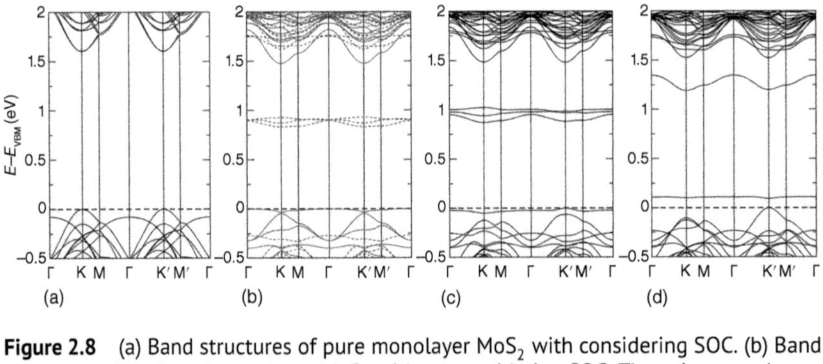

Figure 2.8 (a) Band structures of pure monolayer MoS_2 with considering SOC. (b) Band structures of Cr-doped monolayer MoS_2 without considering SOC. The spin-up and spin-down channels are indicated by the dashed and solid lines, respectively. (c) Band structures of Cr-doped monolayer MoS_2 with considering SOC. (d) Band structures of V-doped monolayer MoS_2 with considering SOC. E_F is set to 0 eV. Source: Singh and Schwingenschlögl [60]/with permission of John Wiley & Sons.

Figure 2.8d shows the band structure of V-doped MoS_2. Compared with the case of Cr-doped case, the valley polarization is preserved. And more importantly, the polarized valleys form the valence band edges, benefiting for practical applications. The valley polarization in V-doped MoS_2 is 120 meV, which is sizeable for further exploiting the valley physics.

For the doping atoms, they can also be adsorbed on the surface, which might induce spin polarization as well and then break the time-reversal symmetry. According to this scheme, Qi et al. investigated the valleytronic properties of monolayer MoS_2 when Sc, Mn, Fe, and Cu atoms are adsorbed on the surface [61]. Figure 2.9 shows the band structures of monolayer MoS_2 adsorbed by Sc, Ti, V, Cr, Mn, Fe, Co, and Cu atoms with considering SOC, where the magnetization orientation is along the out-of-plane direction. For the adsorption of Sc, Mn, Fe, and Cu atoms, one valley is higher than the other valley, leading to the valley polarization. While for the cases of Ti, V, Cr, and Co, the bands from the transition-metal (TM) atom hybridize with the valence band of the host atoms, which makes the valleys indistinguishable. For the cases of Sc and Cu, the valley spin splitting at the $+K$ point is more significant as compared with that at the $-K$ point. That is why the $+K$ valley is higher in energy than the $-K$ valley. For the cases of Mn and Fe, the spin splitting at the $-K$ point is more significant than that at the $+K$ point, resulting in the $+K$ valley lying below the

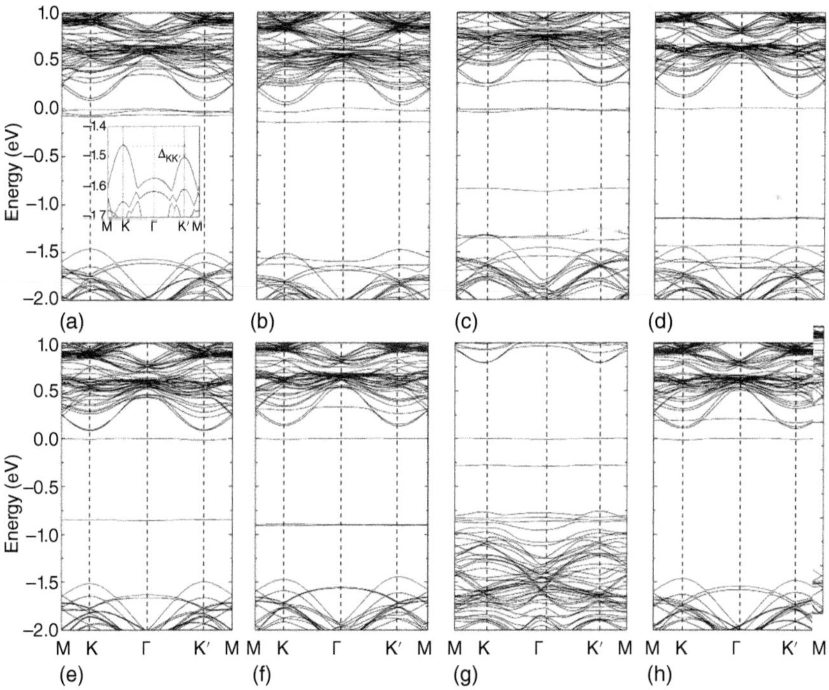

Figure 2.9 Band structures of monolayer MoS_2 adsorbed by Sc (a), Ti (b), V (c), Cr (d), Mn (e), Fe (f), Co (g), and Cu (h) atoms with considering SOC. Insert in (a) is the magnification of the bands around the valleys. E_F is set to 0 eV. Source: Chen et al. [61]. Reproduced with permission of The Royal Society of Chemistry.

$-K$ valley in energy. This discrepancy results from the opposite exchange interaction: it is positive for Mn and Fe cases and negative for Sc and Cu cases. Based on these systems, the anomalous charge and spin and valley Hall effect can be obtained [61].

The n-p codoping, with the coexistence of both n-type and p-type dopants, is shown to be a promising method for engineering the electronic properties. Using first-principles calculations, Yang et al. [62] explored the valleytronic properties of n-p codoped monolayer MoS_2 [62]. For constructing the codoping configuration, one Nb atom (Nb_1) is first introduced into monolayer MoS_2, which tends to substitute the Mo atom. Then, the second Nb atom (Nb_2) is introduced, which favors the top site above the Nb_1 atom. For the adsorption of Nb_2 atom, the adsorption energy is 3.845 eV for the pure monolayer MoS_2, and it is 3.789 eV for Nb_1 doped MoS_2. Before including SOC, the quadratic non-Dirac band dispersions can be observed at the Γ point. In this case, the system is referred to as spin-gapless semiconductor. When including SOC, such band dispersion is deformed, and a band gap is introduced. Such spin polarization combined with the broken inversion symmetry would lead to the valley polarization. It should be noted that a 3×3 supercell is used here in the calculations, which results in the band folding. By unfolding the bands, a valley polarization of 125 meV is obtained.

As in the same group of TMDs, monolayer WS_2 shares many similar properties with MoS_2, including the valley physics. To induce the valley polarization in monolayer WS_2, Zhang et al. introduced the adsorption of 3d TM atoms [63]. To determine the energetically stable configurations, various sites for the adsorption of 3d TM atoms are considered. And the site above the W atom is found to be energetically most stable. The adsorption strength is related to the electronic configuration of 3d TM atoms. For the TM atoms with half-filled and unfilled states, i.e. Sc, Ti, V, Fe, Co, and Ni, they are chemically adsorbed on monolayer WS_2. While for Cr, Mn, Cu, and Zn, the adsorption energy is small. By adsorbing TM atom on monolayer WS_2, the spin polarization would be introduced. For the cases of Sc, Ti, V, and Cr, the total magnetic moment increases from 2.98 to 6 μ_B with increasing atomic number. And because of the strong hybridization with the host atoms, the 4s electrons of the TM are transferred to the 3d orbitals, which increases the magnetic moment. For the cases of Mn, Fe, Co, and Ni, the total magnetic moment decreases from 5 to 0 μ_B with increasing atomic number. For the case of Cu, the 4s electrons are transferred to the 3d orbital of W, forming a magnetic moment of 0.9 μ_B. For the case of Zn, due to the closed 3d and 4s shells, no magnetic moment is obtained.

The spin-polarized band structure of TM adsorbed monolayer WS_2 with considering SOC is shown in Figure 2.10. It can be seen that the local energy levels appear in the band gap. Among these 3d TM atoms, the introduction of Sc, Ti, V, Cr, Mn, Fe, Co, and Cu can be considered as n-type dopants, because of the fact that the partially filled impurity levels are close to the conduction band. This would enhance the n-type mobile carrier density. Except Zn adsorbed case, the large valley spin splitting can be observed in the conduction band. More importantly, the valley spin splitting at the $+K$ and $-K$ valleys are different with considering SOC. In this regard, one valley shifts higher in energy with respect to another valley, forming the valley polarization. The values for the valley polarization in these systems range from 30.86 to

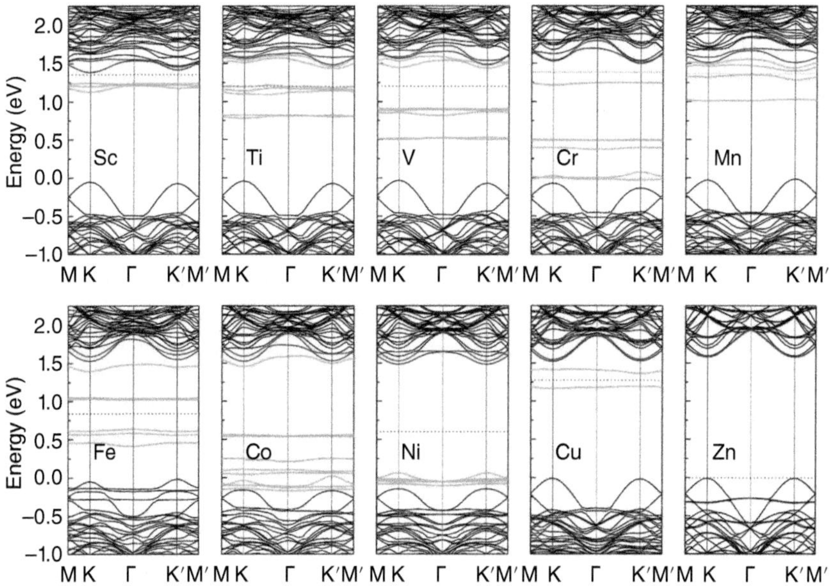

Figure 2.10 Spin-polarized band structures of TM atom doped monolayer WS_2 with including SOC. E_F is set to 0 eV. Source: Guo et al. [63]/with permission of IOP Publishing.

83.78. And the case of Ti exhibits the largest valley polarization of 83.78 meV. Such valley polarization is sought into the joint effects of SOC and time-reversal symmetry breaking.

Recently, monolayer MoSSe is shown to feature the interesting valley physics [64, 65], which are considered as a promising valleytronic material. To incorporate monolayer MoSSe into valleytronic devices, effective approach for achieving the valley polarization is necessary. To this end, using first-principles calculations, Dai et al. proposed that valley polarization can be realized in Janus monolayer MoSSe via atomic doping of V and Cr [66]. Monolayer MoSSe is a semiconductor with a direct band gap of 1.56 eV. The band edges locate at the $+K$ and $-K$ points in both the conduction and valence bands form two pairs of valleys. The valence band edge is mainly from the Mo-d_{x2-y2} and d_{xy} orbitals, while the conduction band edge is dominated by Mo-d_{z2} orbital. For pure monolayer MoSSe, it is not spin polarized. By considering SOC, the band gap decreases to 1.47 eV. And a large valley spin splitting of 169 meV occurs in the valence band, while this value is one 14 meV in the conduction band. Such discrepancy results from the different orbital contribution to the valence and conduction edges.

To break the time-reversal inversion symmetry, Cr and V atoms are selected as the dopants, as their atomic radii are close to that of Mo atom. This would not damage the crystal structure and then the valley feature significantly. The calculated binding energies for Cr and V doping are found to be 2.204 and 2.773 eV, respectively, which suggest the stability of the doped systems. After introducing the dopant, the spin polarization is induced in monolayer MoSSe. For the case of Cr, the total magnetic moment is calculated to be 0 μ_B. But, the doped Cr atom has a local magnetic moment of 2.0 μ_B, which can be attributed to the more localized character of the 3d orbital as compared with the 4d orbitals. And because of the strong hybridization with the Cr atom, the host atoms also have magnetic moments, which are aligned antiferromagnetically to the Cr atom. For the case of V, a total magnetic moment of 1.0 μ_B is obtained, which results from the fact that V has less electron with respect to Mo. And the magnetic moment is mainly localized on the V atom, and the magnetic moment localized on the host atoms are rather small.

For the spin-polarized band structure of Cr-doped monolayer MoSSe without considering SOC, although several defect levels appear in the band gap, the valley feature is preserved. The spin polarization lifts the spin degeneracy of the bands of Cr-doped MoSSe, and the $+K$ and $-K$ valleys show the same spin sign. The states of Cr atom spread in a large energy range, indicating a strong hybridization with the host atoms. Upon considering SOC, the valley degeneracy of the $+K$ and $-K$ valleys are lifted, resulting in a valley polarization of 10 meV. It should be noted that there are some defect levels locating between the valleys and the Fermi level, making it not applicable for valleytronics. For V-doped monolayer MoSSe, similar to the Cr-doped case, the valley feature is preserved when excluding SOC and a large spin splitting in the band structure is observed. There are one spin-up and one spin-down bands locating in the band gap. Importantly, the defect levels lie far away from the Fermi level. Therefore, there is no defect level locating between the valleys and the Fermi level. By considering SOC, the valley spin splitting at the $+K$ valley is weaker than that at the $-K$ valley, resulting in the valley polarization of 59 meV. And importantly, the defect levels lie far away from the valleys and the Fermi level in the presence of SOC, making it an ideal valleytronic material.

In the works mentioned earlier, the time-reversal inversion symmetry is broken by introducing magnetic metal atoms, which has been proposed as an effective way for realizing valley polarization. However, the doped metal atoms in the host materials normally tend to form clusters, which would deform the valley behaviors. In fact, spin polarization can also be induced by introducing nonmetal dopants. Using first-principles calculations, Dai et al. proposed that promising valley polarization in monolayer Tl_2O is realized through doping nonmetal atoms [67]. Monolayer Tl_2O shows a metal-shrouded structure with one O atomic layer sandwiched between two Tl atomic layers. The lattice constant of monolayer Tl_2O is optimized to be 3.51 Å. The space group of monolayer Tl_2O is P62m, and the inversion symmetry is broken. Monolayer Tl_2O is a semiconductor with an indirect band gap of 1 eV. Its valence band maximum lies at the + K and $-K$ point. For its conduction band, although the minimum locates at the Γ point, the lowest band at the $+K$ and $-K$ points also form valleys. When taking SOC into consideration, the valleys in the conduction band show a large valley spin splitting, while it is small in the valence band.

To induce the spin polarization in monolayer Tl_2O, C and N atoms are selected as the dopants because their atomic radii are close to that of O. By introducing the nonmetal dopants, the annoying cluster effect can be avoided. The calculated binding energy for doping C and N are 4.9 and 6.42 eV, respectively, which suggests that the doped systems are experimentally feasible. Due to the differing electronic configurations of C/N from that of O, the spin polarization is expected. And the calculations show that the total magnetic moment for the C-doped case is 2.0 μ_B, agreeing with the fact that C has two less electrons than O atom. While for N-doped case, the magnetic moment is only 0.37 μ_B. Such small values can be attributed to the strong hybridization between N and the host atoms. For both cases, the spin polarization is mainly distributed on the dopants. When excluding SOC, due to the existence of spin polarization, the spin-up and spin-down bands are asymmetric. However, the $+K$ and $-K$ valleys are still degenerate. The Fermi level crosses the valence band for both cases, making them semimetallic. It should be noted that the Fermi level in such systems could be easily tuned by electron doping. When considering SOC, the degeneracy between the $+K$ and $-K$ valleys are lifted for both cases, causing the valley polarization. For the N-doped monolayer Tl_2O, the valley spin splitting induced by SOC is larger than that at the $-K$ point, and the spin signs are opposite at the two valleys. The valley polarization for the N-doped monolayer Tl_2O is 23 meV. In addition, the defect levels introduced by N doping lie below the valleys, indicating that N doping does not affect the band gap of the host material. For C-doped case, a defect level lies above the valleys. But fortunately, it lies by 200 meV away from the valleys, which would not affect the valley features. The valleys polarization for C-doped case is 44 meV. The strain effect on the valley physics of these systems is also investigated, which is shown to be able to modulate the valley polarization in both systems effectively. For more information, please refer to Ref. [67].

2.1.3 Valley Polarization in van der Waals Heterostructures

Aside from foreign atom doping, magnetic proximity effect is also an effective way to break the time-reversal symmetry and thus realize valley polarization in 2D valleytronic materials. In the early stage, the research in this field is mainly focused on the bulk magnetic substrates, such as EuO, EuS, CoO, MnO, and so on [68–84]. However, the interfacial state raised from the dandling bonds of the bulk magnetic substrates usually would inhibit the valley features, which is not favorable for valleytronic applications. The recent rise of 2D magnetic semiconductors offers a promising avenue for overcoming this problem. With respect to the bulk magnets, 2D magnetic semiconductors feature a clean surface without any dangling bonds. When forming interface with 2D valleytronic materials, the weak van der Waals interaction would dominate the interlayer coupling between them. And thus the formation of excessive interface states can be avoided, beneficial for protecting the valley physics. In this section, we will review and discuss the research on the valley polarization in TMDs induced by contacting with 2D magnetic semiconductors.

Using first-principles calculations, Zhou et al. investigated the valley polarization properties of monolayer WS_2 induced by magnetic proximity effect of monolayer MnO_2 [85]. The lattice constants of monolayer t-MnO_2 and WS_2 are optimized to be 3.03 and 3.15 Å, respectively. The lattice mismatch between them is only 3.8%. Six possible stacking configurations of WS_2/MnO_2 are investigated, and the fcc-II stacking pattern, where the Mo atom sites above Mn or O atoms, are found to be most stable. To confirm the stability of the heterostructure, phonon spectra calculations are performed. The absence of imaginary frequency in the whole Brillouin zone indicates its stability. The interlayer distance is found to be 2.71 Å. More importantly, by contacting with monolayer t-MnO_2, spin polarization is induced in monolayer MoS_2, and the largest magnetic moment is 0.065 μ_B. The valley spin splitting at the $+K$ point is enhanced to 420 meV, while it is reduced to 345 meV at the $-K$ point. This suggests the valley polarization occurs in WS_2/MnO_2. For the trigonal crystal field, the d orbitals split into three groups: one singlet a_1 (d_{z2}) and two doublets e_1 (d_{xy}, d_{x2-y2}) and e_2 (d_{xz}, d_{yz}). The conduction band minimum, the O-p_z orbital is strongly hybridized with the Mn-e_1 and -e_2 orbitals. For the valence band maximum, it is dominated by W-e_1 orbitals. Accordingly, the W-e_1 orbital plays an important role in the valley spin splitting and the valley polarization.

Similar to monolayer MnO_2, monolayer h-VN is also a ferromagnetic material, but exhibiting a half-metallic feature [87]. The Curie temperature of monolayer h-VN is predicted to be 768 K. Meanwhile, its lattice constant is found to be 3.23 Å. This value is close to that of monolayer WS_2 (3.19 Å). Considering this point, Kang et al. constructed the WS_2/VN heterostructure and investigated its valleytronic properties on the basis of first-principles calculations [86]. According to the symmetry, six highly symmetric stacking configurations are considered for the WS_2/VN heterostructure. To determine the most stable configuration, the binding energies between monolayer WS_2 and h-VN are calculated. The structure with S atoms siting above V atoms and W siting above N atoms is found to be most stable. So only this configuration is considered. Figure 2.11a and b presents the band structure of WS_2/VN heterostructure. The lowest conduction band is mainly contributed by the orbitals of h-VN, while the highest valence band is mainly from the orbitals of WS_2. This suggests a typical type-II band alignment. Figure 2.11c shows the spin-projected band structure of WS_2/VN heterostructure. It can be seen that the lowest conduction band and highest valence band of WS_2 still exhibit the valley feature. However, the valley spin

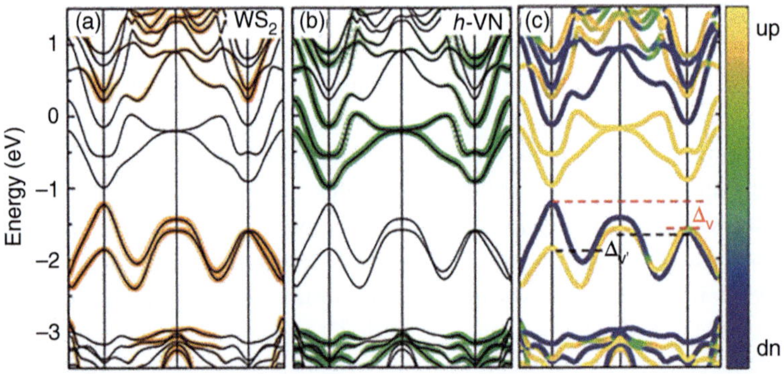

Figure 2.11 (a and b) Orbital resolved band structures of WS_2/VN heterostructure. (c) Spin resolved band structure of WS_2/VN heterostructure. E_F is set to 0 eV. Source: Ke et al. [86]/with permission of American Physical Society.

splitting in the valence band at different valleys is significantly different, which is 627 and 2 meV at $+K$ and $-K$ points. As a result, a valley polarization of 376 meV is produced in the first valence band, and a valley polarization of 249 meV is generated at the second valence band. Moreover, a valley spin splitting of 130 (141) meV is found in the conduction band of WS_2 at the $+K$ ($-K$) point, which results in the valley polarization of 148 (123) meV in the lowest (second) conduction band. As shown in Figure 2.11c, the spins at the $+K$ valley for the lowest conduction band and highest valence band are antiparallel, indicating the bright exciton state. While for the case at the $-K$ point, it is parallel, suggesting the existence of the dark exciton state. Such a feature demonstrates that the spin alignment of carriers in WS_2 can be altered by constructing WS_2/VN heterostructure. And based on these findings, Kang et al. proposed a prototype filter device for both the valley and spin [86].

Recently, the WSe_2/CrI_3 heterostructure has been fabricated through transferring mechanically exfoliated monolayer WSe_2 onto CrI_3 substrate [88, 89]. Such a heterostructure shows distinguishable energy and intensity under left and right circular polarization, which is a hallmark of the valley polarized state. And the valley polarization is estimated to be 3.5 meV. To explore the underlying physics of the valley polarization in WSe_2/CrI_3 heterostructure, Liu et al. performed systematical investigation on its valleytronic properties based on first-principles calculations [90]. To construct the WSe_2/CrI_3 heterostructure, a 2×2 supercell of monolayer WSe_2 is used to match one unit cell of monolayer CrI_3. To determine the most stable structure of WSe_2/CrI_3 heterostructure, three high symmetrical stacking patterns are considered. In the C-1 configuration, one Cr atom sites above one Se atom. In C-2 configuration, one Cr atom sites above one W atom. In C-3 configuration, two Cr atoms site above one Se and one W atoms, respectively. The calculated binding energies show that there is only a small energy barrier between these three structures. Moreover, these three configurations share similar band structures.

Taking C-3 as an example, in WSe_2/CrI_3 heterostructure, the valley features are well preserved. Moreover, the direct band gap features of WSe_2 at the $+K$ and $-K$ points are also preserved. After contacting with the ferromagnetic substrate of CrI_3, the valley degeneracy between the $+K$ and $-K$ valleys is lifted, resulting in the valley polarization. The magnetic moment on the W atom is found to be nearly zero, which suggests the interaction with CrI_3 is rather weak. This is similar to the case where the valley polarization is induced by external magnetic field, which is favorable for practical valleytronic applications. Furthermore, the electrostatic potential gradient is also an important factor for lifting the valley degeneracy. And previous work has shown that under an electrical field, the valley polarization induced by magnetic field can be significantly enhanced [91]. In this heterostructure, there indeed is an interlayer electrostatic potential gradient, generating a vertical electrical field. However, the interlayer electric field is rather weak in WSe_2/CrI_3 heterostructure.

Besides Ref. [90], Ren et al. also investigated the valleytronic properties of WSe_2/CrI_3 heterostructure [92]. A valley polarization of 2 meV is obtained in WSe_2/CrI_3 heterostructure. They also attribute this feature to the joint effect of time-reversal symmetry breaking and inversion symmetry breaking. Moreover, they show that the valley polarization is insensitive to the stacking patterns of WSe_2/CrI_3

heterostructure. By reversing the magnetization in the CrI_3 layer, the valley splitting and polarization at the $+K$ and $-K$ points are switchable.

Besides monolayer WSe_2, monolayer CrI_3 is also found to be a suitable ferromagnetic substrate to induce valley polarization for monolayer $AgBiP_2Se_2$. Using first-principles calculations, Cheng et al. predicted monolayer $AgBiP_2Se_2$ is a 2D valleytronic material with large valley spin splitting in the conduction band, and find that the valley polarization can be induced by contacting with monolayer CrI_3 [93]. Bulk $AgBiP_2Se_6$ exhibits a lamellar structure and is an antiferroelectric semiconductor [94]. By reducing the layer thickness to monolayer, it transforms into a ferroelectric semiconductor with out-of-plane electric polarization. The lattice constant of monolayer $AgBiP_2Se_6$ is found to be 6.754 Å. Monolayer $AgBiP_2Se_6$ presents an indirect band gap of 1.22 eV, with the valence band maximum locating at the $+K$ and $-K$ points and conduction band minimum locating at the Γ point. For the conduction bands at the $+K$ and $-K$ points, there is also an energy extrama. Therefore, there are two pairs of valleys in the valence and conduction bands of monolayer $AgBiP_2Se_6$. Figure 2.12a shows the orbital resolved band structure of monolayer $AgBiP_2Se_6$ without considering SOC. It can be seen that the valence band edges are mainly from the Ag orbitals, while the conduction band edges are dominated by the Bi orbitals. Hence, the SOC strength within the conduction band would be larger than that in the valence band. When taking SOC into consideration, as shown in Figure 2.12b, the valleys in the conduction band experiences a significant spin splitting of 472 meV. And due to the time-reversal symmetry, the spins at $+K$ and $-K$ valleys are opposite.

The $AgBiP_2Se_2/CrI_3$ heterostructure is constructed with the unit cells, resulting in a lattice mismatch of 3.6%. The phonon spectra calculations are carried out, and only tiny imaginary frequency appears at the Γ point, which confirms the stability of $AgBiP_2Se_2/CrI_3$ heterostructure. Figure 2.12c and d shows the band structures of $AgBiP_2Se_2/CrI_3$ heterostructure under two different ferroelectric polarizations. When the ferroelectric polarization pointing to the interface, the spin splitting at

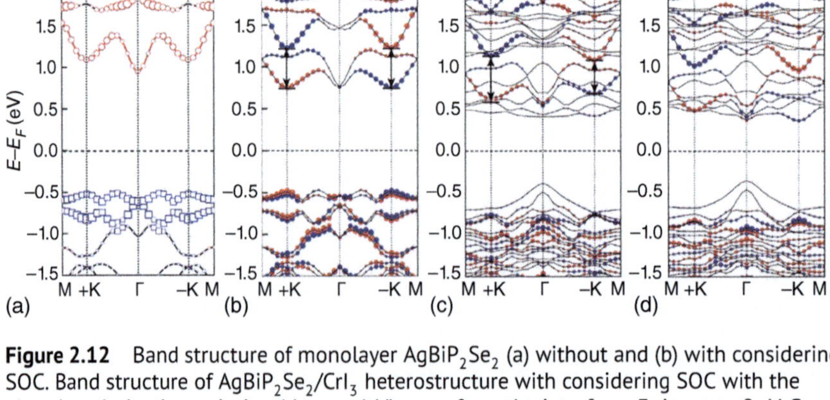

Figure 2.12 Band structure of monolayer $AgBiP_2Se_2$ (a) without and (b) with considering SOC. Band structure of $AgBiP_2Se_2/CrI_3$ heterostructure with considering SOC with the electric polarization pointing (c) to and (d) away from the interface. E_F is set to 0 eV. Source: Pei et al. [93]/with permission of American Chemical Society.

the $+K$ valley in the conduction band is enhanced to 551 meV and the spin splitting at the $-K$ valley in the conduction is reduced to 359 meV. In this regard, the valley polarization occurs in $AgBiP_2Se_2/CrI_3$ heterostructure. Though valley polarization is obtained in this case, the lowest conduction band locates at the Γ point, which would deform the valley physics, making it not applicable. Interestingly, by reversing the ferroelectric polarization, the lowest conduction band transforms from the Γ point to the $-K$ point, see Figure 2.12d. In this case, the valley physics is preserved in the conduction band, enabling the anomalous valley Hall effect. Therefore, the valley polarization in $AgBiP_2Se_2/CrI_3$ heterostructure can be controlled via external electric field [93], which is promising for developing controllable valleytronic devices.

Considering the positive role played by monolayer CrI_3 for realizing valley polarization, Xu et al. systematically investigated the valleytronic properties of the heterostructures constructed by monolayer TMD (MoS_2, $MoSe_2$, $MoTe_2$, WS_2, and WSe_2) and CrI_3 [95]. Moreover, the monolayer $CrBr_3$ is also considered as the ferromagnetic substrate. In total, 10 systems are investigated, including MoS_2/CrI_3, $MoSe_2/CrI_3$, $MoTe_2/CrI_3$, WS_2/CrI_3, WSe_2/CrI_3, $MoS_2/CrBr_3$, $MoSe_2/CrBr_3$, $MoTe_2/CrBr_3$, $WS_2/CrBr_3$, and $WSe_2/CrBr_3$. For constructing the heterostructure, a unit cell of the ferromagnetic substrate is employed to match a 2×2 supercell of 2H-phase TMD monolayer. For all these systems, the ferromagnetic properties monolayer $CrBr_3$ or CrI_3 are well preserved. The calculations on the magnetic anisotropy energy show that the magnetic easy axes of all these systems favor the out-of-plane direction.

WSe_2/CrI_3, $MoTe_2/CrI_3$, $WSe_2/CrBr_3$, and $MoTe_2/CrBr_3$ without considering SOC feature a semiconducting state with a much smaller band gap as compared with those of their respective monolayers. And a similar character is also found for $MoSe_2/CrBr_3$, $MoS_2/CrBr_3$, $WS_2/CrBr_3$, WS_2/CrI_3, MoS_2/CrI_3, and $MoSe_2/CrI_3$ heterostructures. When taking SOC into consideration, the band gaps of WSe_2/CrI_3 and $MoTe_2/CrI_3$ are further decreased. While for $WSe_2/CrBr_3$ and $MoTe_2/CrBr_3$, the introduction of SOC even induces the phase transition from insulating to metallic. In addition, the $+K$ and $-K$ valleys of monolayer TMD are well preserved in the heterostructures, and the valleys are from the Mo or W. For pure monolayer TMD, the $+K$ and $-K$ valleys are degenerate in energy when excluding SOC. After including SOC, the spin degeneracies at the $+K$ and $-K$ valleys are broken, and the highest valence bands at $+K$ and $-K$ points feature opposite spins. By interfacing with the ferromagnetic substrate of monolayer $CrBr_3$ or CrI_3, the spin degeneracies at the $+K$ and $-K$ points are further broken, leading to the valley polarization. Remarkably, the valley polarization of $WSe_2/CrBr_3$ and $MoTe_2/CrBr_3$ are found to be ~15.2 and 28.7 meV, respectively. This suggests that $WSe_2/CrBr_3$ and $MoTe_2/CrBr_3$ are suitable candidates to generate large valley polarization.

Using first-principles calculations, Dai et al. employed monolayer $CrCl_3$ as the ferromagnetic substrate to induce valley polarization in monolayer 2H-CrX_2 (X = S, Se) [96]. Monolayer 2H-CrS_2 ($CrSe_2$) is found to be a 2D valleytronic semiconductor with a direct band gap of 0.93 (0.75) eV locating at the $+K$ and $-K$ points. This results in two pairs of valleys, respectively, in the valence and conduction bands.

For monolayer 2H-CrS_2 ($CrSe_2$), the including of SOC induces a valley spin splitting of 68.5 (90.6) meV in the valence band. While for the conduction band, the valley splitting is extremely small, which is found to be 4.1 and 15.4 meV, respectively. This difference is attributed to the different orbital contributions of the valence and conduction band edges. Given the large valley splitting, monolayer 2H-CrX_2 is considered as a promising 2D valleytronic material. For realizing valley polarization in monolayer 2H-CrX_2, monolayer $CrCl_3$ is selected by considering two facts. First, monolayer $CrCl_3$ is a ferromagnetic semiconductor with large band gap of 1.51 eV. Second, monolayer $CrCl_3$ has a lattice constant close to that of monolayer 2H-CrX_2. When forming the heterostructures, the valley polarization is successfully obtained. Additionally, the valley polarization is found to be dependent on the stacking patterns.

Similar to transition metal trihalides, monolayer TM dihalides are also discovered as 2D ferromagnetic semiconductors, which hold potential to be used as ferromagnetic substrates to realize valley polarization. One typical example is monolayer $NiCl_2$. Monolayer $NiCl_2$ can be fabricated from its layered bulk counterpart. And it is a promising 2D ferromagnetic semiconductor with a band gap of 2.4–2.8 eV [97, 98]. The Curie temperature of monolayer $NiCl_2$ is 120 K [99]. Using first-principles calculations, Teng et al. employed monolayer $NiCl_2$ as the ferromagnetic substrate to induce the valley polarization in WSe_2 [100]. For the $WSe_2/NiCl_2$ heterostructure, six possible configurations are considered. The calculated binding energy shows that the T1, T2, T5, and T6 configurations are lower in energy with respect to the T3 and T4 configurations. The phonon spectra of T1, T2, T5, and T6 configurations present no negative values in the whole Brillouin zone, indicating their dynamical stability. Therefore, only these four configurations are considered for $WSe_2/NiCl_2$ heterostructure.

As these four configurations share similar band structure, only the band structure of T5 configuration for $WSe_2/NiCl_2$ heterostructure is discussed. With interfacing with the ferromagnetic substrate of monolayer $NiCl_2$, the valley polarization is realized. And by reversing the magnetization orientation, the valley polarization is also reversed, but the magnitude of the valley polarization is preserved. The obtained valley polarization for T1, T2, T5, and T6 configurations are 1.17, 9.13, −11.87, and −11.01 meV, respectively. As the valley polarization in T2, T5, and T6 configurations are significantly larger than that of T1 configuration, the interlayer stacking pattern can be used to tune the valley physics in $WSe_2/NiCl_2$ heterostructure. And to further increase the valley polarization in monolayer WSe_2, Teng et al. proposed to increase the heterostructure from two to three layers, namely, a WSe_2 layer is sandwiched between two layers of $NiCl_2$. In this case, the valley polarization indeed is increased significantly.

Using first-principles calculations, Tang et al. [101] and Yin et al. [102] also employed monolayer transition metal dihalides as the ferromagnetic substrates to induce valley polarization in monolayer TMD. Tang et al. [101] showed that the magnitude of the valley polarization in the heterostructures is related to the strength of the magnetic proximity effect, namely depending on the interlayer charge transfer and Coulomb interaction. For a given heterostructure, the type-III configuration features a large valley polarization as compared with that of a type-I or type-II band alignment. Yin et al. [102] found that biaxial strain can induce a transition of the band alignment from type-II to type-I, which in turn influences the valley physics. Besides binary ferromagnetic systems, ternary ferromagnetic systems are also used as substrates to achieve the valley polarization, and similar results are obtained [103–106].

2.2 Rashba Effects

From the preceding sections, we can see that the valley physics is correlated with SOC. In this section, we will discuss another SOC-related phenomena – Rashba effect. Rashba effect was first proposed in seminal works discussing the SOC-induced band splitting in wurtzite semiconductors, such as CdS and CdSe [107, 108]. Later, the Bychkov–Rashba model is reported to describe the motion of electrons in the 2D electron gas system with a potential gradient [109]. Rashba effect generally is aroused from potential gradient induced by structure inversion asymmetry. Under such a potential gradient (E_z), an electron with k and spin σ would experience an effective magnetic field [109, 110], which can be represented by the Rashba Hamiltonian:

$$H_R = \lambda\sigma \cdot (E_Z \times k)$$

Here, λ is the coupling constant. As a result of this effect, the spin degenerated parabolic bands would split into two bands with opposite spin-polarized states. And the corresponding energy values are:

$$E^{\pm}(k) - \frac{\hbar^2 k^2}{2m^*} \pm \alpha_R |k|$$

Here, m^* is the effective mass of electrons, and α_R is the Rashba parameter that indicates the strength of Rashba effect.

The gate tunability of Rashba effect makes it possible to electrically control the spin degree of freedom, without needing the magnetic field or materials [111, 112]. One typical example is the Datta–Das spin field effect transistor, which is proposed by Datta et al. [113]. In such a transistor, the spin precession in the 2D electron gas can be engineered via Rashba effect. Since then, great efforts have been devoted to the research on Rashba effect, and many systems have been demonstrated to be with the Rashba effect, such as semiconductor systems [114–119], perovskite oxides [120–122], the metal surfaces [123–127], and so on. Recently, the discovery of Rashba effect in 2D systems attracts rapidly increasing attention as the strength of Rashba effect is significantly enhanced in 2D systems. The realization of Rashba effect in 2D materials offers opportunities for promising spintronic applications and hosting nontrivial phenomena. In this section, we will discuss the recent theoretical progress of the Rashba effect in 2D materials.

Using first-principles calculations, Freeman et al. found that a remarkable Rashba spin splitting can be produced in monolayer $LaOBiS_2$ [128]. By excluding SOC, the band structure of monolayer $LaOBiS_2$ shows a semiconducting nature with a direct band gap locating at the X point. The bands in the low energy area are dominated by the BiS_2 layers. In this regard, monolayer $LaOBiS_2$ can be considered as a quantum well, in which the La_2O_2 layer behaves like a potential barrier and thus prevents the overlap of the wave functions from the two BiS_2 layers. This results in that the two highest valence bands (HVB-1, 2) and two lowest conduction bands (LCB-1, 2) are twofold degenerate, and the degenerated states are separate from different BiS_2 layers.

In fact, the La_2O_2 layer plays an important role for producing the Rashba effect in monolayer $LaOBiS_2$. In monolayer $LaOBiS_2$, the bonding between La_2O_2 layer and BiS_2 layers is ionic, which enables the BiS_2 layers under polar electric field with opposite directions. This makes the two BiS_2 layers individually exhibit a Rashba effect. There are two polar fields with opposite directions. The Rashba spin splitting is attributed to the two polar fields. Therefore, there are two sets of Rashba states localized at the top and bottom BiS_2 layers. And the two sets of state are degenerated in energy because of the inversion symmetry of monolayer $LaOBiS_2$. More importantly, the spin splitting energy of the HVBs is found to be 119 meV, resulting in a significant large Rashba parameter of 4.78 eV Å. Such large value is significantly desirable for practical applications. In addition, under an external electric field perpendicular along the z direction, the degenerate Rashba bands split, which is attributed to the Stark effect. This further confirms that the states from top and bottom BiS_2 layers are separated not only spatially but also energetically [128].

As BiS_2-based layered compounds usually feature similar properties, the Rashba effect might be expected in other BiS_2-based systems. To this end, Dai et al. investigated the electronic properties of monolayer $SrFBiS_2$ and $BiOBiS_2$ [129]. Different from $LaOBiS_2$, $SrFBiS_2$, and $BiOBiS_2$ are two new members of BiS_2-based compounds. They also exhibit a layered structure with the space group of P4/nmm. In each slab of $SrFBiS_2$ ($BiOBiS_2$), the interaction between Bi_2O_2 and BiS_2 (Sr_2F_2 and BiS_2) layers is ionic. The band structures of monolayer $SrFBiS_2$ and $BiOBiS_2$ are shown in Figure 2.13. When excluding SOC, as shown in Figure 2.13a and c, monolayers $SrFBiS_2$ and $BiOBiS_2$ are semiconductors with a direct gap of 0.89 and 1.00 eV, respectively. The lowest conduction band and the highest valence band around the X point are referred to as α and β, respectively. For both systems, the bands in the low energy area form the BiS_2 layers, and the SrF and BiO layers have almost no contribution. Accordingly, the SrF and BiO layers in monolayer $SrFBiS_2$ and $BiOBiS_2$, respectively, behave like the potential barrier preventing the overlap between the wave functions from the opposite BiS_2 layers. Therefore, the bands α and β are twofold degenerate without considering SOC. When taking SOC into consideration, as shown in Figure 2.13b and d, the band gaps of monolayers $SrFBiS_2$ and $BiOBiS_2$ are reduced to 0.48 and 0.50 eV, respectively. This is sought into the strong SOC strength within the constituent atoms in both systems. Besides the band gap narrowing, a spin splitting occurs for the bands α and β for both systems, indicating a Rashba-like effect.

As compared with the cases with inversion symmetry for whole structure, the identification of Rashba effect in systems with inversion asymmetry is more

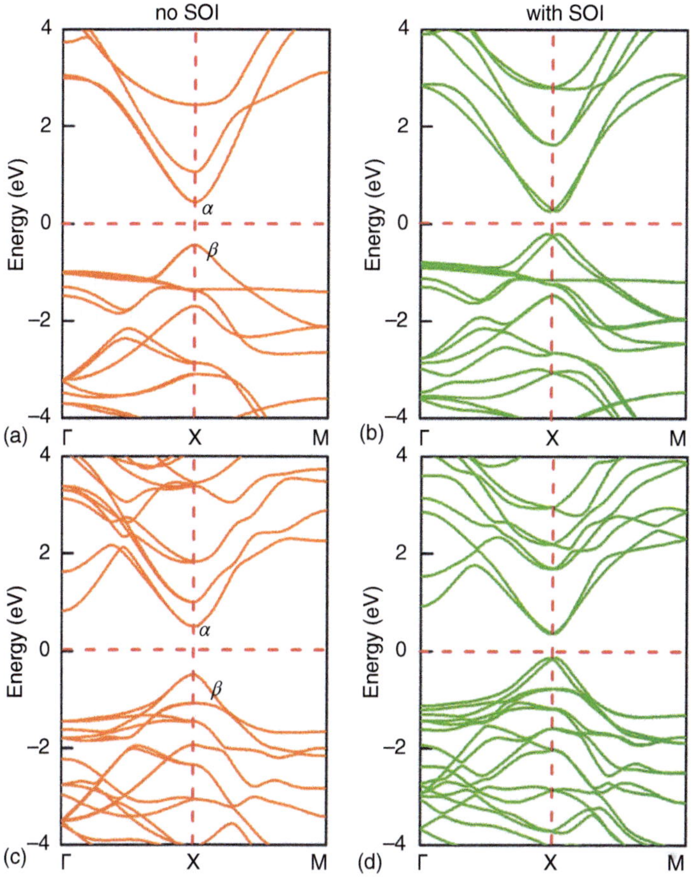

Figure 2.13 Band structures of monolayer $SrFBiS_2$ (a) without and (b) with considering SOC. Band structures of monolayer $BiOBiS_2$ (c) without and (d) with considering SOC. E_F is set to 0 eV. Source: Ma et al. [129]. Reproduced with permission of The Royal Society of Chemistry.

straightforward. One typical example is monolayer BiTeX (X = Br and I). Using first-principles calculations, Dai et al. investigated the Rashba effect in monolayer BiTeX [130]. Bulk BiTeX displays a layered structure with the P3m1 space group [131, 132]. In each slab, one Te atomic layer and one X atomic layer sandwich one Bi atomic layer, resulting in a triple layer. Each Bi atom coordinates with three Te and three X atoms, which form a distorted octahedron. Different from the strong intralayer interaction, the interlayer coupling is the van der Waals interaction. Such weak interlayer coupling offers a natural cleaving plane, and thus monolayer BiTeX can be readily obtained experimentally. Because of the broken inversion symmetry, the charge in monolayer BiTeX would distribute unevenly along the out-of-plane direction, which would lead to the potential gradient and thus electric polarization. Then the magnitude of the dipole moment in monolayer BiTeX is examined. The dipole moment in monolayer BiTeBr is significantly larger than that of the monolayer BiTeI. This is sought into the specific crystal structure of monolayer BiTeX. In detail, as compared with the Bi-I bond in monolayer BiTeI, the bong length of Bi-Br in monolayer BiTeBr is smaller. Based on the Debye equation, the dipole moments of the Bi-X bonds increase in the order of Bi-Br < Bi-I. While for the Bi-Te bonds in these two systems, the bond lengths are comparable with only a slight difference. As the Bi-X bond behaves like cancellers to the net dipole moments, the net dipole moments would increase in the order of BiTeI < BiTeBr.

Figure 2.14 shows the band structures of monolayer BiTeX. When excluding SOC, a semiconducting character with an indirect band gap is observed for both systems. For the three conduction bands shown in Figure 2.14, they mainly form the Bi-p orbitals. While for the six valence bands, they are dominated by the Te/X-p orbitals. Considering the constituent atoms of monolayer BiTeX are heavy atoms, strong SOC

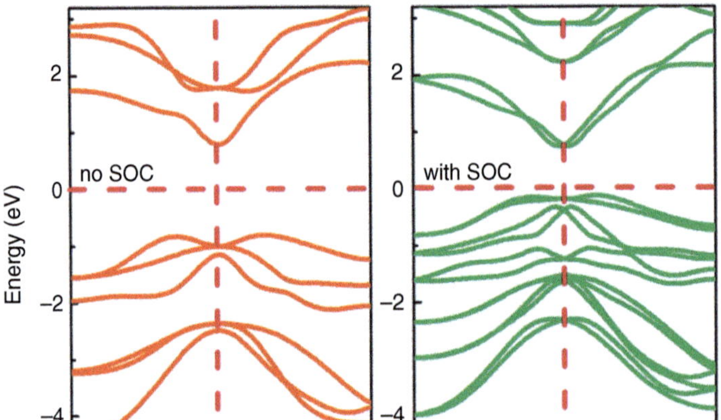

Figure 2.14 Band structures of monolayer BiTeX. The upper panels show the band structure of monolayer BiTeBr with and without considering SOC. The lower panels show the band structures of monolayer BiTeI without and with considering SOC. E_F is set to 0 eV. Source: Ma et al. [130]. Reproduced with permission of The Royal Society of Chemistry.

strength is expected in both systems. As shown in Figure 2.14, when including SOC, the band gaps are significantly reduced, which is related to the energy splitting under the SOC. In addition, as compared with the band structures without considering SOC, the conduction band minimum shift slight off the Γ point, giving rise to the Rashba spin splitting. To quantitatively describe the strength of the Rashba effect in monolayer BiTeX, the Rashba parameter is calculated. The Rashba parameter is found to be 1.31 and 1.86 meV, respectively, for monolayer BiTeBr and BiTeI eV Å. After estimating the Rashba effect in monolayer BiTeX [130], Liu et al. employed uniaxial stress to engineer the Rashba effect in monolayer BiTeI on the basis of first-principles calculations and model analysis [131]. And using first-principles calculations, Li et al. studied the effect of biaxial tensile strain on the Rashba effect of monolayer BiTeBr [132]. These two works demonstrate that applying strain is an effective way to modulate the Rashba effect of monolayer BiTeX.

Following the strategy of material discovery by substituting the chemical element Bi in monolayer BiTeI with Sb in the same group, Kent et al. obtained monolayer SbTeI [133]. In monolayer SbTeI, the atoms are stacked in the sequence of I-Sb-Te, forming a hexagonal lattice with the space group P3m1. To examine the stability of monolayer SbTeI, the formation energy with respect to the bulk structure is calculated. The obtained formation energy is found to be small, indicating that monolayer SbTeI can be synthesized. In addition to the formation energy, the phonon spectra of monolayer SbTeI is also calculated. All branches of the phonon dispersions are real, suggesting the dynamical stability. Without considering SOC, monolayer SbTeI is a semiconductor with an indirect band gap. The valence band maximum locates slightly off the Γ point, while the conduction band minimum sites at the Γ point. The orbitals in the low energy area are from the p orbital of all these three species. Upon taking SOC into consideration, as shown in Figure 2.15a, the band structure of monolayer SbTeI changes significantly. The enlarged view of the bands near the conduction band minimum are shown in Figure 2.15b and c. Obviously, the initially degenerate bands split into two, forming the Rashba spin splitting. To characterize the Rashba spin splitting in monolayer SbTeI, the Rashba parameter is calculated. Interestingly, the Rashba parameter is found to be 1.39 eV Å, which is comparable with that of BiTeI [130].

Different from the giant Rashba effect in monolayer BiTeX, the Rashba effect in monolayer MXY (M = Mo, W; X/Y = S, Se, Te) is rather weak, although it also exhibits an intrinsic polar electric field induced by its broken inversion symmetry. Using first-principles calculations, Cheng et al. predicted that the Rashba parameters in monolayer MXY are around 0.01 eV Å [64]. Apparently, this is not applicable for practical applications. To enhance the Rashba effect in monolayer MoXY, Chu et al. proposed that in-plane strain is an effective way [65]. Using first-principles calculations, Chu et al. investigated the influence of biaxial strain on the Rashba effect of monolayer MoXY. Considering the fact that the physics of monolayer MXY are essentially the same, we take monolayer WSeTe as an example to demonstrate the tunability of the Rashba effect. Figure 2.16b plots the band structure of monolayer WSeTe without applying strain. For the highest valence bands around the Γ point, they are mainly from the Se-p_z and W-d_{z2} orbitals. In order to clarify the orbital

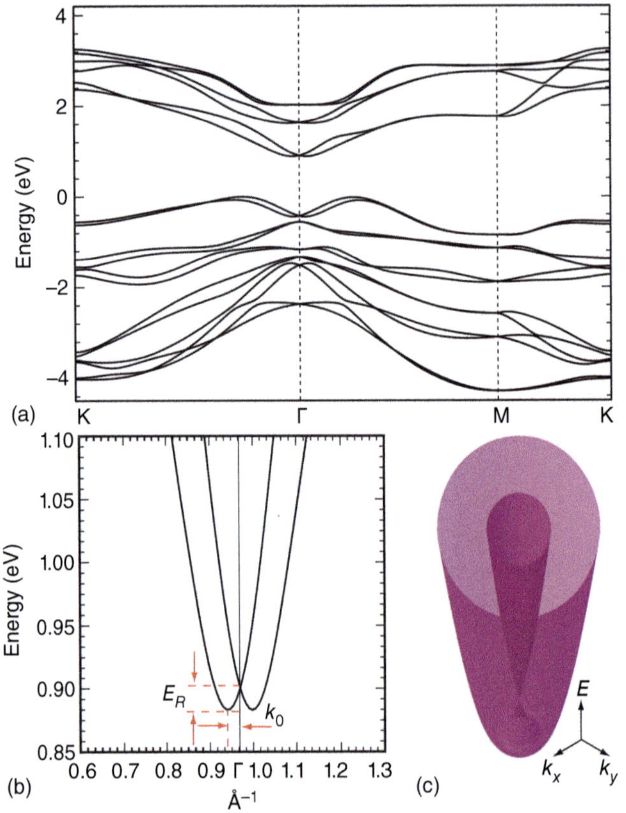

Figure 2.15 Band structures of monolayer SbTeI with considering SOC. (b) Enlarged view of the bands near the conduction band minimum. (c) 3D plot of the bands shown in (b). E_F is set to 0 eV. Source: Zhuang et al. [133]/with permission of American Physical Society.

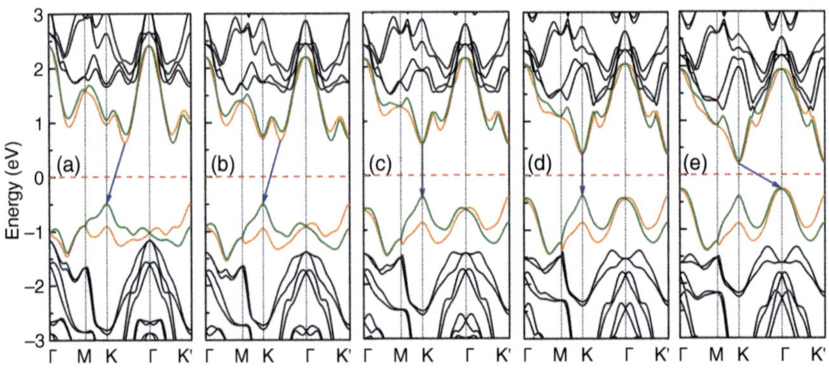

Figure 2.16 Band structures of monolayer MoSeTe with considering SOC under various biaxial strains. The fundamental band gaps are marked by the arrows. E_F is set to 0 eV. (a) −2%, (b) 0%, (c) 2%, (d) 4%, and (e) 6%. Source: Yao et al. [65]/with permission of American Physical Society.

dependence of the Rashba effect in monolayer WSeTe, the partial SOC is artificially switched on or off. When the SOC of both Se-p_z and W-d_{z2} orbitals are switched off, the Rashba spin splitting at around the Γ point can be neglected, suggesting the important role of these two orbitals. By further switching of the SOC of Se-p_z or W-d_{z2}, it is found that the latter one contributes more to the Rashba spin splitting. In fact, this agrees with fact that the SOC strength within d orbitals is stronger than that of the p orbitals.

Then, biaxial strain ranging from −2 to 6% is applied on monolayer WSeTe, and the corresponding band structures are shown in Figure 2.16. Under the tensile strain, the highest valence bands around the Γ point are shifted up, and the Rashba spin splitting energy is decreased. While for the compressive strain, the highest valence bands around the Γ point are shifted down, and the Rashba spin splitting energy is increased. Such changes are strongly related to local electric field change induced by the change of the orbital overlap between atoms. To confirm this point, the work function change under various strains is studied. Indeed, the out-of-plane dipole of monolayer WSeTe decreases with increasing tensile strain and increases with increasing compressive strain. This suggests that the overlap between the Se-p_z and W-d_{z2} orbitals can influence the local electric field, thus modulating the Rashba effect in monolayer WSeTe. Moreover, by increasing the compressive strain or decreasing the tensile strain, the Rashba parameter of monolayer WSeTe can be increased. For instance, under 2% compressive strain, the Rashba parameter can be increased to 50%, large enough for modulating the spin states. Besides the strain effect, the external electric field is also applied on monolayer WSeTe to modulate the Rashba spin splitting. However, it is found to be with a low efficiency as compared with strain [65]. In addition to Ref. [65], Ren et al. also investigated the influence of external electric field and in-plane biaxial strain on the Rashba spin splitting of monolayer MXY on the basis of first-principles calculations, and similar results are found [134]. Moreover, they show that the anisotropic character of the Rashba spin splitting of monolayer MXY can be significantly enhanced under compressive strain [134].

Based on Refs. [65, 134], the Rashba effect in monolayer MXY can be effectively modulated by external electric field and in-plane strain. However, these approaches suffer from the constant energy consumption. To address this issue, Yang et al. proposed charge doping is a new and effective way to engineer the Rashba effect in monolayer MXY on the basis of first-principles calculations [135]. And here, monolayer WSeTe is also taken as an example. Due to the difference in the electronegativity values of M, X, and Y atoms, the M atom features a positive charge, while X and Y atoms show a negative charge. And the X atom has more electrons than the Y atom. This gives rise to a local electric field in the out-of-plane direction pointing from Y to X atoms. To investigate the effect of charge doping, the charge ranging from −0.3 e to 0.3 e is doped in monolayer WSeTe. Here, the positive and negative charges correspond to dop electrons and hole, respectively. It is found that the charge doping can engineer the Rashba effect for the valence band at the Γ point and for the conduction band at the M point. For the electron doping, the Rashba effect can be enhanced. Taking the case with electron doping of 0.3 e as an example, the positive charge on

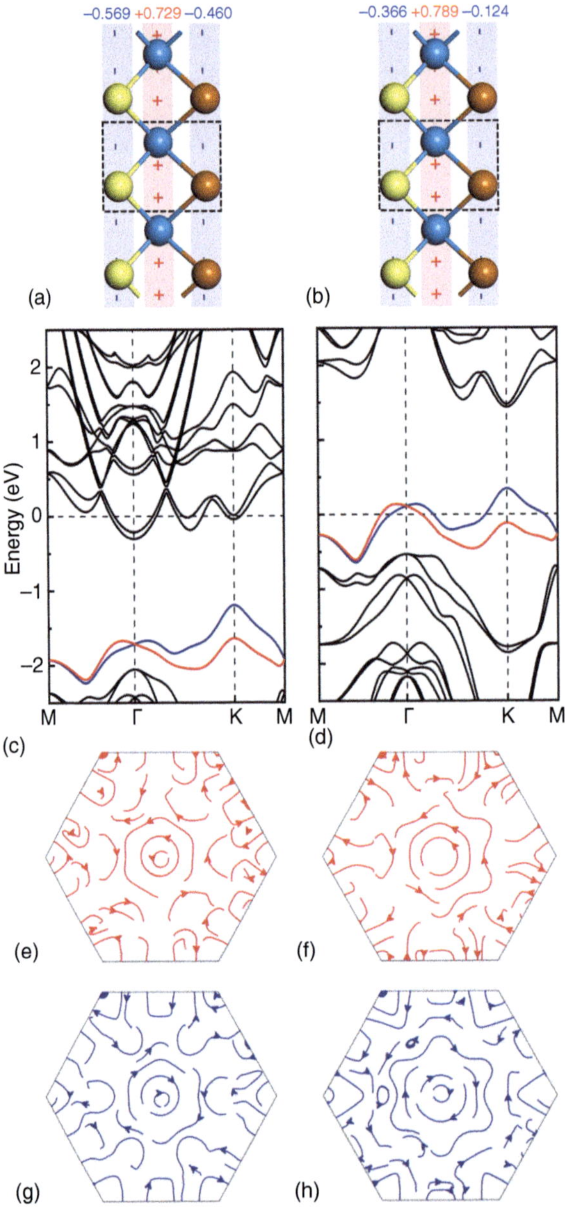

Figure 2.17 Charges on W, Se, and Te atoms for monolayer WSeTe with doping (a) electron of +0.3 e and (b) hole of −0.3 e. Band structures of monolayer WSeTe with doping (c) electron of +0.3 e and (d) hole of −0.3 e. Spin textures around the Γ point of monolayer WSeTe with doping (e and f) electron of +0.3 e and (g and h) hole of −0.3 e. E_F is set to 0 eV. Source: Chen et al. [135]. Reproduced with permission of The Royal Society of Chemistry.

W atom decreases and the negative charges on the Se and Te atoms increase with respect to the case without charge doping, see Figure 2.17a. Furthermore, the electron doping shifts the Fermi level up and the Rashba parameter is increased by 10.5%. And the Rashba spin splitting in monolayer WSeTe with electron doping of 0.3 e is further confirmed by the spin textures shown in Figure 2.17e and g. Different from electron doping, hole doping would weaken the Rashba effect for both the valence and conduction bands. Taking hole doping of −0.3 e as an example, as shown in

Figure 2.17b, the W atom has more positive charge, while the Se and Te atoms have less negative charge as compared with the neutral case. From Figure 2.17d, it can be seen that in this case the Fermi level shifts down, and the Rashba parameter is reduced by 6.1%. And the corresponding spin texture suggesting the Rashba spin splitting is shown in Figure 2.17f and h. Therefore, charge doping can effectively engineer the Rashba effect in monolayer WSeTe.

Recently, monolayer III–VI chalcogenide NX (N = Ga, In; X = S, Se, Te) has received great interest [136–139]. Monolayer NX is shown to be semiconducting with an indirect band gap. Importantly, monolayer NX has a super high in-plane stiffness value, even comparable with that of graphene [140, 141]. The field effect transistors based on monolayer NX have been fabricated by several experimental groups, and the on/off ratio and electron mobility are shown to be remarkably high [136, 142]. Similar to monolayer TMD, monolayer NX also has mirror symmetry, and the inversion symmetry is broken. Upon imposing external electric field in the out-of-plane direction, the mirror symmetry of monolayer NX has a possibility to be broken, and the Rashba spin splitting might be obtained. To verify this expectation, Ju et al. comprehensively studied the Rashba effect in monolayer NX on the basis of first-principles calculations [143]. The band structures of six systems share similar band structure, and the SOC has slight influence on the band structure. For monolayer GaTe, the conduction band minimum locates at the M point, while for other systems, the conduction band minimum sites at the Γ point. The valence band maximum of monolayer NX locates between the Γ and M point. As the inversion symmetry of monolayer NX is broken, there is a spin splitting at the K point when including SOC.

Figure 2.18 shows the band structures of monolayer NX with SOC under an external electric field of 0.9 V Å^{-1}. The valence bands at the Γ point are enlarged. It should be noted that such an electric field can be realized by using ionic liquid in

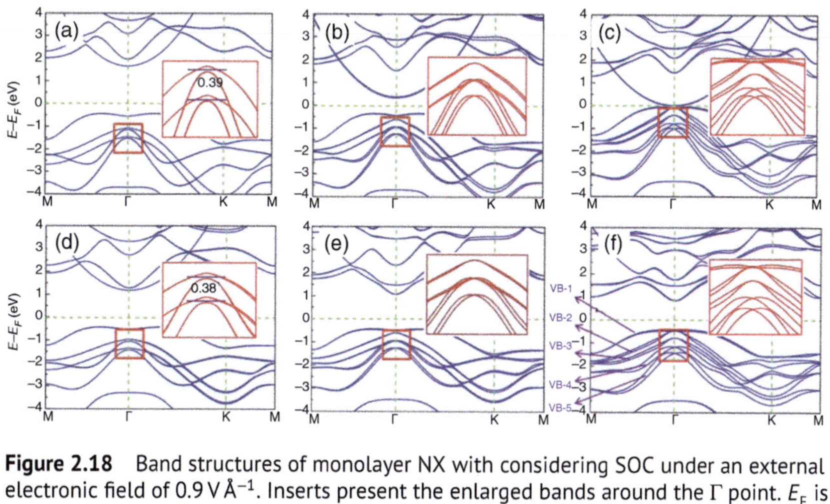

Figure 2.18 Band structures of monolayer NX with considering SOC under an external electronic field of 0.9 V Å^{-1}. Inserts present the enlarged bands around the Γ point. E_F is set to 0 eV. (a) GaS, (b) GaSe, (c) GaTe, (d) InS, (e) InSe, and (f) InTe. Source: Ju et al. [143]/with permission of IOP Publishing.

experiment. For monolayer GaS and InS, the Rashba spin splitting is not significant because of the weak SOC strength, see Figure 2.18a and d. Different from these two cases, significant Rashba spin splitting occurs in the other four systems. From the inserts in Figure 2.18, the typical Rashba spin splitting is obtained in VB-4 and VB-5 bands. And the Rashba strength of NTe is remarkably larger than that of NSe. Therefore, the external electric field indeed can induce the Rashba effect in monolayer NX. By further examining the Rashba parameters, it is found that the Rashba parameter is strongly related to the anions. For InS and GaS under the external electric field of 0.9 V Å^{-1}, the Rashba parameters are similar, almost zero. Upon further increasing the electric field, the Rashba parameters gradually increased. For GaSe and InSe under the external electric field of 0.9 V Å^{-1}, the Rashba parameters are 0.54 and 0.58 eV Å, respectively. While for GaTe and InTe under the external electric field of 0.9 V Å^{-1}, the values are 1.37 and 1.70 eV Å, respectively. The Rashba spin splitting in these systems is also confirmed by the calculated spin texture.

Although Rashba effect is successfully obtained in monolayer NX, the Rashba parameters are relatively weak. To enhance the Rashba parameters in monolayer NX, constructing a Junus structure by substitution is shown to be an effective way [144]. Using first-principles calculations, Ju et al. systematically investigated the Rashba effect in Janus monolayer group III–VI chalcogenides [144]. Monolayer NX consists of four atomic layers that are stacked in the sequence of X-N-N-X. By replacing one X layer by Y layer, the polar XAAY (A = Ga, In; X a Y S, Se, Te) ternary compounds can be obtained. And by replacing one X and one N layer, the polar XABY (A, B = Ga, In; X ≠ Y = S, Se, Te) quaternary compounds can be realized. To confirm the stability of these systems, the phonon spectra calculations are performed. And the absence of negative frequency or tiny negative frequency around the Γ point confirm the stability of these systems. Furthermore, the thermal stabilities of these systems are also confirmed by carrying out the molecular dynamic simulations. As the physical mechanisms of these systems are similar, only monolayer SeInGaTe and SInGaSe are taken as examples to discuss the Rashba effect.

Figure 2.19a and b presents the band structures without considering SOC. It can be seen that the spin degeneracy is preserved for both systems, and no spin polarization is generated in these systems. The band gaps of monolayer SeInGaTe and SInGaSe are found to be 1.57 and 1.75 eV, respectively. Upon considering SOC, as shown in Figure 2.19a and b, the bands around the Fermi level experience a spin splitting. And the Rashba spin splitting is obtained around the conduction band minimum at the Γ point. This is attributed to the broken inversion symmetry of monolayer SeInGaTe and SInGaSe. The local electric field induced by the broken inversion symmetry can be confirmed by the work function changes. As presented in Figure 2.19c and d, the work function changes, which is directly related to dipole moment, between the opposite surfaces are estimated to be 1.34 and 1.36 eV, respectively, for monolayer SeInGaTe and SInGaSe. This results in the nonzero potential gradient in the out-of-plane direction, and thus the Rashba effect in monolayer SeInGaTe and SInGaSe. The influence of external electric field on the Rashba spin splitting in the conduction bands of monolayer SeInGaTe and SInGaSe are also investigated. By applying a positive electric field (pointing from Te to Se or from

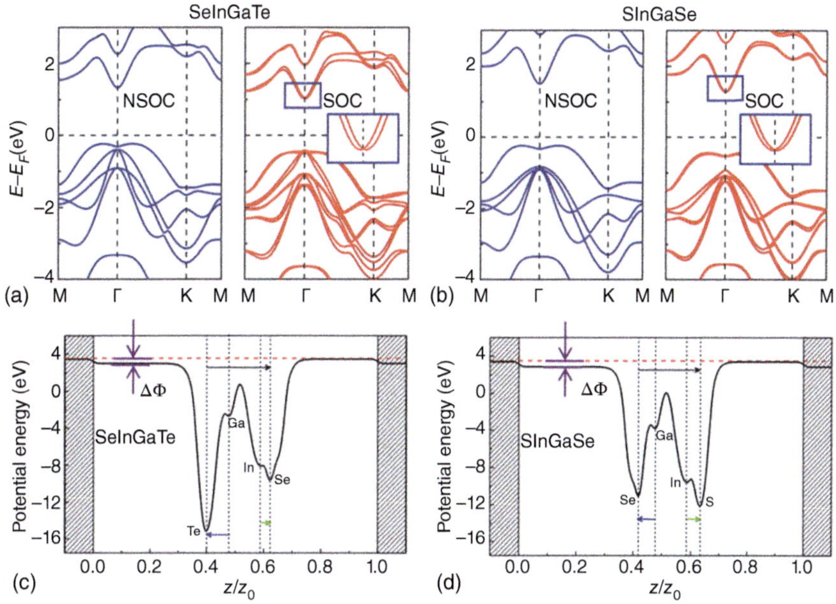

Figure 2.19 Band structures of monolayer (a) SeInGaTe and (b) SInGaSe without (blue lines) and with (red lines) considering SOC. Insets in (a and b) show the enlarged conduction bands around the Γ point. Planar average of the electrostatic potential energy of monolayer (c) SeInGaTe and (d) SInGaSe. E_F is set to 0 eV. Source: Ju et al. [144]. Reproduced with permission of The Royal Society of Chemistry.

Se to S), the Rashba spin splitting can be significantly enhanced. While for applying a negative electric field, the Rashba spin splitting can be weakened. For example, the Rashba parameter for monolayer SeInGaTe is 0.67 eV Å, which increases to 0.96 eV Å under the external electric field of 0.9 V Å^{-1} and decreases to 0.21 eV Å^{-1} under the external electric field of −0.9 V Å^{-1}. Therefore, Janus monolayer group III–VI chalcogenides hold great promises for future spintronic applications.

Another typical example for realizing Rashba spin splitting by constructing Janus structure is monolayer MA_2Z_4 family. Recently, 2D layered $MoSi_2N_4$ and WSi_2N_4 are synthesized in experiment via chemical vapor deposition (CVD), providing a new family of 2D materials [28]. And monolayer $MoSi_2N_4$ and WSi_2N_4 can be readily obtained in experiment. More importantly, this synthesis method is applicable for other similar systems with a formula of MA_2Z_4 (M = W, V, Nb, Ta, Ti, Zr, Hf, or Cr; A = Si or Ge, and Z = N, P, or As). Inspired by the synthesized method of MSi_2N_4 (M = Mo and W), Ren et al. proposed that monolayer $MSiGeN_4$ (M = Mo and W) can also be achieved by introducing Si and Ge during the CVD growth [145]. Using first-principles calculations, Ren et al. systematically studied the Rashba effect in monolayer $MSiGeN_4$. Monolayer $MSiGeN_4$ can be considered as a Si-N layer and a Ge-N layer sandwiching a MN_2 layer. Monolayer $MSiGeN_4$ exhibits a hexagonal lattice with the space group of No. 156. Therefore, the mirror symmetry is broken for monolayer $MSiGeN_4$. The optimized lattice constants of monolayer $MoSiGeN_4$ and $WSiGeN_4$ are found to be $a = b = 2.963$ and 2.964 Å,

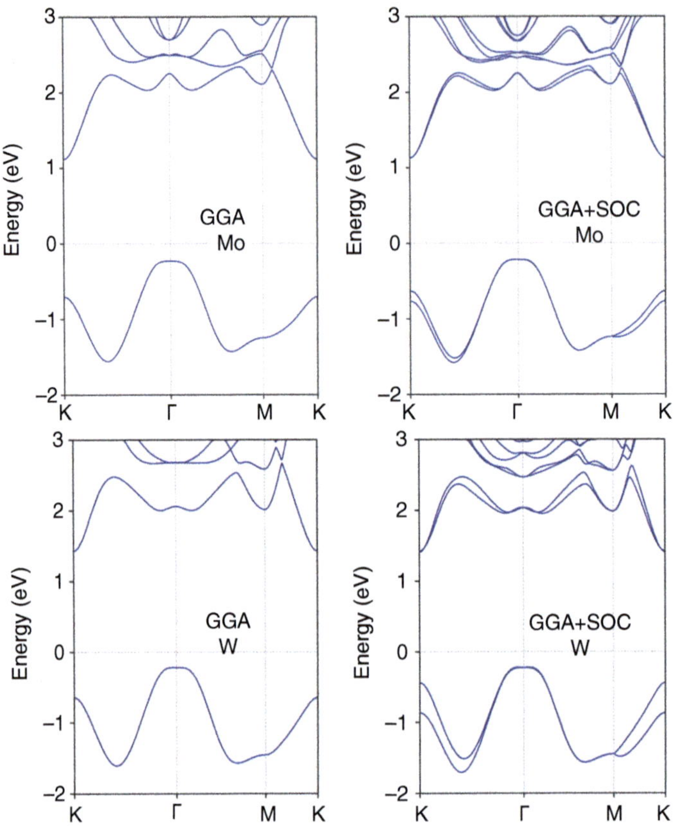

Figure 2.20 Band structures of monolayer $MSiGeN_4$ without and with considering SOC. E_F is set to 0 eV. Source: Guo et al. [145]. Reproduced with permission of The Royal Society of Chemistry.

respectively. And the dynamical, thermal, and mechanical stabilities of monolayer $MSiGeN_4$ are confirmed by the phonon spectra calculations, molecular dynamical simulations, and the Born criteria, respectively. Therefore, it is possible to realize monolayer $MSiGeN_4$ in experiment.

Figure 2.20 shows the band structure of monolayer $MSiGeN_4$. Regardless of SOC, monolayer $MSiGeN_4$ exhibits an indirect band gap. The conduction band minimum for monolayer $MSiGeN_4$ lies at the K point, and the valence band maximum locates slightly away from the Γ point. Because of the intrinsic out-of-plane electric field induced by the broken inversion symmetry, Rashba spin splitting occurs for the bands around the Γ point when taking SOC into consideration. The band gap values for monolayer $MSiGeN_4$ without and with considering SOC are very close, which are estimated to be around 1.3 and 1.6 eV for monolayer $MoSiGeN_4$ and $WSiGeN_4$, respectively. Moreover, there is a Zeeman-type spin splitting in the valence bands at the K point. The Rashba parameters for monolayer $MoSiGeN_4$ and $WSiGeN_4$ are found to be 0.033 and 0.111 eV Å, respectively. Obviously, these values are rather small. For practical application, further modulation based on external methods is needed.

References

1 Rycerz, A., Tworzydło, J., and Beenakker, C.W.J. (2007). Valley filter and valley valve in graphene. *Nat. Phys.* 3: 172–175.

2 Shkolnikov, Y., De Poortere, E., Tutuc, E., and Shayegan, M. (2002). Valley splitting of AlAs two-dimensional electrons in a perpendicular magnetic field. *Phys. Rev. Lett.* 89: 226805.

3 Gunawan, O., Shkolnikov, Y.P., Vakili, K. et al. (2006). Valley susceptibility of an interacting two-dimensional electron system. *Phys. Rev. Lett.* 97: 186404.

4 Ohkawa, F.J. and Uemura, Y. (1977). Theory of valley splitting in an N-channel (100) inversion layer of Si III. Enhancement of splittings by many-body effects. *J. Phys. Soc. Jpn.* 43: 925–932.

5 Sham, L., Allen, S., Kamgar, A., and Tsui, D. (1978). Valley–valley splitting in inversion layers on a high-index surface of silicon. *Phys. Rev. Lett.* 40: 472–475.

6 Bloss, W., Sham, L., and Vinter, V. (1979). Interaction-induced transition at low densities in silicon inversion layer. *Phys. Rev. Lett.* 43: 1529–1532.

7 Schaibley, J.R., Yu, H., Clark, G. et al. (2016). Valleytronics in 2D materials. *Nat. Rev. Mater.* 1: 16055.

8 Takashina, K., Ono, Y., Fujiwara, A. et al. (2006). Valley polarization in Si(100) at zero magnetic field. *Phys. Rev. Lett.* 96: 236801.

9 Karch, J., Tarasenko, S.A., Ivchenko, E.L. et al. (2011). Photoexcitation of valley-orbit currents in (111)-oriented silicon metal-oxide-semiconductor field-effect transistors. *Phys. Rev. B* 83: 121312.

10 Isberg, J., Gabrysch, M., Hammersberg, J. et al. (2013). Generation, transport and detection of valley-polarized electrons in diamond. *Nat. Mater.* 12: 760–764.

11 Zhu, Z., Collaudin, A., FauquØ, B. et al. (2011). Field-induced polarization of Dirac valleys in bismuth. *Nat. Phys.* 8: 89–94.

12 Novoselov, K.S., Geim, A.K., Morozov, S.V. et al. (2004). Electric field effect in atomically thin carbon films. *Science* 306: 666–669.

13 Castro Neto, A.H., Guinea, F., Peres, N.M.R. et al. (2009). The electronic properties of graphene. *Rev. Mod. Phys.* 81: 109–162.

14 Zeng, M., Xiao, Y., Liu, J. et al. (2018). Exploring two-dimensional materials toward the next-generation circuits: from monomer design to assembly control. *Chem. Rev.* 118: 6236–6296.

15 Vitale, S.A., Nezich, D., Varghese, J.O. et al. (2018). *Small* 14: 1801483.

16 Novoselov, K.S., Jiang, D., Schedin, F. et al. (2005). Two-dimensional atomic crystals. *Proc. Natl. Acad. Sci. U. S. A.* 102: 10451–10453.

17 Gunlycke, D. and White, C.T. (2011). Graphene valley filter using a line defect. *Phys. Rev. Lett.* 106: 136806.

18 Xu, X., Yao, W., Xiao, D., and Heinz, T.F. (2014). Spin and pseudospins in layered transition metal dichalcogenides. *Nat. Phys.* 10: 343–350.

19 Xiao, D., Liu, G.-B., Feng, W. et al. (2012). Coupled spin and valley physics in monolayers of MoS_2 and other group-VI dichalcogenides. *Phys. Rev. Lett.* 108: 196802.

20 Mak, K.F., He, K.L., Shan, J., and Heinz, T.F. (2012). Control of valley polarization in monolayer MoS_2 by optical helicity. *Nat. Nanotechnol.* 7: 494–498.

21 Cao, T., Wang, G., Han, W. et al. (2012). Valley-selective circular dichroism of monolayer molybdenum disulphide. *Nat. Commun.* 3: 1–5.

22 Zhu, Z.Y., Cheng, Y.C., and Schwingenschlögl, U. (2011). Giant spin-orbit-induced spin splitting in two-dimensional transition-metal dichalcogenide semiconductors. *Phys. Rev. B* 84: 153402.

23 Ma, Y., Kou, L., Du, A. et al. (2018). Conduction-band valley spin splitting in single-layer H-Tl_2O. *Phys. Rev. B* 97: 035444.

24 Dou, K., Ma, Y., Peng, R. et al. (2020). Promising valleytronic materials with strong spin-valley coupling in two-dimensional MN_2X_2 (M = Mo, W; X= F, H). *Appl. Phys. Lett.* 117: 172405.

25 Yang, C., Song, Z., Sun, X., and Lu, J. (2021). Valley pseudospin in monolayer $MoSi_2N_4$ and $MoSi_2As_4$. *Phys. Rev. B* 103: 035308.

26 Chang, M.C., Ho, P.H., Tseng, M.F. et al. (2020). Fast growth of large-grain and continuous MoS_2 films through a self-capping vapor-liquid-solid method. *Nat. Commun.* 11: 1–9.

27 Radisavljevic, B., Radenovic, A., Brivio, J. et al. (2011). Single-layer MoS_2 transistors. *Nat. Nanotechnol.* 6: 147–150.

28 Hong, Y.L., Liu, Z., Wang, L. et al. (2020). Chemical vapor deposition of layered two-dimensional $MoSi_2N_4$ materials. *Science* 369: 670–674.

29 Cai, Y., Zhang, G., and Zhang, Y.W. (2014). Polarity-reversed robust carrier mobility in monolayer MoS_2 nanoribbons. *J. Am. Chem. Soc.* 136: 6269–6275.

30 Zeng, H., Dai, J., Yao, W. et al. (2012). Valley polarization in MoS_2 monolayers by optical pumping. *Nat. Nanotechnol.* 7: 490–493.

31 Li, Y., Ludwig, J., Low, T. et al. (2014). Valley splitting and polarization by the Zeeman effect in monolayer $MoSe_2$. *Phys. Rev. Lett.* 113: 266804.

32 Aivazian, G., Gong, Z., Jones, A.M. et al. (2015). Magnetic control of valley pseudospin in monolayer WSe_2. *Nat. Phys.* 11: 148–152.

33 Li, X., Cao, T., Niu, Q. et al. (2013). Coupling the valley degree of freedom to antiferromagnetic order. *Proc. Natl. Acad. Sci. U. S. A.* 110: 3738–3742.

34 Tong, W.Y., Gong, S.J., Wan, X., and Duan, C.G. (2016). Concepts of ferrovalley material and anomalous valley Hall effect. *Nat. Commun.* 7: 1–7.

35 Song, Z., Sun, X., Zheng, J. et al. (2018). Spontaneous valley splitting and valley pseudospin field effect transistors of monolayer $VAgP_2Se_6$. *Nanoscale* 10: 13986–13993.

36 Zhang, X., Yu, Z., Wang, S.S. et al. (2016). Theoretical prediction of MoN_2 monolayer as a high capacity electrode material for metal ion batteries. *J. Mater. Chem. A* 4: 15224–15231.

37 Zang, Y., Ma, Y., Peng, R. et al. (2021). Large valley-polarized state in single-layer NbX_2(X= S, Se): theoretical prediction. *Nano Res.* 14: 834–839.

38 Doran, N.J., Titterington, D.J., Ricco, B., and Wexler, G. (1978). A tight binding fit to the bandstructure of 2H-$NbSe_2$ and NbS_2. *J. Phys. C Solid State Phys.* 11: 685–698.

39 Wang, H., Huang, X., Lin, J. et al. (2017). High-quality monolayer superconductor $NbSe_2$ grown by chemical vapour deposition. *Nat. Commun.* 8: 1–8.

40 Zhu, X., Guo, Y., Cheng, H. et al. (2016). Signature of coexistence of superconductivity and ferromagnetism in two-dimensional $NbSe_2$ triggered by surface molecular adsorption. *Nat. Commun.* 7: 1–8.

41 Lian, C.S., Si, C., and Duan, W. (2018). Unveiling charge-density wave, superconductivity, and their competitive nature in two-dimensional $NbSe_2$. *Nano Lett.* 18: 2924–2929.

42 Yan, R., Khalsa, G., Schaefer, B.T. et al. (2019). Thickness dependence of superconductivity in ultrathin NbS_2. *Appl. Phys. Express* 12: 023008.

43 Lu, A.Y., Zhu, H., Xiao, J. et al. (2017). Janus monolayers of transition metal dichalcogenides. *Nat. Nanotechnol.* 12: 744–749.

44 Zhang, C., Nie, Y., Sanvito, S., and Du, A. (2019). First-principles prediction of a room-temperature ferromagnetic Janus VSSe monolayer with piezoelectricity, ferroelasticity, and large valley polarization. *Nano Lett.* 19: 1366–1370.

45 Guan, Z. and Ni, S. (2020). Predicted 2D ferromagnetic Janus VSeTe monolayer with high Curie temperature, large valley polarization and magnetic crystal anisotropy. *Nanoscale* 12: 22735–22742.

46 Zhao, Y.F., Shen, Y.H., Hu, H. et al. (2021). Combined piezoelectricity and ferrovalley properties in Janus monolayer VClBr. *Phys. Rev. B* 103: 115124.

47 Du, W., Ma, Y., Peng, R. et al. (2020). Prediction of single-layer $TiVI_6$ as a promising two-dimensional valleytronic semiconductor with spontaneous valley polarization. *J. Mater. Chem. C* 8: 13220–13225.

48 Huang, B., Clark, G., Navarro-Moratalla, E. et al. (2017). Layer-dependent ferromagnetism in a van der Waals crystal down to the monolayer limit. *Nature* 546: 270–273.

49 Zhao, P., Ma, Y., Lei, C. et al. (2019). Single-layer $LaBr_2$: two-dimensional valleytronic semiconductor with spontaneous spin and valley polarizations. *Appl. Phys. Lett.* 115: 261605.

50 Krämer, K., Schleid, T., Schulze, M. et al. (1989). Three bromides of lanthanum: $LaBr_2$, La_2Br_5, and $LaBr_3$. *Z. Anorg. Allg. Chem.* 575: 61–70.

51 Schäfer, H. and Schnering, H.V. (1964). Metal-Metall-Bindungen bei niederen Halogeniden, Oxyden und Oxydhalogeniden schwerer Übergangsmetalle Thermochemische und strukturelle Prinzipien. *Angew. Chem.* 76: 833–849.

52 Magonov, S.N., Zönnchen, P., Rotter, H. et al. (1993). Scanning tunneling and atomic force microscopy study of layered transition metal halides Nb_3X_8 (X = Cl, Br, I). *J. Am. Chem. Soc.* 115: 2495–2503.

53 Peng, R., Ma, Y., Xu, X. et al. (2020). Intrinsic anomalous valley Hall effect in single-layer Nb_3I_8. *Phys. Rev. B* 102: 035412.

54 Cui, Q., Zhu, Y., Liang, J. et al. (2021). Spin-valley coupling in a two-dimensional VSi_2N_4 monolayer. *Phys. Rev. B* 103: 085421.

55 Cheng, H.-X., Zhou, J., Ji, W. et al. (2021). Two-dimensional intrinsic ferrovalley GdI_2 with large valley polarization. *Phys. Rev. B* 103: 125121.

56 Kasten, C., Müller, P.H., and Schienle, M. (1984). Magnetic ordering in GdI_2. *Solid State Commun.* 51: 919–921.

57 Felser, C., Ahn, K., Kremer, R.K. et al. (1999). Giant negative magnetoresistance in GdI_2: prediction and realization. *J. Solid State Chem.* 147: 19–25.

58 Ahn, K., Felser, C., Seshadri, R. et al. (2000). Giant negative magnetoresistance in GdI_2. *J. Alloys Compd.* 303: 252–256.

59 Cheng, Y.C., Zhang, Q.Y., and Schwingenschlögl, U. (2014). Valley polarization in magnetically doped single-layer transition-metal dichalcogenides. *Phys. Rev. B* 89: 155429.

60 Singh, N. and Schwingenschlögl, U. (2017). A route to permanent valley polarization in monolayer MoS_2. *Adv. Mater.* 29: 1600970.

61 Chen, X., Zhong, L., Li, X., and Qi, J. (2017). Valley splitting in the transition-metal dichalcogenide monolayer via atom adsorption. *Nanoscale* 9: 2188–2194.

62 Wei, X., Zhang, J., Zhao, B., and Yang, Z. (2020). Coexistence of valley polarization and Chern insulating states in MoS_2 monolayers with n-p cooping. *Sci. Rep.* 10: 9851.

63 Guo, S., Wang, Y., and Zhang, J. (2020). Realization of valley polarization in monolayer WS_2 via 3d transition metal atom adsorption. *J. Phys. D. Appl. Phys.* 53: 384001.

64 Cheng, Y.C., Zhu, Z.Y., Tahir, M., and Schwingenschlögl, U. (2013). Spin orbit–induced spin splittings in polar transition metal dichalcogenide monolayers. *Europhys. Lett.* 102: 57001.

65 Yao, Q.-F., Cai, J., Tong, W.-Y. et al. (2017). Manipulation of the large Rashba spin splitting in polar two-dimensional transition-metal dichalcogenides. *Phys. Rev. B* 95: 165401.

66 Peng, R., Ma, Y., Zhang, S. et al. (2018). Valley polarization in Janus single-layer MoSSe via magnetic doping. *J. Phys. Chem. Lett.* 9: 3612–3617.

67 Xu, X., Ma, Y., Zhang, T. et al. (2019). Nonmetal-atom-doping-induced valley polarization in single-layer Tl_2O. *J. Phys. Chem. Lett.* 10: 4535–4541.

68 Zhang, Q., Yang, S.A., Mi, W. et al. (2016). Large spin-valley polarization in monolayer $MoTe_2$ on top of EuO(111). *Adv. Mater.* 28: 959–966.

69 Qi, J., Li, X., Niu, Q., and Feng, J. (2015). Giant and tunable valley degeneracy splitting in $MoTe_2$. *Phys. Rev. B* 92: 121403(R).

70 Song, Y., Zhang, Q., Mi, W., and Wang, X. (2016). Valley polarization and p-/n-type doping of monolayer WTe_2 on top of Fe_3O_4(111). *Phys. Chem. Chem. Phys.* 18: 15039–15045.

71 Song, Y., Wang, X., and Mi, W. (2017). Ferroelectricity tailored valley splitting in monolayer $WTe_2/YMnO_3$ heterostructures: a route toward electrically controlled valleytronics. *Adv. Electron. Mater.* 3: 1700245.

72 Liang, X., Deng, L., Huang, F. et al. (2017). The magnetic proximity effect and electrical field tunable valley degeneracy in MoS_2/EuS van der Waals heterojunctions. *Nanoscale* 9: 9502–9509.

73 Su, S., Barlas, Y., Li, J. et al. (2017). Effect of intervalley interaction on band topology of commensurate graphene/EuO heterostructures. *Phys. Rev. B* 95: 075418.

74 Li, N., Zhang, J., Xue, Y. et al. (2018). Large valley polarization in monolayer $MoTe_2$ on a magnetic substrate. *Phys. Chem. Chem. Phys.* 20: 3805–3812.

75 Yang, G., Li, J., Ma, H. et al. (2018). Induced valley splitting in monolayer MoS_2 by an antiferromagnetic insulating CoO(111) substrate. *Phys. Rev. B* 98: 235419.

76 Xu, L., Yang, M., Shen, L. et al. (2018). Large valley splitting in monolayer WS_2 by proximity coupling to an insulating antiferromagnetic substrate. *Phys. Rev. B* 97: 041405(R).

77 Zhang, K., Wang, L., and Wu, X. (2019). Spin polarization and tunable valley degeneracy in a MoS_2 monolayer via proximity coupling to a Cr_2O_3 substrate. *Nanoscale* 11: 19536–19542.

78 Zhang, F., Mi, W., and Wang, X. (2019). Tunable valley and spin splitting in 2H-VSe_2/$BiFeO_3$(111) triferroic heterostructures. *Nanoscale* 11: 10329–10338.

79 Pei, Q. and Mi, W. (2019). Electrical control of magnetic behavior and valley polarization of monolayer antiferromagnetic $MnPSe_3$ on an insulating ferroelectric substrate from first principles. *Phys. Rev. Appl.* 11: 014011.

80 Xue, X., Wang, X., and Mi, W. (2019). Valley and spin splitting in monolayer TX_2/antiferromagnetic MnO (T = Mo, W; X = S, Se) van der Waals heterostructures. *J. Phys. D. Appl. Phys.* 52: 115303.

81 Zollner, K., Junior, P.E.F., and Fabian, J. (2020). Giant proximity exchange and valley splitting in transition metal dichalcogenide/hBN/(Co, Ni) heterostructures. *Phys. Rev. B* 101: 085112.

82 Li, J., Gu, L., and Wu, R. (2020). Possible realization and protection of valley-polarized quantum Hall effect in Mn/WS_2. *Phys. Rev. B* 101: 024412.

83 Zhou, W., Yang, Z., Li, A. et al. (2020). Spin and valley splittings in Janus monolayer WSSe on a MnO(111) surface: large effective Zeeman field and opening of a helical gap. *Phys. Rev. B* 101: 045113.

84 Zhang, C., Zhang, S., Lin, Y. et al. (2021). Strong valley splitting in d0 two-dimensional SnO induced by magnetic proximity effect. *Nanotechnology* 32: 225201.

85 Zhou, B., Li, Z., Wang, J. et al. (2019). Tunable valley splitting and an anomalous valley Hall effect in hole-doped WS_2 by proximity coupling with a ferromagnetic MnO_2 monolayer. *Nanoscale* 11: 13567–13575.

86 Ke, C., Wu, Y., Yang, W. et al. (2019). Large and controllable spin-valley splitting in two-dimensional WS_2/h–VN heterostructure. *Phys. Rev. B* 100: 195435.

87 Kuklin, A.V., Shostak, S.A., and Kuzubov, A.A. (2018). Two-dimensional lattices of VN: emergence of ferromagnetism and half-metallicity on nanoscale. *J. Phys. Chem. Lett.* 9: 1422–1428.

88 Seyler, K.L., Zhong, D., Huang, B. et al. (2018). Valley manipulation by optically tuning the magnetic proximity effect in WSe_2/CrI_3 heterostructures. *Nano Lett.* 18: 3823–3828.

89 Zhong, D., Seyler, K.L., Linpeng, X. et al. (2017). van der Waals engineering of ferromagnetic semiconductor heterostructures for spin and valleytronics. *Sci. Adv.* 3: e1603113.

90 Zhang, Z., Ni, X., Huang, H. et al. (2019). Valley splitting in the van der Waals heterostructure WSe_2/CrI_3: the role of atom superposition. *Phys. Rev. B* 99: 115441.

91 Mirzaei, M., Vazifehshenas, T., Salavati-fard, T. et al. (2018). Plasmon-phonon coupling in a valley-spin-polarized two-dimensional electron system: a theoretical study on monolayer silicene. *Phys. Rev. B* 98: 045429.

92 Hu, T., Zhao, G., Gao, H. et al. (2020). Manipulation of valley pseudospin in WSe_2/CrI_3 heterostructures by the magnetic proximity effect. *Phys. Rev. B* 101: 125401.

93 Pei, Q., Zhou, B., Mi, W., and Cheng, Y. (2019). Triferroic material and electrical control of valley degree of freedom. *ACS Appl. Mater. Interfaces* 11: 12675–12682.

94 Gave, M.A., Bilc, D., Mahanti, S.D. et al. (2005). On the lamellar compounds $CuBiP_2Se_6$, $AgBiP_2Se_6$ and $AgBiP_2S_6$. Antiferroelectric phase transitions due to cooperative Cu+ and Bi3+ ion motion. *Inorg. Chem.* 44: 5293–5303.

95 Zhang, H., Yang, W., Ning, Y., and Xu, X. (2020). Abundant valley-polarized states in two-dimensional ferromagnetic van der Waals heterostructures. *Phys. Rev. B* 101: 205404.

96 Lei, C., Ma, Y., Zhang, T. et al. (2020). Valley polarization in monolayer CrX_2 (X = S, Se) with magnetically doping and proximity coupling. *New J. Phys.* 22: 033002.

97 Hacohen, Y.R., Popovitz-Biro, R., Grunbaum, E. et al. (2002). Vapor-liquid-solid growth of $NiCl_2$ nanotubesvia reactive gas laser ablation. *Adv. Mater.* 14: 1075–1078.

98 Ackerman, J., Fouassier, C., Holt, E.M., and Holt, S.L. (1972). The 5° crystal spectra of nickel(II) chloride and nickel(II) bromide. *Inorg. Chem.* 11: 3118–3122.

99 Lu, M., Yao, Q., Xiao, C. et al. (2019). Mechanical, electronic, and magnetic properties of NiX_2 (X = Cl, Br, I) layers. *ACS Omega* 4: 5714–5721.

100 Teng, S., Mao, X., Liu, Z. et al. (2020). Manipulation of valley splitting for the $WSe_2/NiCl_2$ heterostructure by adjusting the interlayer spacing and constructing a $NiCl_2/WSe_2/NiCl_2$ heterojunction. *New J. Phys.* 22: 103061.

101 Li, Q., Chen, K.-Q., and Tang, L.-M. (2020). Large valley splitting in van der Waals heterostructures with type-III band alignment. *Phys. Rev. Appl.* 13: 014064.

102 Mao, X., Li, J., Li, C. et al. (2019). Biaxial strain induced band transition and valley–spin coupling in the ferromagnetic semiconducting $WSe_2/1T\text{-}FeCl_2$ heterostructure. *J. Mater. Chem. C* 7: 9398–9405.

103 Sun, Y.-Y., Shang, L., Ju, W. et al. (2019). Tuning valley polarization in two-dimensional ferromagnetic heterostructures. *J. Mater. Chem. C* 7: 14932–14937.

104 Hu, H., Tong, W.-Y., Shen, Y.-H., and Duan, C.-G. (2020). Electrical control of the valley degree of freedom in 2D ferroelectric/antiferromagnetic heterostructures. *J. Mater. Chem. C* 8: 8098–8106.

105 Marfoua, B. and Hong, J. (2020). Electric filed dependent valley polarization in 2D $WSe_2/CrGeTe_3$ heterostructure. *Nanotechnology* 31: 425702.

106 Khan, I., Marfoua, B., and Hong, J. (2021). Electric field induced giant valley polarization in two dimensional ferromagnetic $WSe_2/CrSnSe_3$ heterostructure. *npj 2D Mater. Appl.* 5: 10.

107 Rashba, E.I. (1960). Properties of semiconductors with an extremum loop. 1. Cyclotron and combinational resonance in a magnetic field perpendicular to the plane of the loop. *Sov. Phys. Solid State* 2: 1109–1122.

108 Casella, R.C. (1960). Toroidal energy surfaces in crystals with wurtzite symmetry. *Phys. Rev. Lett.* 5: 371–373.

109 Bychkov, Y.A. and Rashba, E.I. (1984). Properties of a 2D electron gas with lifted spectral degeneracy. *JETP Lett.* 39: 78–81.

110 Ishizaka, K., Bahramy, M.S., Murakawa, H. et al. (2011). Giant Rashba-type spin splitting in bulk BiTeI. *Nat. Mater.* 10: 521–526.

111 Manchon, A., Koo, H.C., Nitta, J. et al. (2015). New perspectives for Rashba spin–orbit coupling. *Nat. Mater.* 14: 871.

112 Chuang, P., Ho, S.-C., Smith, L. et al. (2015). All-electric all-semiconductor spin field-effect transistors. *Nat. Nanotechnol.* 10: 35.

113 Datta, S. and Das, B. (1990). Electronic analog of the electro-optic modulator. *Appl. Phys. Lett.* 56: 665.

114 Nitta, J., Akazaki, T., Takayanagi, H., and Enoki, T. (1997). Gate control of spin-orbit interaction in an inverted $In_{0.53}Ga_{0.47}As/In_{0.52}Al_{0.48}As$ heterostructure. *Phys. Rev. Lett.* 78: 1335.

115 Sato, Y., Gozu, S.-i., Kita, T., and Yamada, S. (2001). An investigation of tunable spin–orbit interactions in front-gated $In_{0.75}Ga_{0.25}As/In_{0.75}Al_{0.25}As$ heterojunctions. *Phys. E.* 10: 77–80.

116 Koga, T., Nitta, J., Akazaki, T., and Takayanagi, H. (2002). Rashba spin-orbit coupling probed by the weak antilocalization analysis in InAlAs/InGaAs/InAlAs quantum wells as a function of quantum well asymmetry. *Phys. Rev. Lett.* 89: 046801.

117 Li, S.-S. and Xia, J.-B. (2008). Linear rashba model of a hydrogenic donor impurity in GaAs/GaAlAs quantum wells. *Nanoscale Res. Lett.* 4: 178–180.

118 Toloza Sandoval, M.A., Ferreira da Silva, A., deAndrada e Silva, E.A., and La Rocca, G.C. (2009). *J. Supercond. Nov. Magn.* 23: 171.

119 Becker, C.R., Zhang, X.C., Pfeuffer-Jeschke, A. et al. (2003). Very large rashba spin–orbit splitting in HgTe quantum wells. *J. Supercond.* 16: 625–634.

120 Tao, L.L. and Wang, J. (2016). Strain-tunable ferroelectricity and its control of Rashba effect in $KTaO_3$. *Appl. Phys.* 120: 234101.

121 Shanavas, K.V. and Satpathy, S. (2014). Electric field tuning of the Rashba effect in the polar perovskite structures. *Phys. Rev. Lett.* 112: 086802.

122 da Silveira, L.G.D., Barone, P., and Picozzi, S. (2016). Rashba-Dresselhaus spin-splitting in the bulk ferroelectric oxide $BiAlO_3$. *Phys. Rev. B Condens. Matter* 93: 245159.

123 Lee, H. and Choi, H.J. (2012). Role of d orbitals in the Rashba-type spin splitting for noble-metal surfaces. *Phys. Rev. B Condens. Matter* 86: 045437.

124 Gong, S.-J., Duan, C.-G., Zhu, Y. et al. (2013). Controlling Rashba spin splitting in Au (111) surface states through electric field. *Phys. Rev. B Condens. Matter* 87: 035403.

125 Gong, S.-J., Cai, J., Yao, Q.-F. et al. (2016). Orbital control of Rashba spin orbit coupling in noble metal surfaces. *J. Appl. Phys.* 119: 125310.

126 Krupin, O., Bihlmayer, G., Starke, K. et al. (2005). Rashba effect at magnetic metal surfaces. *Phys. Rev. B Condens. Matter* 71: 201403.

127 Ast, C.R., Henk, J., Ernst, A. et al. (2007). Giant spin splitting through surface alloying. *Phys. Rev. Lett.* 98: 186807.

128 Liu, Q., Guo, Y., and Freeman, A.J. (2013). Tunable Rashba effect in two-dimensional $LaOBiS_2$ films: ultrathin candidates for spin field effect transistors. *Nano Lett.* 13: 5264–5270.

129 Ma, Y., Dai, Y., Yin, N. et al. (2014). Ideal two-dimensional systems with a gain Rashba-type spin splitting: $SrFBiS_2$ and $BiOBiS_2$ nanosheets. *J. Mater. Chem. C* 2: 8539–8545.

130 Ma, Y., Dai, Y., Wei, W. et al. (2014). Emergence of electric polarity in BiTeX (X = Br and I) monolayers and the giant Rashba spin splitting. *Phys. Chem. Chem. Phys.* 16: 17603–17609.

131 Zhang, S.-H. and Liu, B.-G. (2019). Anisotropic Rashba effect and charge and spin currents in monolayer BiTeI by controlling symmetry. *Phys. Rev. B* 100: 165429.

132 Yang, W., Guan, Z., Wang, H., and Li, J. (2021). Ideal strength and strain engineering of the Rashba effect in two-dimensional BiTeBr. *Phys. Chem. Chem. Phys.* 23: 6552–6560.

133 Zhuang, H.L., Cooper, V.R., Xu, H. et al. (2015). Rashba effect in single-layer antimony telluroiodide SbTeI. *Phys. Rev. B* 92: 115302.

134 Hu, T., Jia, F., Zhao, G. et al. (2018). Intrinsic and anisotropic Rashba spin splitting in Janus transition-metal dichalcogenide monolayers. *Phys. Rev. B* 97: 235404.

135 Chen, J., Wu, K., Ma, H. et al. (2020). Tunable Rashba spin splitting in Janus transition-metal dichalcogenide monolayers via charge doping. *RSC Adv.* 10: 6388–6394.

136 Bandurin, D.A., Tyurnina, A.V., Yu, G.L. et al. (2017). High electron mobility, quantum Hall effect and anomalous optical response in atomically thin InSe. *Nat. Nanotechnol.* 12: 223–227.

137 Debbichi, L., Eriksson, O., and Lebègue, S. (2015). Two-dimensional indium selenides compounds: an ab initio study. *J. Phys. Chem. Lett.* 6: 3098–3103.

138 Sun, Y., Luo, S., Zhao, X.-G. et al. (2018). InSe: a two-dimensional material with strong interlayer coupling. *Nanoscale* 10: 7991–7998.

139 Bahuguna, B.P., Saini, L.K., Sharma, R.O., and Tiwari, B. (2018). Hybrid functional calculations of electronic and thermoelectric properties of GaS, GaSe, and GaTe monolayers. *Phys. Chem. Chem. Phys.* 20: 28575–28582.

140 Huang, W., Gan, L., Li, H. et al. (2016). 2D layered group IIIA metal chalcogenides: synthesis, properties and applications in electronics and optoelectronics. *CrystEngComm* 18: 3968–3984.

141 Demirci, S., Avazlı, N., Durgun, E., and Cahangirov, S. (2017). Structural and electronic properties of monolayer group III monochalcogenides. *Phys. Rev. B* 95: 115409.

142 Sucharitakul, S., Goble, N.J., Kumar, U.R. et al. (2015). Intrinsic electron mobility exceeding 103 cm^2/ (V s) in multilayer InSe FETs. *Nano Lett.* 15: 3815–3819.

143 Ju, W., Wang, D., Li, T. et al. (2020). Electric field control of Rashba spin splitting in 2D NIIIXVI (N = Ga, In; X = S, Se, Te) monolayer. *J. Phys. Condens. Matter* 32: 175503.

144 Ju, W., Wang, D., Li, T. et al. (2020). Remarkable Rashba spin splitting induced by an asymmetrical internal electric field in polar III–VI chalcogenides. *Phys. Chem. Chem. Phys.* 22: 9148–9156.

145 Guo, S.-D., Mu, W.-Q., Zhu, Y.-T. et al. (2021). Predicted septuple-atomic-layer Janus $MSiGeN_4$ (M = Mo and W) monolayers with Rashba spin splitting and high electron carrier mobilities. *J. Mater. Chem. C* 9: 2464–2473.

3

Ferromagnetic Order in Two- and One-Dimensional Materials

The research on intrinsic ferromagnetic order in 2D material dates back to 2012. Using first-principles calculations, Ma et al. have predicted the monolayer VS_2 and VSe_2 should be ferromagnetic [1]. Based on spin-polarized calculations, it is interesting to find that monolayer VX_2 (X = S, Se) do harbor spin polarization. In monolayer VS_2, the V atom possesses a magnetic moment of 0.486 μ_B, and the neighboring S atom takes a magnetic moment of −0.026 μ_B. In monolayer VSe_2, the magnetic moment on V atom is 0.680 μ_B, and the value on Se atom is −0.048 μ_B. As the existence of magnetic moment does not guarantee the magnetic coupling of the magnetic moments, the magnetic coupling of monolayer VX_2 is investigated. One ferromagnetic state and one antiferromagnetic (AFM) state are considered for both systems. For both cases, the ground state is identified as the ferromagnetic state, which is lower than the AFM state by 205 and 377 meV, respectively, in energy.

Moreover, Ma et al. proposed that strain can deliberately modulate the magnetic properties of monolayer VX_2 [1]. Here, the tensile and compressive strain are uniformly applied along both the armchair and zigzag directions, and the strain ranges from −5 to 5%. The magnetic moments on V and X atoms are found to increase monotonically with increasing strain from −5 to 5%. As compared with the pure case, the magnetic moment on V atom is increased by 87 and 135%, respectively, for monolayer VS_2 and VSe_2 under 5% strain, and the magnetic moment on X atom is increased by 215 and 144%. Under compressive strain of 5%, the magnetic moment on V atom is decreased to 0.307 and 0.330 μ_B, respectively, for monolayer VS_2 and VSe_2. Therefore, the spin polarization in monolayer VX_2 is robust against strain. In addition to the magnetic moment, the variation of magnetic coupling as a function of strain is also investigated. It is found that the ferromagnetic state is preserved to be the ground state under strain from −5 to 5%. The energy difference between the AFM and ferromagnetic states increases with increasing strain from −5 to 5%. That means, the ferromagnetic state is weakened under compressive strain and is enhanced under tensile strain. To understand the change of magnetic moment under strain, the joint effect of ionic and covalent bonding interactions is proposed. And for the change of the energy difference between AFM and ferromagnetic states, the combined effect of through-bond and through-space interactions is proposed.

Calculations and Simulations of Low-Dimensional Materials: Tailoring Properties for Applications, First Edition. Ying Dai, Wei Wei, Yandong Ma, and Chengwang Niu.

These results open up a new direction in 2D magnetism. Soon after this theoretical work [1], two experimental groups separately demonstrate the magnetism in monolayer VX_2 [2, 3].

In 2017, Xu et al. proposed the magnetism in monolayer CrI_3 [4]. They show that the magnetic anisotropy is possible to remove the restriction of the Mermin–Wanger theorem [5], which enables the occurrence of 2D magnetism. Using magneto-optical Kerr effect microscopy, monolayer CrI_3 is demonstrated to be an Ising ferromagnet with easy magnetization axis along out-of-plane direction. The Curie temperature is estimated to be 45 K. This value is slightly smaller than that of the bulk case, suggesting the weak interlayer interaction. Moreover, CrI_3 exhibits a layer-dependent magnetic property. This work provides new insight into the magnetism in 2D materials. Nearly at the same time, based on the scanning magneto-optic Kerr microscopy, Zhang et al. reported the intrinsic long-range ferromagnetic state in monolayer $Cr_2Ge_2Te_6$ [6]. Interestingly, the transition temperature between the ferromagnetic and paramagnetic states can be obtained by using the small electric field. Such feature is different from the case of three-dimensional materials wherein the transition temperature is insensitive to the magnetic field. Moreover, an effective anisotropy can be induced by the applied field. Therefore, monolayer $Cr_2Ge_2Te_6$ is concluded as an ideal 2D Heisenberg ferromagnet and is promising for investigating fundamental spin behaviors [6]. Following the discovery of monolayer CrI_3 and $Cr_2Ge_2Te_6$, Zhang et al. reported a device fabrication technique and exfoliated monolayer Fe_3GeTe_2 from the layered metallic magnet [7]. Importantly, the ferromagnetic order is preserved in Fe_3GeTe_2 when the layer thickness thins down to monolayer. The out-of-plane magnetization easy axis enables its long-range ferromagnetic order. Compared with the bulk counterpart, the Curie temperature is suppressed. However, by using ionic gate, the Curie temperature can be raised to room temperature.

Based on these works, extensive efforts have been devoted to exploring intrinsic ferromagnetic orders in 2D materials. In Section 3.1, we will discuss the recent theoretical progress on the intrinsic ferromagnetic order in 2D materials. And in Section 3.2, we will briefly discuss the recent theoretical progress on the intrinsic ferromagnetic order in 1D molecular nanowires.

3.1 Intrinsic Ferromagnetic Order in 2D Materials

Similar to CrI_3, $CrCl_3$ and $CrBr_2$ also exhibit a layered structure and are expected to show similar physical properties. In fact, before the experimental work on monolayer CrI_3 [4], Lam et al. already predicted the intrinsic ferromagnetism in monolayer CrX_3 (X = F, Cl, Br, and I) [8]. They found that monolayer CrX_3 can be easily exfoliated in monolayer form due to a low cleavage energy and a high in-plane stiffness. Through phonon spectra calculations and molecular dynamic simulations, these monolayer systems are shown to be stable. The calculations show that the magnetic moment is found to be 3 μ_B per unit cell. To examine the magnetic ground states of these systems, four different magnetic configurations are considered. For all these systems, the ferromagnetic state is found to be most stable. This suggests that monolayer CrX_3 exhibits intrinsic magnetic state. To establish the ferromagnetic state, magnetic anisotropy energy is a key factor. For systems with small magnetic anisotropy energy, the coupling between magnetic moments tends to form paramagnetic feature. For all these systems, the easy axis is found to be along the out-of-plane direction for all these four systems. The corresponding magnetic anisotropy energies for CrF_3, $CrCl_3$, $CrBr_3$, and CrI_3 are estimated to be 120, 32, 186, and 686 μeV per Cr atom, respectively, indicating the robust intrinsic ferromagnetism.

All monolayer systems exhibit a semiconducting character with an indirect band gap. With increasing the atom number of X, the band gap decreases from 4.68 to 1.53 eV monotonically. The states around the Fermi level are fully spin polarized and dominated by one spin channel. Interestingly, Cr-3d orbitals in the spin-up direction are occupied, while Cr-3d orbitals in the spin-down direction are unoccupied. This is sought into the crystal field theory. Under the octahedral field, the d orbitals of Cr split into two groups: three low-lying t_{2g} orbitals and two high-lying e_2 orbitals. Based on the Hund's rule, the electronic configuration of Cr^{3+} is $t_{2g}3e_g0$. Under such condition, all the occupied d orbitals should be in the same direction. Therefore, monolayer CrX_3 exhibits an intrinsic ferromagnetic semiconducting state.

Considering the important role of magnetic anisotropy energy in establishing the ferromagnetic state in monolayer CrX_3, Yan et al. investigated the influence of strain on the magnetic anisotropy energy on the basis of first-principles calculations [9]. Interestingly, the magnetic anisotropy energy on monolayer CrX_3 shows a strain dependence. For monolayer CrI_3, it increases with increasing compressive strain, and an opposite trend is found in monolayer $CrCl_3$ and $CrBr_3$. In addition to the magnetic anisotropy energy, the magnetic ground state also experiences a phase transition. For more detail, please refer to Ref. [9].

Interestingly, although Refs. [8, 9] confirm the out-of-plane easy axis of monolayer $CrCl_3$, recent experiments show that monolayer $CrCl_3$ favors the in-plane magnetic easy axis [10, 11]. To solve this puzzle, using first-principles calculations, Wu et al. systematically investigated the magnetic properties of monolayer CrX_3 (X = Cl, Br, I) and found that the inclusion of magnetic shape anisotropy is important for estimating the magnetic easy axis [12]. The magnetic anisotropy energy consists of two parts: the SOC-induced magnetocrystalline anisotropy (MCA) energy and the dipole-dipole interaction induced magnetic shape anisotropy (MSA) energy. For all these three systems, the values of MCA are positive. This suggests that the MCA part favors the out-of-plane direction, which is consistent with previous works [8, 9]. Moreover, the MCA increases monotonically with increasing atom number from Cl, Br, to I, which can be attributed to the increase in SOC strength from Cl, Br, to I. Different from the case of MCA, the values of MSA for all these three systems are found to be negative. Therefore, the MSA favors in-plane magnetic easy axis. The resultant magnetic anisotropy energy is found to be negative for monolayer $CrCl_3$ and positive for monolayer $CrBr_3$ and CrI_3. As a consequence, monolayer favors in-plane magnetic easy axis, while monolayer $CrBr_3$ and CrI_3 favor out-of-plane magnetic easy axis. With these results in hand, the aforementioned puzzle could be easily understood.

Another typical class of metal trihalides that receive great attention is monolayer MnX_3 (X = F, Cl, Br, I). Using first-principles calculations, Sun et al. investigated the electronic and magnetic properties of monolayer MnX_3 [13]. Monolayer MnX_3 shares similar structure as monolayer CrX_3, and each unit cell contains two Mn and six X atoms. The Mn atom coordinates with six X atoms. The spin-polarized calculations show that monolayer MnX_3 is spin polarized. To reveal the coupling among the magnetic moments in monolayer MnX_3, the ferromagnetic state, antiferromagnetic-Néel (AFM-N), antiferromagnetic-zigzag (AFM-ZZ), antiferromagnetic-stripy (AFM-SR), and the mixed AFM-N-ST configurations are considered. For all these systems, the ferromagnetic configuration is found to be the ground state. Interestingly, the magnetic moment per Mn atom increases from 3.92 μ_B in MnF_3 to 4.27 μ_B in MnI_3. These values agree well with the +3 oxidation state of Mn atom, which favors the high-spin state of $t_{2g}3e_g1$. The magnetic anisotropy energy of monolayer MnX_3 is calculated, which indicates that all these systems favor the in-plane magnetic easy axis.

Figure 3.1a and c presents the spin-polarized band structures of monolayer MnF_3 and MnI_3. As monolayer $MnCl_3$ and $MnBr_3$ share similar band structure, they are not shown here. For monolayer MnF_3, $MnCl_3$, and $MnBr_3$, the minority-spin channel shows a large band gap, while the majority-spin channel presents a Dirac point at the *K* point at the Fermi level. For monolayer MnI_3, the Dirac point locates above the Fermi level based on Perdew–Burke–Ernzerhof (PBE) functional and locates at the Fermi level based on Heyd-Scuseria-Ernzerhof (HSE06) functional. Consequently, monolayer MnX_3 is a Dirac half-metal. Figure 3.1b and d shows the 3d plot of the majority bands around the Fermi level for MnF_3 and MnI_3. As compared with monolayer MnF_3, the Dirac bands become slightly flatter for monolayer MnI_3, which can be attributed to the stronger SOC strength. The corresponding projections of the Dirac bands on 2D plane are shown in Figure 3.1b and d. It should be noted that the bands shown in Figure 3.1a and c do not include SOC. Upon including SOC, a small band gap of 3–10 meV opens at the Dirac point. This indicates that the feature of the Dirac half-metal will be preserved in monolayer MnX_3 in the presence of SOC. Furthermore, the Curie temperature for the ferromagnetic state of monolayer MnX_3 is investigated via Matropolis Monte Carlo simulations. The obtained Curie temperature for monolayer MnX_3 is found to be ranging from 450 to 720 K, beneficial for practical applications.

Similar to CrI_3, bulk VI_3 also exhibits a layered structure wherein the slabs are stacked in the *c* direction. It has a band gap of 0.6 and 0.7 eV [14, 15]. Moreover, VI_3 is spin polarized with a ferromagnetic state. Therefore, spin polarization is also expected in monolayer VI_3. Using first-principles calculations, Nachtigall et al. demonstrated that monolayer VI_3 has intrinsic ferromagnetism and exhibits half-metallicity [16]. Importantly, the half-metallic Dirac point is observed around the Fermi level. Different from the Dirac point of graphene, the Dirac point is dominated by the d orbitals from V atom. Upon including SOC, a band gap of 12 meV is opened at the Dirac point. Using the Ising model, the Curie temperature for the ferromagnetic state of monolayer VI_3 is estimated to be 98 K. Fortunately, carrier doping is demonstrated to be able to increase the Curie temperature. Recently,

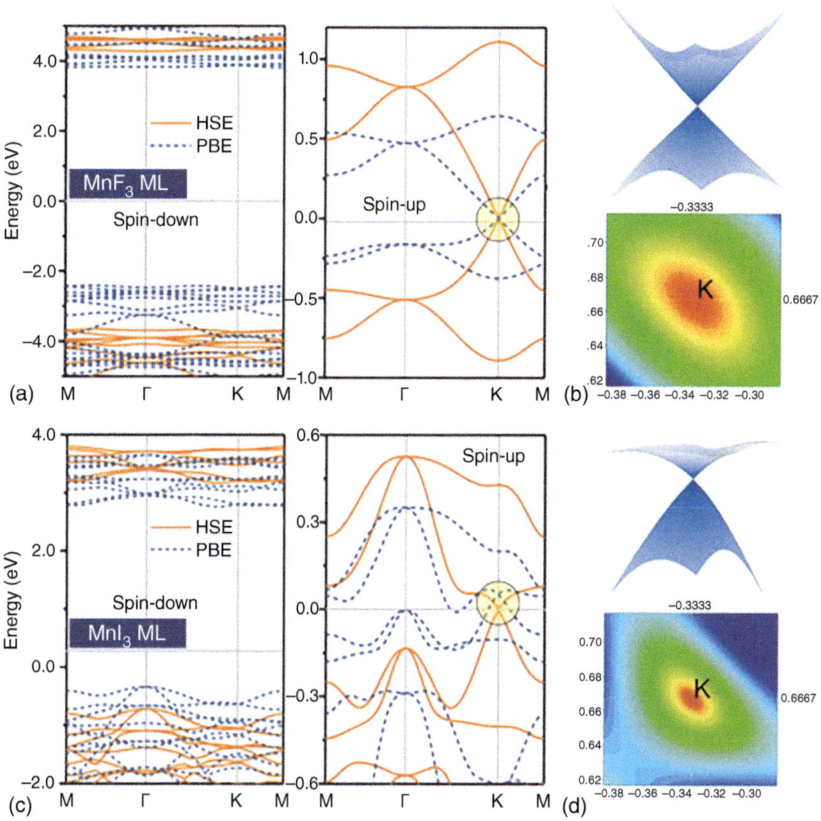

Figure 3.1 Spin-polarized band structures of monolayer (a) MnF_3 and (c) MnI_3 based on PBE and HSE06 functionals. 3D plot of the Dirac bands and the corresponding projections of Dirac bands on 2D plane for monolayer (b) MnF_3 and (d) MnI_3. E_F is set to 0 eV. Source: Sun and Kioussis [13]/with permission of American Physical Society.

on the other hand, Long et al. proposed that the half-metallic state identified in monolayer VI_3 is a metastable state [17]. The Mott-insulator state is more stable than the half-metallic state by 0.2–0.3 eV per VI_3. And the magnetic easy axis is along the out-of-plane direction. Later, Wu et al. showed that the monolayer VI_3 is an Ising-like ferromagnetic material [18]. The orbital moment on each V atom is 1 μ_B, which is antiparallel to its spin moment of 2 μ_B. Therefore, the net magnetic moment on each V atom is 1 μ_B for monolayer VI_3. Using first-principles calculations, Ren et al. determined the magnetic anisotropy energy for VI_3 via examining the spin orientations in terms of selection rules [19]. They show that monolayer VI_3 in ferromagnetic state is uniaxial, which is different from CrI_3. Except for these works, there are many other works devoted to monolayer metal trihalides [20–30].

A family of materials related to monolayer transition metal trihalides that hold equal promise for exhibiting ferromagnetic order are monolayer transition metal dihalides. Similar to transition metal trihalides, transition metal dihalides present a layered structure, which are studied decades ago [31]. Using first-principles calculations, Hennig et al. systematically studied the electronic and magnetic properties of monolayer iron dihalides FeX_2 (X = Cl, Br, I) [32]. Monolayer FeX_2 share the typical 1T structure of transition metal dichalcogenides. In monolayer FeX_2, the Fe atom is in an octahedral configuration coordinated with six X atoms. The exfoliation energies for monolayer $FeCl_2$, $FeBr_2$, and FeI_2 are found to be 70, 75, and 83 meV atom^{-1}, respectively, indicating the high experimental feasibility of these materials. The stability of these materials is demonstrated by the phonon spectra calculations. The lattice constants for monolayer $FeCl_2$, $FeBr_2$, and FeI_2 are found to be 3.49, 3.70, and 3.98 Å, respectively. For all these three systems, the magnetic moment per Fe atom is found to be 4 μ_B. This agrees with the electronic configuration of d6 for Fe^{2+} with four unpaired electrons. The coupling between the magnetic moments on Fe atoms is ferromagnetic for all these three systems. The ferromagnetic state in these systems can be attributed to the Goodenough–Kanamori rules, as the Fe-X-Fe angle is nearly 90°. All these three systems favor in-plane magnetic easy axis.

Monolayer FeX_2 all exhibit a half-metallic character with one spin channel crossing the Fermi level. Based on the orbital-resolved band structure, the two bands crossing the Fermi level are dominated by the Fe-d_{xz}, d_{yz}, d_{xy}, and d_{x2-y2} orbitals. This results from the hybridization between the d orbitals of Fe atom and the p orbital of Cl atom. While for the Fe-d_{z2} orbital, its out-of-plane character makes it not contribute to the hybridization and leads to a flat band at ~2 eV above the Fermi level. For the other two systems, they share similar feature. Furthermore, for the bands crossing the Fermi level, the Fermi velocities are found to between 1.2 × 105 and 3.4 × 105 m s^{-1}, suggesting the high carrier mobility in these materials.

For transition metal dihalides, the transition metal can also be 4f rare-earth elements. One example is GdI_2. Bulk GdI_2 is shown to be a ferromagnetic material with room temperature Tc in experiment [33–35]. It features a layered structure, similar to 2H-MoS_2. This is different from the case of FeX_2. Using first-principles calculations, Wang et al. investigated the electronic and magnetic properties of monolayer GdI_2 [36]. The exfoliation energy for monolayer GdI_2 is 0.26 J m^{-2}, indicating that monolayer GdI_2 can be readily obtained in experiment. The stability of monolayer GdI_2 is confirmed by performing phonon spectra calculations and molecular dynamic simulations. The calculation shows that the ferromagnetic configuration is found to be the ground state, which is lower than the AFM configuration by 139 meV per Gd atom in energy. The magnetic moment is mainly distributed on the Gd atom. The magnetic anisotropy energy indicates that monolayer GdI_2 favors the in-plane magnetic easy axis. Based on the Monte Carlo simulations, the Curie temperature is 745 K.

Monolayer GdI_2 is a ferromagnetic semiconductor. With including SOC, the band structure experiences only a slight change. Both the conduction and valence band edges are dominated by the Gd atoms. The valence and conduction band edges are from opposite spin channels. Therefore, monolayer GdI_2 is a bipolar magnetic semiconductor. Simply by shifting the Fermi level via gate voltage, completely spin-polarized currents with reversible spin polarization can be realized and tuned.

The preceding two works clearly show a high potential of monolayer transition metal dihalides for achieving intrinsic ferromagnetism in 2D limit. Up to now, a series of monolayer transition metal dihalides have been identified as 2D ferromagnetic materials, including NiX_2, CoX_2, MnX_2, VX_2, and so on [37–42]. And in fact, the intrinsic ferromagnetic order is also identified in other transition metal based monolayer binary compound beyond the dihalides. For example, Shenoy et al. performed a systematical first-principles study on the magnetic behaviors in non-van der Waals systems and proposed monolayer Cr_2O_3 has an insulating ferromagnetic phase [43]. Wang et al. reported a class of monolayer Cr_3X_4 (X = S, Se, Te), wherein the two oxidation states of Cr lead to the double exchange interaction and hence enhance ferromagnetic order significantly and result in Curie temperature up to 370 K for Cr_3Se_4 and 460 K for Cr_3Te_4, while for monolayer Cr_3Se_4 and Cr_3Te_4, they are identified as ferromagnetic half-metals with 100% spin-polarized currents [44]. Ghergherehchi et al. explore the magnetic properties of monolayer FeX (X = S, Se, Te) and show that while monolayer FeS and FeSe are nonmagnetic, monolayer FeTe is a good candidate for spintronic applications [45]. Zhang et al. predict monolayer $CrTe_2$ as a ferromagnetic material and hold a Curie temperature above 300 K [46].

Except for binary systems, ternary monolayer systems are also shown to exhibit intriguing ferromagnetic behaviors [47–56]. For example, Sun et al. reported the intrinsic ferromagnetism monolayer CrOX (CrOCl and CrOBr) on the basis of first-principles calculations [57]. Bulk MOX (M = Cr/V/Ti, X = Cl/Br), crystallizing in an orthorhombic structure, are known as AFM semiconductors with spin-Peierls features [58–62]. Each M atom coordinates with 4 O and 2 X atoms, forming a distorted octahedral structure of MO_4Cl_2. To fabricate monolayer MOX, mechanical cleavage and liquid exfoliation can be employed. The exfoliation energies for monolayer CrOCl and CrOBr are much smaller than that of graphene, indicating that they can be easily obtained in experiment. The dynamical and thermal stabilities of monolayer CrOCl and CrOBr are confirmed by carrying out the phonon spectra calculations and molecular dynamic simulations.

To study the magnetic coupling between the magnetic moments of monolayer CrOCl and CrOBr, six magnetic configurations are considered. The ground state for monolayer CrOCl and CrOBr is found to be ferromagnetic. To identify the magnetic easy axis, the magnetic anisotropy energy is calculated. For both systems, the out-of-plane direction is shown to be the magnetic easy axis. Therefore, the spins in monolayer CrOCl and CrOBr are along the out-of-plane direction. Figure 3.2 displays the band structure and density of states of monolayer CrOCl. And monolayer CrOBr shares similar features. As shown in Figure 3.2, monolayer CrOCl is a ferromagnetic semiconductor, exhibiting a band gap of 2.38 eV. Its conduction and valence band edges are from the same spin channel. The conduction band edges are dominated by the Cr-3d orbitals, while the valence band edges are mainly from the Cl-2p orbitals. To investigate the spin dynamics, Monte Carlo simulations are carried out. The Curie temperature is estimated to be 160 K for monolayer CrOCl. While for monolayer CrOBr, it shows a smaller Curie temperature. The identification of intrinsic in monolayer CrOX provides a new avenue to explore 2D semiconducting intrinsic ferromagnets from bulk crystals.

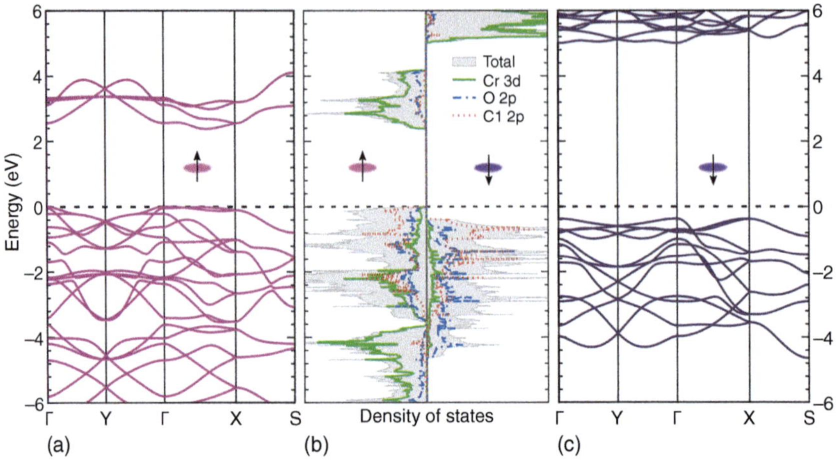

Figure 3.2 Band structures of monolayer CrOCl with (a) majority and (c) minority spins. (b) Projected density of states of monolayer CrOCl. E_F is set to 0 eV. Source: Miao et al. [57]/with permission of American Chemical Society.

Similar to MOX, CrSX (X = Cl, Br, I) also crystallizes in an orthorhombic structure. It exhibits a layered structure, suggesting the experimental feasibility of monolayer structure. Using first-principles calculations, Wang et al. explored the electronic and magnetic properties of monolayer CrSX [63]. Monolayer CrSX is spin polarized. The magnetic moment is found to be 3 μ_B per f.u., which is mainly contributed by the Cr atom. After considering one ferromagnetic and four AFM configurations, ferromagnetic configuration is found to be the ground state with the lowest energy. The intrinsic ferromagnetic state in monolayer CrSX can be understood by the Goodenough–Kanamori–Anderson rules, as the cation–anion–cation bond angles approximate 90°.

All these systems exhibit a semiconducting feature. Monolayer CrSCl shows a direct band gap of 0.856 eV, while monolayer CrSBr and CrSI harbor an indirect band gap of 0.757 and 0.473 eV, respectively. The decreasing of band gap with increasing atomic number of X can be attributed to weakening of the CrS_4X_2 octahedral crystal field. The valence and conduction band edges are from the same spin channel. For all these systems, the conduction band minimum is dominated by Cr-3p orbitals, while the valence band maximum is mainly from X- and S-p orbitals. Moreover, Wang et al. [63] stress that as the valence and conduction bands near the Fermi level are full spin polarized, the 100% spin-polarized current can be realized in monolayer CrSX under either electron or hole doping. Therefore, monolayer CrSX is a promising ferromagnetic semiconductor.

In addition to the d/f-electron based monolayer ferromagnets, magnetic properties are also achieved in materials without d or f electrons, namely, d0 magnetism [64, 65]. Compared with the former, d0 magnetism holds advantages in high-speed and long-distance polarized transport. The underlying physics for realizing d0 magnetism is to partially occupy the p orbitals with localized character. Inspired by the recent breakthroughs in the synthesis of 2D nonstoichiometric compounds [66–68], such as Na_2Cl and Na_3Cl, Zhao et al. propose that intrinsic d0 magnetism could be obtained in the nonstoichiometric compounds [65]. Using the high-throughput first-principles calculations, Zhao et al. screen out three types of 2D alkali metal subnitrides, i.e. Na_2N, K_2N, and Rb_2N with $\overline{P}6m2$ and $\overline{P}3m1$ space group (H phase and T phase), and tetragonal I phase with P4/mmm space group are stable. Taking K_2N as an example, their electronic and magnetic properties are investigated. In monolayer K_2N, each N atom accepts two s electrons from the bonding K atoms, leaving one unpaired electron. Therefore, a magnetic moment of 1 μ_B per N atom is expected for all these three systems. The calculations indeed show that these three phases of monolayer K_2N are all spin polarized, which are mainly resulted from the $p_{x/y}$ orbitals of N atoms. The magnetic moment on each N atom is calculated to be 0.81, 0.72, and 0.79 μ_B, respectively, for monolayer H-K_2N, T-K_2N, and I-K_2N. The magnetic ground state is found to be ferromagnetism, which is lower than the AFM state by 0.11, 0.05, and 0.05 eV, respectively, for H-K_2N, T-K_2N, and I-K_2N. The Curie temperature is estimated to be 1180, 513, and 484 K, respectively, for monolayer H-K_2N, T-K_2N, and I-K_2N. Based on the spin-polarized band structures of these systems, monolayer H- and I-K_2N are ferromagnetic metals, while T-K_2N is half-metallic. This discrepancy is mainly from the different occupancies of the K-s orbital around the Fermi level. For T-K_2N, the s orbital is occupied in the spin-down channel, while it is partially occupied for the other two phases. These exciting results of course will inspire further research on d0 magnetism in 2D limit.

Besides the aforementioned works, there is a lot of theoretical research devoted to intrinsic ferromagnetism in 2D lattice. For more detail, please refer to Refs. [69–79].

3.2 Intrinsic Ferromagnetic Order in 1D Molecular Nanowires

With the increasing demand for device miniaturization, developing molecular nanowires as building blocks for spintronic devices has attracted great attention. Molecular nanowire ferromagnets are promising alternatives to monolayer ferromagnets and hold a wide range of applications from high-capacity storage devices to quantum computers [80–82]. To realize molecular nanowire ferromagnets, it is important to choose suitable functional molecules. Up to now, several molecular nanowires have been predicated to exhibit the magnetic properties [83–92]. In the following, we will discuss the several typical examples of molecular nanowires with ferromagnetism.

Using first-principles calculations, Chen et al. investigated the electronic and magnetic properties of a series of vanadium naphthalene (V_n–1Np$_n$) sandwich clusters (SWCs) and the VNp sandwich nanowire (SWN) [89]. On the basis of the energetically stable structure of V_n–1Np$_n$ SWCs, the infinite VNp SWN is constructed. Figure 3.3a shows the crystal structure of VNp SWN. Each unit cell contains two Np rings and two V atoms. The magnetic moment is calculated to be 1 μ_B per V atom. By considering different magnetic configurations, AFM configuration is found to be the ground state, which is lower by 58 meV than the ferromagnetic state. In the

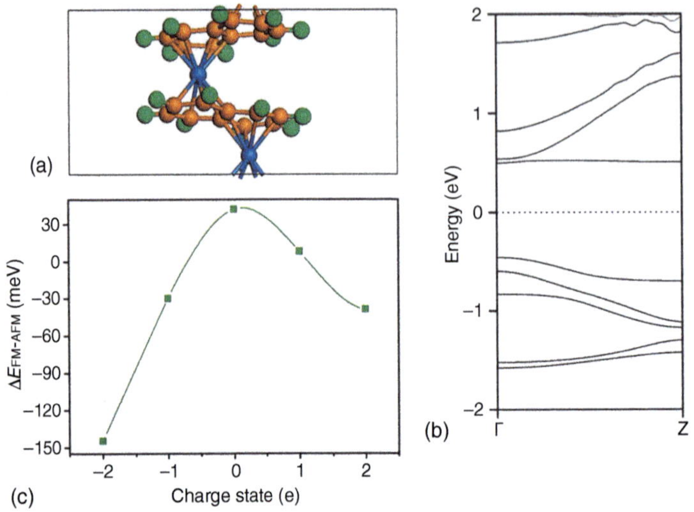

Figure 3.3 (a) Crystal structure of VNp SWN. (b) Band structure of VNp SWN. (c) Energy difference between ferromagnetic and antiferromagnetic states as a function of carrier doping. E_F is set to 0 eV. Source: Li et al. [89]/with permission of American Chemical Society.

AFM state, the lattice constant is found to be 7.27 Å. From the band structure shown in Figure 3.3b, it can be seen that VNp SWN is semiconducting with a band gap of 1.1 eV. By looking at the structure of VNp SWN, the Peierls deformation might exist. To confirm this point, the VNp SWN with a rotation angle of 0° is also investigated. Interestingly, the magnetic ground state is found to be ferromagnetic, which is lower than the AFM state by 76 meV. And there is a majority band crossing the Fermi level, which is the hallmark of the Peierls instability. Compared with the rotated case, VNp SWN with a rotation angle of 0° is higher in energy by 400 eV. Such energy gain is related to the electronic contribution because of the band gap opening.

Furthermore, the magnetic coupling in VNp SWN can be modulated under electron or hole doping. Under the doping of one electron per unit cell, the magnetic ground state of VNp SWN transfers from AFM coupling to ferromagnetic coupling, the ferromagnetic coupling is lower in energy than the AFM coupling by 40 meV, see Figure 3.3c. Upon further increasing electron doping, the ferromagnetic state can be enhanced. Under one hole doping, the AFM coupling is preserved in VNp SWN. While for doping of two holes, the magnetic coupling of VNp SWN can be switched to ferromagnetic state. Considering these merits, VNp SWN has potential applications in nanoelectronics and spintronics [89].

Another typical class of molecular nanowire ferromagnets is transition metal phthalocyanine (M-Pc, M = Cr, Mn, Co, Ni, Cu, and Zn) nanowire (M-PcNW). Using first-principles calculations, Dai et al. investigated the electronic and magnetic properties of M-PcNW [91]. The results show that Cr-PcNW, Mn-PcNW, Co-PcNW, and Cu-PcNW are spin polarized, while Ni-PcNW and Zn-PcNW are not spin polarized. The total magnetic moment per unit cell is found to be 4, 3, 1, and 1 μ_B, respectively, for Cr-PcNW, Mn-PcNW, Co-PcNW, and Cu-PcNW. The magnetic moment is mainly localized on the M atom, while the neighboring N atoms are slightly spin polarized. Considering all the possible magnetic configurations, the magnetic couplings in Cr-PcNW and Cu-PcNW are found to be AFM, Co-PcNW shows paramagnetic coupling, and Mn-PcNW exhibits a ferromagnetic state. The ferromagnetic state in Mn-PcNW is 84 meV lower in energy than the AFM state. More interestingly, the electronic structure calculation shows that Mn-PcNW displays half-metallic property. These results provide ideal platforms for future molecular spintronics.

It should be noted that although extensive efforts have been devoted to molecular nanowires with ferromagnets, the experimental realization of the spintronic properties in molecular nanowires is still under exploration. It is thus of great significance to explore new and suitable molecular nanowires in future.

References

1 Ma, Y., Dai, Y., Guo, M. et al. (2012). Evidence of the existence of magnetism in pristine VX_2 monolayers (X = S, Se) and their strain-induced tunable magnetic properties. *ACS Nano* 6: 1695–1701.

2 Gao, D., Xue, Q., Mao, X. et al. (2013). Ferromagnetism in ultrathin VS_2 nanosheets. *J. Mater. Chem. C* 1: 5909–5916.

3 Xu, K., Chen, P., Li, X. et al. (2013). Ultrathin nanosheets of vanadium diselenide: a metallic two-dimensional material with ferromagnetic charge-density-wave behavior. *Angew. Chem. Int. Ed.* 52: 10477–10481.

4 Huang, B., Clark, G., Navarro-Moratalla, E. et al. (2017). Layer-dependent ferromagnetism in a van der Waals crystal down to the monolayer limit. *Nature* 546: 270–273.

5 Mermin, N.D. and Wagner, H. (1966). Absence of ferromagnetism or antiferromagnetism in one- or two-dimensional isotropic Heisenberg models. *Phys. Rev. Lett.* 17: 1133–1136.

6 Gong, C., Li, L., Li, Z. et al. (2017). Discovery of intrinsic ferromagnetism in two-dimensional van der Waals crystals. *Nature* 546: 265.

7 Deng, Y., Yu, Y., Song, Y. et al. (2018). Gate-tunable room-temperature ferromagnetism in two-dimensional Fe_3GeTe_2. *Nature* 563: 94–99.

8 Zhang, W., Qu, Q., Zhu, P., and Lam, C. (2015). Robust intrinsic ferromagnetism and half semiconductivity in stable two-dimensional single-layer chromium trihalides. *J. Mater. Chem. C* 3: 12457–12468.

9 Webster, L. and Yan, J.A. (2018). Strain-tunable magnetic anisotropy in monolayer $CrCl_3$, $CrBr_3$, and CrI_3. *Phys. Rev. B* 98: 144411.

10 Cai, X., Song, T., Wilson, N.P. et al. (2019). Atomically thin CrCl3: an in-plane layered antiferromagnetic insulator. *Nano Lett.* 19: 3993–3998.

11 Kim, H., Yang, B., Li, S. et al. (2019). Evolution of interlayer and intralayer magnetism in three atomically thin chromium trihalides. *Proc. Natl. Acad. Sci. U. S. A.* 116: 11131.

12 Xue, F., Hou, Y., Wang, Z., and Wu, R. (2019). Two-dimensional ferromagnetic van der Waals $CrCl_3$ monolayers with enhanced anisotropy and Curie temperature. *Phys. Rev. B* 100: 224429.

13 Sun, Q. and Kioussis, N. (2018). Prediction of manganese trihalides as two-dimensional dirac half-metals. *Phys. Rev. B* 97: 094408.

14 Son, S., Coak, M.J., Lee, N. et al. (2019). Bulk properties of the van der Waals hard ferromagnet VI_3. *Phys. Rev.* B 99: 041402(R).

15 Kong, T., Stolze, K., Timmons, E.I. et al. (2019). VI_3 – a new layered ferromagnetic semiconductor. *Adv. Mater.* 31: 1808074.

16 He, J., Ma, S., Lyu, P., and Nachtigall, P. (2016). Unusual Dirac halfmetallicity with intrinsic ferromagnetism in vanadium trihalide monolayers. *J. Mater. Chem. C* 4: 2518.

17 Wang, Y.-P. and Long, M.-Q. (2020). Electronic and magnetic properties of van der Waals ferromagnetic semiconductor VI_3. *Phys. Rev. B* 101: 024411.

18 Yang, K., Fan, F., Wang, H. et al. (2020). VI_3: a two-dimensional Ising ferromagnet. *Phys. Rev. B* 101: 100402(R).

19 Zhao, G., Liu, X., Hu, T. et al. (2021). Difference in magnetic anisotropy of the ferromagnetic monolayers VI_3 and CrI_3. *Phys. Rev. B* 103: 014438.

20 Lado, J.L. and Fernández-Rossier, J. (2017). On the origin of magnetic anisotropy in two dimensional CrI_3. *2D Mater.* 4: 035002.

21 Wang, Y., Li, S., Zhang, C.W., and Wang, P. (2018). High-temperature Dirac half-metal $PdCl_3$: a promising candidate for realizing quantum anomalous hall effect. *J. Mater. Chem. C* 6: 10284–10291.

22 Tian, Y., Gao, W., Henriksen, E.A. et al. (2019). Optically driven magnetic phase transition of monolayer $RuCl_3$. *Nano Lett.* 19: 7673–7680.

23 Subhan, F., Khan, I., and Hong, J. (2019). Pressure-induced ferromagnetism and enhanced perpendicular magnetic anisotropy of bilayer CrI_3. *J. Phys. Condens. Matter* 31: 355001.

24 Mukherjee, T., Chowdhury, S., Jana, D., and Voon, L. (2019). Strain induced electronic and magnetic properties of 2D magnet CrI_3: a DFT approach. *J. Phys. Condens. Matter* 31: 335802.

25 Li, H., Xu, Y.K., Lai, K., and Zhang, W.B. (2019). The enhanced ferromagnetism of single-layer CrX_3 (X = Br and I) via van der Waals engineering. *Phys. Chem. Chem. Phys.* 21: 11949–11955.

26 Pizzochero, M., Yadav, R., and Yazyev, O. (2020). Magnetic exchange interactions in monolayer CrI_3 from many-body wavefunction calculations. *2D Mater.* 7: 35005.

27 Kashin, I.V., Mazurenko, V.V., Katsnelson, M.I., and Rudenko, A.N. (2020). Orbitally-resolved ferromagnetism of monolayer CrI_3. *2D Mater.* 7: 025036.

28 Geng, J., Chan, I.N., Ai, H. et al. (2020). Magnetic and electronic properties of 2D TiX_3 (X = F, Cl, Br and I). *Phys. Chem. Chem. Phys.* 22: 17632–17638.

29 Xu, M., Huang, C., Li, Y. et al. (2020). Electrical control of magnetic phase transition in a type-I multiferroic double-metal trihalide monolayer. *Phys. Rev. Lett.* 124: 067602.

30 Kurumaji, T., Seki, S., Ishiwata, S. et al. (2013). Magnetoelectric responses induced by domain rearrangement and spin structural change in triangular-lattice helimagnets NiI_2 and CoI_2. *Phys. Rev. B* 87: 014429.

31 McGuire, M. (2017). Crystal and magnetic structures in layered, transition metal dihalides and trihalides. *Crystals* 7: 121.

32 Ashton, M., Gluhovic, D., Sinnott, S.B. et al. (2017). Two-dimensional intrinsic half-metals with large spin gaps. *Nano Lett.* 17: 5251–5257.

33 Taraphder, A., Laad, M.S., Craco, L., and Yaresko, A.N. (2008). GdI_2: a new ferromagnetic excitonic solid? *Phys. Rev. Lett.* 101: 136410.

34 Ryazanov, M., Simon, A., and Kremer, R.K. (2008). Magnetic freezing and spin frustration in the triangular lattice magnets GdI_2H_x ($0 \leq x < 1$). *Phys. Rev. B Condens. Matter Mater. Phys.* 77: 104423.

35 Kasten, A., Moller, P.H., and Schienle, M. (1984). Magnetic ordering in GdI_2. *Solid State Commun.* 51: 919–921.

36 Wang, B., Zhang, X., Zhang, Y. et al. (2020). Prediction of a two-dimensional high-TC f-electron ferromagnetic semiconductor. *Mater. Horiz.* 7: 1623–1630.

37 Kulish, V. and Huang, W. (2017). Single-layer metal halides MX_2 (X = Cl, Br, I): stability and tunable magnetism from first principles and Monte Carlo simulations. *J. Mater. Chem. C* 5: 8734–8741.

38 Wu, X., Cai, Y., Xie, Q. et al. (2012). Magnetic ordering and multiferroicity in MnI_2. *Phys. Rev. B: Condens. Matter Mater. Phys.* 86: 134413.

39 Farooq, M., Khan, I., Moaied, M., and Hong, J. (2017). Hydrogen functionalization induced two-dimensional ferromagnetic semiconductor in Mn di-halide systems. *Phys. Chem. Chem. Phys.* 19: 29516–29524.

40 Lu, M., Yao, Q., Xiao, C. et al. (2019). Mechanical, electronic, and magnetic properties of NiX_2 (X = Cl, Br, I) layers. *ACS Omega* 4: 5714–5721.

41 Yekta, Y., Hadipour, H., Şaşıoğlu, E. et al. (2021). Strength of effective Coulomb interaction in two-dimensional transition-metal halides MX_2 and MX_3 (M=Ti, V, Cr, Mn, Fe, Co, Ni; X=Cl, Br, I). *Phys. Rev. Mater.* 5: 034001.

42 Xiao, H., Wang, X., Wang, R. et al. (2019). Intrinsic magnetism and biaxial strain tuning in two-dimensional metal halides V_3X_8 (X = F, Cl, Br, I) from first principles and Monte Carlo simulation. *Phys. Chem. Chem. Phys.* 21: 11731–11739.

43 Bandyopadhyay, A., Frey, N.C., Jariwala, D., and Shenoy, V.B. (2019). Engineering magnetic phases in two-dimensional non-van der Waals transition-metal oxides. *Nano Lett.* 19: 7793–7800.

44 Zhang, X., Wang, B., Guo, Y. et al. (2019). High Curie temperature and intrinsic ferromagnetic half-metallicity in two-dimensional Cr_3X_4 (X = S, Se, Te) nanosheets. *Nanoscale Horiz.* 4: 859.

45 Bafekry, A., Sarsari, I.A., Faraji, M. et al. (2021). Electronic and magnetic properties of two-dimensional of FeX (X = S, Se, Te) monolayers crystallize in the orthorhombic structures. *Appl. Phys. Lett.* 118: 143102.

46 Sun, X., Li, W., Wang, X. et al. (2020). Room temperature ferromagnetism in ultra-thin van der Waals crystals of 1T-$CrTe_2$. *Nano Res.* 13: 3358–3363.

47 Xu, B., Li, S., Jiang, K. et al. (2020). Switching of the magnetic anisotropy via strain in two dimensional multiferroic materials: CrSX (X = Cl, Br, I). *Appl. Phys. Lett.* 116: 052403.

48 Hu, T., Wan, W., Li, Y. et al. (2002). Large Magnetic anisotropy energy and robust half-metallic ferromagnetism in 2D $MnXSe_4$ (X = As, Sb). *Ann. Phys.* 532: 2000365.

49 Feng, X., Liu, J., Ma, X., and Zhao, M. (2020). Ferroelectricity and multiferroicity in two-dimensional $Sc_2P_2Se_6$ and $ScCrP_2Se_6$ monolayers. *Phys. Chem. Chem. Phys.* 22: 7489–7496.

50 Zhang, F., Zhang, H., Mi, W., and Wang, X. (2020). Electronic structure, magnetic anisotropy and Dzyaloshinskii–Moriya interaction in Janus $Cr_2I_3X_3$ (X = Br, Cl) bilayers. *Phys. Chem. Chem. Phys.* 22: 8647–8657.

51 Wang, C., Zhou, X., Zhou, L. et al. (2019). A family of high-temperature ferromagnetic monolayers with locked spin-dichroism-mobility anisotropy: MnNX and CrCX (X = Cl, Br, I; C = S, Se, Te). *Sci. Bull.* 64: 293–300.

52 Mahajan, A. and Bhowmick, S. (2021). Decoupled strain response of ferroic properties in a multiferroic $VOCl_2$ monolayer. *Phys. Rev. B* 103: 075436.

53 Ding, N., Chen, J., Dong, S., and Stroppa, A. (2020). Ferroelectricity and ferromagnetism in a VOI_2 monolayer: role of the Dzyaloshinskii-Moriya interaction. *Phys. Rev. B* 102: 165129.

54 Xu, C., Zhang, J., Guo, Z. et al. (2021). A first-principles study on the electronic property and magnetic anisotropy of ferromagnetic CrOF and CrOCl monolayers. *J. Phys. Condens. Matter* 33: 195804.

55 Song, G., Li, D., Zhou, H. et al. (2021). Intrinsic room-temperature ferromagnetic semiconductor $InCrTe_3$ monolayers with large magnetic anisotropy and large piezoelectricity. *Appl. Phys. Lett.* 118: 123102.

56 Wu, D., Zhuo, Z., Lv, H., and Wu, X. (2021). Two-dimensional $Cr_2X_3S_3$ (X = Br, I) Janus semiconductor with intrinsic room-temperature magnetism. *J. Phys. Chem. Lett.* 11: 2905–2911.

57 Miao, N., Xu, B., Zhu, L. et al. (2018). 2D intrinsic ferromagnets from van der Waals antiferromagnets. *J. Am. Chem. Soc.* 140: 2417–2420.

58 Coïc, L., Spiesser, M., Palvadeau, P., and Rouxel, J. (1981). Chromium(III) oxyhalides: magnetic and optical properties, lithium intercalation. *J. Mater. Res. Bull.* 16: 229–236.

59 Maule, C., Tothill, J., Strange, P., and Wilson, J. (1988). An optical investigation into the 3d1 and 3d2 transition-metal halides and oxyhalides, compounds near to delocalisation. *J. Phys. C Solid State Phys.* 21: 2153–2179.

60 Seidel, A., Marianetti, C., Chou, F. et al. (2003). S = 1/2 chains and spin-Peierls transition in TiOCl. *Phys. Rev. B* 67: 020405.

61 Glawion, S., Scholz, M., Zhang, Y. et al. (2009). Electronic structure of the two-dimensional Heisenberg antiferromagnet VOCl: a multiorbital Mott insulator. *Phys. Rev. B* 80: 155119.

62 Zhang, Y., Foyevtsova, K., Jeschke, H. et al. (2010). Can the Mott insulator TiOCl be metallized by doping? A first-principles study. *Phys. Rev. Lett.* 104: 146402.

63 Guo, Y., Zhang, Y., Yuan, S. et al. (2018). Chromium sulfide halide monolayers: intrinsic ferromagnetic semiconductors with large spin polarization and high carrier mobility. *Nanoscale* 10: 18036–18042.

64 Zhang, L., Pan, J., Yan, Q., and Hu, Z. (2020). Computational study of a novel 2D Ferromagnetic Metal: the Ce_2C Monolayer. *Phys. Status Solidi (RRL)* 14: 2000324.

65 Jiang, X., Liu, Q., Xing, J., and Zhao, J. (2019). Two-dimensional AXenes: a new family of room-temperature d0 ferromagnets and their structural phase transitions. *J. Phys. Chem. Lett.* 10 (24): 7753–7759.

66 Shi, G., Chen, L., Yang, Y. et al. (2018). Two-dimensional Na–Cl crystals of unconventional stoichiometries on graphene surface from dilute solution at ambient conditions. *Nat. Chem.* 10: 776–779.

67 Oganov, A. (2018). 2D materials worth their salt. *Nat. Chem.* 10: 694–695.

68 Zhang, L., SHi, G., Peng, B. et al. (2018). Two-dimensional Ca-Cl crystals under ambient conditions observed directly by cryo-electron microscopy. *arXiv*:1812.07195v1.

69 Zhang, W., Wong, P., Zhu, R., and Wee, A. (2019). Van der Waals magnets: wonder building blocks for two-dimensional spintronics? *InfoMat* 1: 479–495.

70 Wei, S., Tang, X., Liao, X. et al. (2019). Recent progress of spintronics based on emerging 2D materials: CrI_3 and Xenes. *Mater. Res. Express* 6: 122004.

71 Tang, X. and Kou, L. (2019). Two-dimensional ferroics and multiferroics: platforms for new physics and applications. *J. Phys. Chem. Lett.* 10: 6634–6649.

72 Mak, K., Shan, J., and Ralph, D. (2019). Probing and controlling magnetic states in 2D layered magnetic materials. *Nat. Rev. Phys.* 1: 646–661.

73 Zhai, B., Du, J., Li, X. et al. (2019). Two-dimensional ferromagnetic materials and related van der Waals heterostructures: a first-principle study. *J. Semicond.* 40: 081509.

74 Sethulakshmi, N., Mishra, A., Ajayan, P. et al. (2019). Magnetism in two-dimensional materials beyond graphene. *Mater. Today* 27: 107–122.

75 Li, H., Ruan, S., and Zeng, Y. (2019). Intrinsic van der Waals magnetic materials from bulk to the 2D limit: new frontiers of spintronics. *Adv. Mater.* 10: 1900065.

76 Gbertini, M., Koperski, M., Morpurgo, A., and Novoselov, K. (2019). Magnetic 2D materials and heterostructures. *Nat. Nanotechnol.* 14: 408–419.

77 Gong, C. and Zhang, X. (2019). Two-dimensional magnetic crystals and emergent heterostructure devices. *Science* 363: 4450.

78 Ma, Y., Kuc, A., Jing, Y. et al. (2017). Two-dimensional haeckelite NbS_2: a diamagnetic high-mobility semiconductor with Nb^{4+} ions. *Angew. Chem. Int. Ed.* 56: 1–6.

79 Ma, Y., Dai, Y., Li, X. et al. (2014). Prediction of two-dimensional materials with half-metallic Dirac cones: $Ni_2C_{18}H_{12}$ and $Co_2C_{18}H_{12}$. *Carbon* 17: 382–388.

80 Leuenberger, M. and Loss, D. (2001). Quantum computing in molecular magnets. *Nature* 410: 789–793.

81 Wu, M. and Zeng, X. (2011). Transition-metal-molecular sandwich nanowires as magnetic on/off switch. *Appl. Phys. Lett.* 99: 053121.

82 Lakshmi, S., Roche, S., and Cuniberti, G. (2009). Spin-valve effect in zigzag graphene nanoribbons by defect engineering. *Phys. Rev. B* 80: 193404.

83 Maslyuk, V., Bagrets, A., Meded, V. et al. (2006). Organometallic benzene-vanadium wire: a one-dimensional half-metallic ferromagnet. *Phys. Rev. Lett.* 97: 097201.

84 Wang, L., Cai, Z., Wang, J. et al. (2008). Novel one-dimensional organometallic half metals: vanadium-cyclopentadienyl, vanadium-cyclopentadienyl-benzene, and vanadium-anthracene wires. *Nano Lett.* 8: 3640–3644.

85 Zhou, L., Yang, S.-W., Ng, M.-F. et al. (2008). One-dimensional iron–cyclopentadienyl sandwich molecular wire with half metallic, negative differential resistance and high-spin filter efficiency properties. *J. Am. Chem. Soc.* 130: 4023–4027.

86 Wu, X. and Zeng, X. (2009). Double metallocene nanowires. *J. Am. Chem. Soc.* 131: 14246.

87 Cho, W., Cho, Y., Min, S. et al. (2011). Chromium porphyrin arrays as spintronic devices. *J. Am. Chem. Soc.* 133: 9364.

88 Li, L., Li, J., and Wu, L. (2012). Polymeric fused-ring type iron phthalocyanine nanosheet and its derivative ribbons and tubes. *J. Phys. Chem. C* 116: 9235–9242.

89 Li, Y., Zhou, Z., and Chen, Z. (2012). From anadium naphthalene ($V_{n-1}Np_n$) sandwich clusters to VNp sandwich nanowire: structural, energetic, electronic, and magnetic properties. *J. Phys. Chem. A* 116: 1648–1654.

90 Zhang, X., Tian, Z., Yang, S., and Wang, J. (2011). Magnetic manipulation and half-metal prediction of one-dimensional bimetallic organic sandwich molecular wires $[CpTM_1CpTM_2]\infty$ (TM_1 = Ti, Cr, Fe; TM_2 = Sc-Co). *J. Phys. Chem. C* 115: 2948–2953.

91 Ma, Y., Dai, Y., Zhang, Z. et al. (2012). Magnetic properties of phthalocyanine-based organometallic nanowire. *Appl. Phys. Lett.* 101: 062405.

92 Ma, Y., Dai, Y., Wei, W., and Huang, B. (2013). Engineering intriguing electronic and magnetic properties in novel one-dimensional staircase-like metallocene wires. *J. Mater. Chem. C* 1: 941–946.

4

Two-Dimensional Topological States

The discovery of topological states of matter, as one of the most transformative breakthroughs, has greatly enriched fundamental knowledge in condensed matter physics and materials science [1–6]. According to the respective electronic structures, topologically nontrivial materials have been found in both insulators and semimetals, which host a plethora of exotic phenomena such as the quantized Hall and magnetoelectric effects, edge/surface states, and Fermi arcs, and remarkably they are topologically protected by fundamental symmetries of the bulks. It is generally believed that a material becomes topologically nontrivial if the conduction and valence bands are inverted after including the spin-orbit coupling (SOC), where topological invariants can be calculated based on the wave functions in momentum pace and identify the topological states accurately. Notably, advances in topological band theory using symmetry indicators and topological quantum chemistry provide a convenient way to probe and classify the band topology with the band representations at high-symmetry momenta [7–11]. In fact, for the prediction of topological material candidates, first-principles calculations play a key role and very often guide the experiments. For example, past decade has witnessed the emergence of topological insulators (TIs) in Bi_2Se_3 family [12, 13], topological crystalline insulators (TCIs) in SnTe family [14–18], Dirac semimetals in Na_3Bi and Cd_3As_2 [19–24], Weyl semimetals in TaAs family [25–29], and antiferromagnetic (AFM) TIs in $MnBi_{2n}Te_{3n+1}$ ($n = 1, 2, 3$) family [30–39], which were first proposed theoretically before experimental verification. Recently, guided by the symmetry indicators and topological quantum chemistry, the high-throughput first-principles calculations predicted that at least a quarter of all known materials are topologically nontrivial [40–42].

Topological states started from the two-dimensional (2D) systems. In 1980, Klaus von Klitzing experimentally discovered the quantum Hall plateau in 2D electron gas, i.e. quantum Hall effect. Under strong magnetic field and low temperature, the Hall conductance is quantized in units of e^2/h, where h is the Planck's constant and e is the charge of an electron [43]. It is now realized that the quantum Hall effect is the first experimental realization of topological states with the topologically nontrivial electronic structure characterized by a finite Chern number C and C number of chiral edge states. Interestingly, Haldane proposed that the similar topologically nontrivial characters can be achieved even without external magnetic field, leading

Calculations and Simulations of Low-Dimensional Materials: Tailoring Properties for Applications, First Edition. Ying Dai, Wei Wei, Yandong Ma, and Chengwang Niu.

to the quantum anomalous Hall (QAH) effect [44]. Another important milestone in 2D topological states is the quantum spin Hall (QSH) effect protected by the time-reversal symmetry (T), also called the 2D TI [45, 46], that can be regarded as a combination of two QAH effects with opposite chirality [47, 48]. When T-symmetry is broken usually by magnetic impurity or substrate, QSH insulator can be switched into the QAH insulator, and based on this theoretical prediction, the QAH effect is indeed demonstrated in Cr-doped $(Bi,Sb)_2Te_3$ experimentally [49, 50]. On the other hand, when the role of T-symmetry is replaced by the crystalline mirror and/or rotation symmetries, the TCIs can be obtained. Moreover, research on topological states also started to reach out to 2D magnets, and the quantized spin Hall conductance has been experimentally demonstrated in Mn-doped 2D HgTe, pushing the first-principles calculations to pay more attention to this significant burgeoning research field that has reshaped our understanding of physics and materials [51–53].

4.1 Topological Insulators

Topological insulators are new quantum states of matter protected by the time-reversal symmetry. Distinct from ordinary insulators and semiconductors with strictly separated valence and conduction bands by trivial band gaps in both the bulk and edge/surface, the nontrivial TIs exhibit metallic states on the edge/surface, while the bulk remains insulating [1, 2]. A material with large SOC is usually found to become nontrivial if its valence and conduction bands are inverted. In reality, the band inversion is not new to physics, and it can be traced back to 1939 when William Shockley, using a spinless model, revealed that band inversion induces unique electronic states on the surface [54]. In the absence of SOC, the band inversion can lead to gapless bulk phase. Interestingly, the SOC opens a full gap, giving rise to the QSH insulator as predicted first in graphene with Kane–Mele model in the 2D honeycomb lattice [55]. On the other hand, band inversion is typically induced by the SOC at certain high-symmetry k-points of the Brillouin zone. It refers to the order change between the highest occupied and lowest unoccupied bands, for example the exchange of opposite parities for a material with space inversion symmetry, providing another viewpoint to understand the QSH insulator as realized in the HgTe/CdTe quantum wells with Bernevig–Hughes–Zhang (BHZ) model in the 2D square lattice [56]. These two seminal proposals work out the concept of QSH insulator. However, it is necessary to specify that the opening of band gap or the band inversion driven by SOC is only an important signature but cannot guarantee the emergence of the QSH effect, and further investigation of the topological invariants and gapless edge states are required.

4.1.1 Graphene

Graphene is the single atomic layer of carbon atoms that crystallizes in the honeycomb lattice, and is often regarded as a prototypical 2D material with unique electronic, optical, mechanical, and transport properties [57, 58]. Remarkably, graphene is the first realization of a material where the QSH effect is proposed. In 2005, Kane and Mele described the topology of the single plane graphene with the effect of SOC, and the Hamiltonian can be expressed as [45, 55]:

$$H = t \sum_{\langle i,j \rangle \alpha} c_{i\alpha}^{+}(c_{j\alpha}) + i\lambda \sum_{\langle\langle ij \rangle\rangle \alpha\beta} v_{ij} c_{i\alpha}^{+} s_{\alpha\beta}^{z} c_{j\beta},$$

where $c_{i\alpha}^{+}(c_{j\alpha})$ denotes the creation (annihilation) of an electron with the spin polarization α at site i. $\langle i,j \rangle$ and $\langle\langle i,j \rangle\rangle$ restrict the sums to nearest and next-nearest neighbors, respectively. The first term is the usual nearest-neighbor hopping. The second term represents the intrinsic SOC with $v_{ij} = \pm 1$ chosen as the counterclockwise and clockwise with respect to the positive z-axis for the next-nearest-neighboring hopping.

In the absence of SOC, the system is gapless with two linear Dirac-like band structures centered at the Brillouin zone corners K and K', usually called as Dirac semimetal. Switching on SOC leads to a band gap at the original Dirac points in which the intrinsic SOC plays the corresponding role of the alternative magnetic flux in Haldane's model, and thus harbors two copies of QAH effect with opposite spins and chiralities [44, 45, 55]. As shown in Figure 4.1, the two copies of edge modes with opposite spins are protected by the time-reversal symmetry and related to each other by the Kramers degeneracy theorem, leading to the most important feature of 2D TIs, i.e. the elastic backscattering between spin-helical gapless edge states is forbidden. However, the band gap of graphene is quite small due to the extremely weak intrinsic SOC, making the 2D TIs unrealistic in pristine graphene by employing the current experimental techniques.

4.1.2 HgTe/CdTe Quantum Wells

Parallel to the proposal of 2D TIs in graphene with half-filled bands, could be written as a sum of split elementary band representations (EBRs) in topological quantum

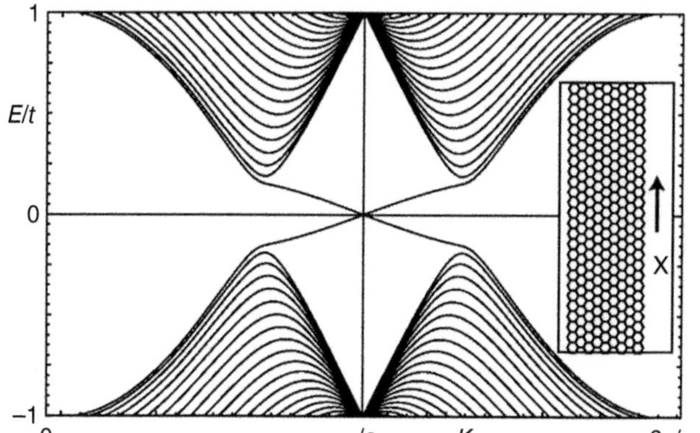

Figure 4.1 Band structure for one-dimensional nanoribbon of graphene. The bands crossing the gap are spin filtered edge states. Source: Kane and Mele [55]/with permission of American Physical Society.

chemistry, Bernevig et al. suggested another route to obtain the 2D TI by utilizing a semiconductor with fully filled valence bands, which are denoted more generically "no linear combination" of EBRs, as proposed in the HgTe/CdTe quantum wells [46, 59]. Remarkably, one year after the theoretical proposal, electrical conductance due to the edge states is indeed observed in transport experiments of the inverted HgTe/CdTe quantum wells, the first signature of 2D TI phase [60].

Similar to other conventional semiconductors composed of light elements, the filled valence bands of CdTe are gapped from the conduction bands with the s-type Γ_6 band lying above the p-type Γ_8 band. By contrast, in HgTe, the p-type Γ_8 band rises above the s-type Γ_6 band, leading to an inverted band structure. Bernevig et al. considered a quantum well structure where HgTe is sandwiched between layers of CdTe. When the thickness of HgTe layer is $d < d_c$ ($d_c = 6.3$ nm), the HgTe/CdTe quantum wells host normal band order because of the dominating contribution from CdTe. However, for $d > d_c$, a band inversion occurs with increasing contribution from HgTe. The inversion of bands versus the increasing thickness d reveals a topological phase transition from the trivial insulator to 2D TIs. Around the critical point, the general form of the effective Hamiltonian can be written as [59, 61]:

$$H_{\text{eff}}(k) = \varepsilon(k) + \begin{bmatrix} m(k) & A\pi^+ & & \\ A^*\pi & -m(k) & & \\ & & m(-k) & -A^*\pi \\ & & -A\pi^+ & -m(-k) \end{bmatrix},$$

where $\varepsilon(k) = C - D\left(k_x^2 + k_y^2\right)$ and $m(k) = M - B\left(k_x^2 + k_y^2\right)$ represent the symmetric and asymmetric mass terms, respectively. The two decoupled blocks are related by the time-reversal operation. M is the mass or gap parameter defined as the energy difference between the E1 and H1 levels at the Γ point. When M equals to zero, the gapless Dirac dispersion appears. As varying the thickness d of the HgTe/CdTe well, the E1 and H1 bands must cross at the d_c, and M changes sign between the two sides of the topological phase transition. The QSH effect appears also in InAs/GaSb quantum wells, but a small bulk band gap leads to the experimental observation requiring extreme conditions, such as the ultralow temperature and precisely controlled molecular-beam epitaxy, which greatly obstructs further experimental investigations and potential room temperature applications [62].

4.1.3 Z_2 Invariant and Spin Chern Number

For a time-reversal symmetric insulator, the nontrivial topology can be accurately characterized by the Z_2 invariant, given by [45, 63]:

$$Z_2 = \frac{1}{2\pi}\left[\oint_{d\tau} A(k)dl - \int_{\tau} F(k)d\tau\right] \text{mod}(2),$$

where $A(k) = i\sum_{n=1}^{N}\left\langle u_n(k) \mid \nabla_k \mid u_n(k)\right\rangle$ represents the Berry connection over all of the occupied states, and $F(k) = \nabla_k \times A(k)$ represents the corresponding Berry curvature [64]. The integrals are over half of the 2D Brillouin zone surface τ and its boundary $d\tau$. There are only two values for the Z_2 invariant due to the presence of the mod(2) term, i.e. "0" and "1," suggesting the binary classification of time-reversal symmetric insulators with topologically trivial and nontrivial phases, respectively. With inversion symmetry, Fu and Kane provide a convenient way to calculate the Z_2 invariant, which requires only the parity of the occupied Bloch wave functions at the time-reversal invariant momentum (TRIM) points in the Brillouin zone [65]. There are four TRIM in the 2D Brillouin zone, and the time-reversal symmetry yields one unique Z_2 invariant ν with:

$$(-1)^{\nu} = \prod_{i=1}^{n} \delta_i, \quad \delta_i = \prod_{m} \xi_m(\Lambda_i)$$

where n is 4 in 2D system. δ_i is the product of parity eigenvalues at the TRIM points Λ_i. $\xi_m(\Lambda_i)$ are parity eigenvalues with $\xi_m(\Lambda_i) = \pm 1$ and m is the number of all occupied bands. According to the Z_2 classification, $\nu = 1$ identifies a topologically nontrivial 2D TI where the system is robust against weak time-reversal invariant perturbations, and $\nu = 0$ characterizes a topologically trivial phase.

This Fu–Kane approach can also be used to probe the three-dimensional (3D) TIs that involve eight TRIM points in the 3D Brillouin zone. In addition, there are four distinct Z_2 invariants $(\nu_0; \nu_1\nu_2\nu_3)$ in three dimensions. Here, $\nu_0 = 1$ characterizes the strong TIs, for which the time-reversal symmetry protects metallic surface bands on all of the surfaces. On the other hand, weak TIs are obtained when $\nu_0 = 0$ but at least one of the indices ν_1, ν_2, or ν_3 is nonzero. The weak TIs can be viewed as a stacking of 2D TIs and display protected metallicity only at surfaces with a certain orientation while other surfaces do not contain topologically protected surface states. When all four invariants are zero, a topologically trivial insulator is obtained.

In 2D systems, spin Chern number C_S provides an equivalent characterization to the Z_2 invariant, i.e. the nonzero C_S is equivalent to the nonzero Z_2 invariant, that is given as $C_S = (C_+ - C_-)/2$ [66, 67]. Here, C_+ and C_- are Chern numbers for the spin-up and spin-down channels [68, 69],

$$C = \frac{1}{2\pi} \int_{BZ} \Omega(k) d^2k,$$

and $\Omega(k)$ is the Berry curvature over all the occupied states,

$$\Omega(k) = \sum_{n<E_F} \sum_{m \neq n} 2\mathrm{Im} \frac{\langle \psi_{nk} | \upsilon_x | \psi_{mk} \rangle \langle \psi_{mk} | \upsilon_y | \psi_{nk} \rangle}{(\varepsilon_{mk} - \varepsilon_{nk})^2}.$$

where m,n are band indices, $\psi_{m/nk}$ and $\varepsilon_{m/nk}$ are the Bloch wave functions and corresponding eigenenergies of band m/n, respectively, and $\upsilon_{x/y}$ are the velocity operators. To distinguish the spin-up and spin-down channels, a projected spin operator P_z, $P_z = PS_zP$, where S_z is the spin operator and P is the projector operator of the occupied states below the Fermi level, needs to be constructed and diagonalized.

Although the SOC causes the spin degree of freedom to no longer be a good quantum number, the eigenvalues of P_z are then not necessarily ± 1. But as long as the spectrum of P_z is gapped, the occupied states can still be divided into the spin-up and spin-down manifolds [67]. The obtained even values of C_S correspond to a topologically trivial insulator state, while odd values of C_S indicate the emergence of a 2D TI phase.

4.1.4 Large Gap Quantum Spin Hall Insulators

Similar to graphene, first-principles calculations show that silicene, germanene, and stanene, atomically thin crystals of Si, Ge, and Sn, with topologically nontrivial electronic structures can realize the QSH effect. They are found to be buckled monolayers and not flat like graphene, and remarkably, larger band gaps are obtained because of the stronger SOC, reaching as much as 23.9 meV and 100 eV for germanene and stanene, respectively [70, 71]. Moreover, the band gap can be effectively tuned by the chemical functionalization, for example, an insulating gap as large as 300 meV is predicted in I-decorated Sn [71]. A large energy gap is essential for the practical utilization of 2D TIs to stabilize the boundary current against the influence of thermally activated bulk carriers.

As the heaviest atom with strong SOC, bismuth is an important ingredient for 2D TIs. The low-buckled GaBi, InBi, and TlBi are predicted to be 2D TIs [73], and the Bi monolayer has attracted great attention due to a relatively large band gap of 0.2 eV [74–76] with the edge states observed experimentally [77]. However, (111)-oriented Bi monolayer is unstable on weakly interacting substrates and grows in the black phosphorous (A17) structure that resembles ($\bar{1}10$) layers of the bulk Bi (A7) structure and turns out to be topologically trivial [78–80]. First-principles calculations indicate that the crystal, electronic, and topological properties of ultrathin Bi films can be drastically modified when decorated by H [72]. Figure 4.2 presents the crystal structures of H-decorated Bi (111) and H-decorated Bi ($\bar{1}10$). In contrast to Bi (111), H-decorated Bi (111) has a quasiplanar geometry and the hexagonal lattice parameter increases from 4.54 to 5.49 Å. H-Bi ($\bar{1}10$) forms a structure with an AB stacking of the pseudosquare layers, and the inversion symmetry is obtained. Moreover, the structure is dynamically stable as shown in Figure. 4.2c.

In the case without SOC, as shown in Figure 4.3a and c, their band structures are gapless and show a semimetallic character with the Dirac band crossing exactly at the K point for the H-decorated Bi (111) and slightly away from the Y point for the case of H-decorated Bi ($\bar{1}10$). This is different from other Bi-based TIs, such as the Bi (111) bilayer [74–76] and the Bi_2Se_3 family [12], but is quite similar to graphene [55]. Taking SOC into account, a band gap opens as illustrated in Figure 4.3b and d. Different from both the Bi-based TIs and graphene, the Dirac-related bands have contributions mainly from the p_x and p_y orbitals while the p_z orbital is removed away from the Fermi level by H, resulting in the large band gap of 1.01 eV. A similar mechanism was reported as well for a Bi/Si system [81]. Their band topology is explicitly confirmed by the Z_2 invariant calculated based on the wave function parities at four TRIM points, i.e. the Γ and three M points for H-decorated Bi (111) and Γ, X, Y, and M for H-decorated Bi ($\bar{1}10$), as displayed in the insets of Figure 4.3b and d. Similarly, the QSH insulators emerge in the F-, Cl-, Br-, and I-decorated Bi (111), with gigantic energy gaps of 1.10, 0.93, 0.88, and 0.87 eV, respectively [72], and Cl-decorated GaBi with a nontrivial band gap of 0.65 eV [82]. In addition, the methyl group is reported

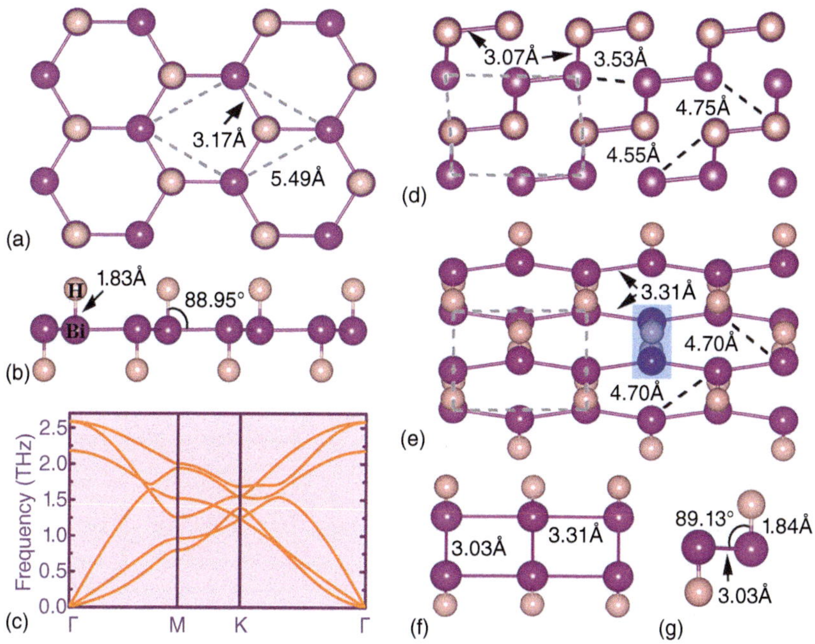

Figure 4.2 (a) Top and (b) side view of optimized structures for H-decorated Bi (111). The corresponding phonon band structure is shown in (c). The top view of the unrelaxed and relaxed crystal structure of H-decorated Bi ($\bar{1}10$) is shown in (d) and (e), respectively. (f) Side view of the relaxed crystal structure of H-decorated Bi ($\bar{1}10$). (g) Zoomed-in view of the highlighted areas in (e). The unit cells are indicated by gray dashed lines. Source: Niu et al. [72]/with permission of American Physical Society.

to be suitable for surface passivation such as the methyl-functionalized surfaces are resilient toward oxidation [83, 84]. The methyl-functionalized Bi, Sb, and Pb are also theoretically proposed to be in the QSH phase with the large nontrivial bulk gaps, and their topologically nontrivial properties are robust against the mechanical deformation [85].

For device applications, it is important to make sure that the predicted large band gap and topological properties are preserved [81]. $\sqrt{3} \times \sqrt{3}$ MoS_2 fits H-decorated Bi (111) nicely both in the lattice constant and in the alignment of the band gaps. First-principles calculations confirm that both the large band gap and nontrivial topological properties are unchanged [72]. Remarkably, the high-temperature QSH effect is experimentally realized with Bi on top of the insulating SiC substrate [86].

A good layered material can be easily made even by the scotch-tape method like graphene and easily obtain the chemically stable 2D system. Therefore, good layered materials and large 2D bulk band gaps are the most important criteria to realize the QSH effect at high temperatures. The previously known thermoelectric materials $ZrTe_5$ and $HfTe_5$ are interlayer weakly bonded binary compounds comparable to graphite. As illustrated in Figure 4.4a, they crystallize in the orthorhombic layered structure with space group *Cmcm* [88]. The interlayer binding energy of $ZrTe_5$ and/or $HfTe_5$ is as weak as that of graphite and is much smaller than that of Bi_2Se_3

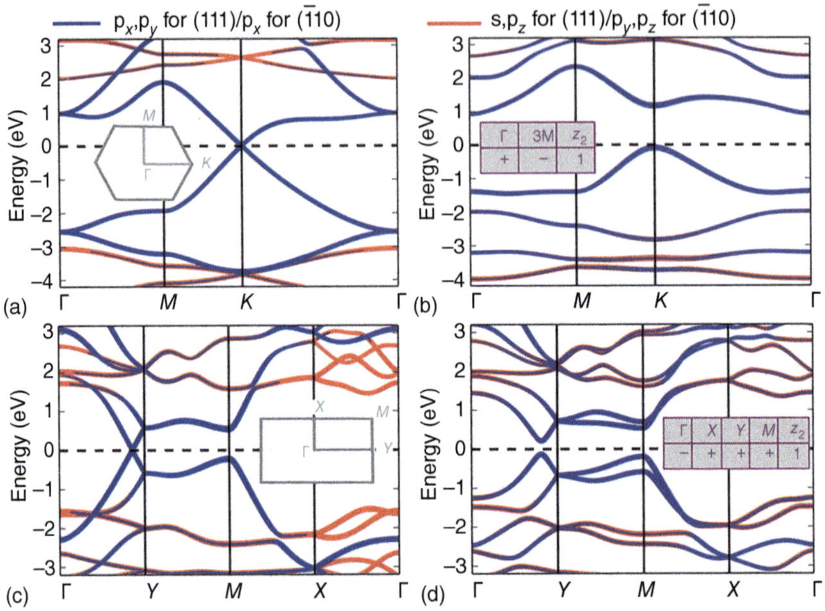

Figure 4.3 Orbitally resolved band structures for (a), (b) H-decorated Bi (111) and (c), (d) H-decorated Bi ($\bar{1}10$), without ((a), (c)) and with ((b), (d)) SOC, weighted with the s, p_x, p_y, and p_z characters. Dark (blue) colors mark states that contribute to the fundamental band gap. The Fermi level is indicated by the dashed line. The insets in (a) and (c) show the 2D Brillouin zone, and those in (b) and (d) show the products of the parities of all occupied bands at the time-reversal invariant momenta and the Z_2 invariant. Source: Niu et al. [72]/with permission of American Physical Society.

and Bi (111), suggesting that experimental fabrication of single-layer $ZrTe_5$ and/or $HfTe_5$ is possible simply by exfoliation from their layered bulk [87]. In the absence of SOC, the system is a semimetal, and one Dirac band crossing appears along the Γ-X direction that is unavoidable because the Dirac-related bands belong to different representations distinguished by the mirror symmetry m_{xz}. The inclusion of SOC will mix them and open up a gap, leading to the QSH insulator with a large band gap of 0.4 eV for $ZrTe_5$ and 0.1 eV for $HfTe_5$ [87]. The 2D nontrivial insulating state is characterized by nonzero Z_2 invariant with $Z_2 = 1$, and support topologically protected gapless edge states as shown in Figure 4.4d.

A SOC-induced band inversion is obtained in single-layer Bi_4Br_4 with a nonzero Z_2 invariant of $Z_2 = 1$, verifying that the single-layer Bi_4Br_4 is a QSH insulator [89]. Bulk Bi_4Br_4 crystallizes in the monoclinic space group *C2/m* that can be regarded as a combined packing of the normal and mirror-reflected single layer along the *z*-axis [90]. The middle Bi_{in} atoms form a zigzag chain, and the Br atoms are tightly attached to Bi_{ex} along the edges of the molecular chain. The complex intrachain and interchain coupling leads to the splitting of $p_x^{\pm}$ orbitals from two adjacent chains within the 2D cell into bonding and antibonding states bonding, and the band order

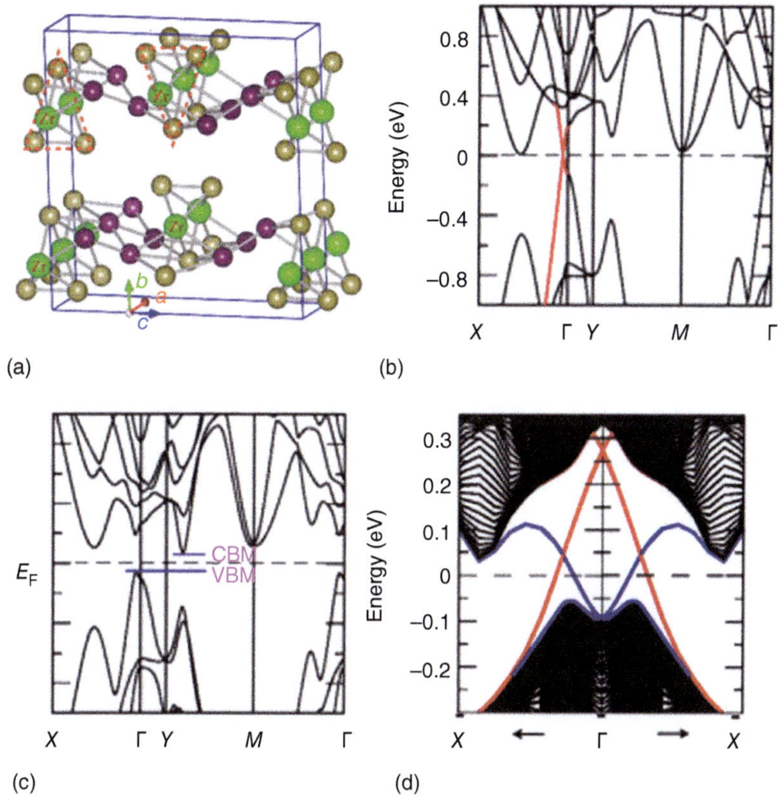

Figure 4.4 (a) The layered crystal structure of 3D bulk $ZrTe_5$ in space group *Cmcm*. The band structures of single-layer $ZrTe_5$ (b) without and (c) with SOC. (d) Calculated gapless edge states for $ZrTe_5$ nanoribbon. Source: Weng et al. [87]/American Physical Society/CC BY 3.0.

is inverted after including SOC as illustrated in Figure 4.5a. One can clearly see the SOC-induced band inversion from the Bi_{in}-p_x and Bi_{ex}-p_x orbitals projected bands in Figure 4.5c and d. In the case without SOC, the CBM at the Γ point is dominated by the Bi_{ex}-p_x orbital with a positive parity, while the VBM is dominated by the Bi_{in}-p_x orbital with a negative parity. Taking SOC into account, both the orbital character and the parity of CBM and VBM are inverted.

Another major milestone in 2D TIs is the large gap QSH insulators in 2D transition metal dichalcogenides with 1T′ structure, namely, 1T′-MX_2 with M = (W, Mo) and X = (Te, Se, S) [91]. An intrinsic band inversion between X-p and M-d orbitals takes place even without including SOC due to the period doubling of the metal chain in the 1T′ structure, which lowers the M-d orbital below X-p orbital around the Γ point. In the absence of SOC, this intrinsic band inversion results in the appearance of two Dirac cones located at finite momenta along Y-Γ-Y in the 2D Brillouin zone. Then, SOC opens up a gap at the Dirac points, leading to the QSH insulator

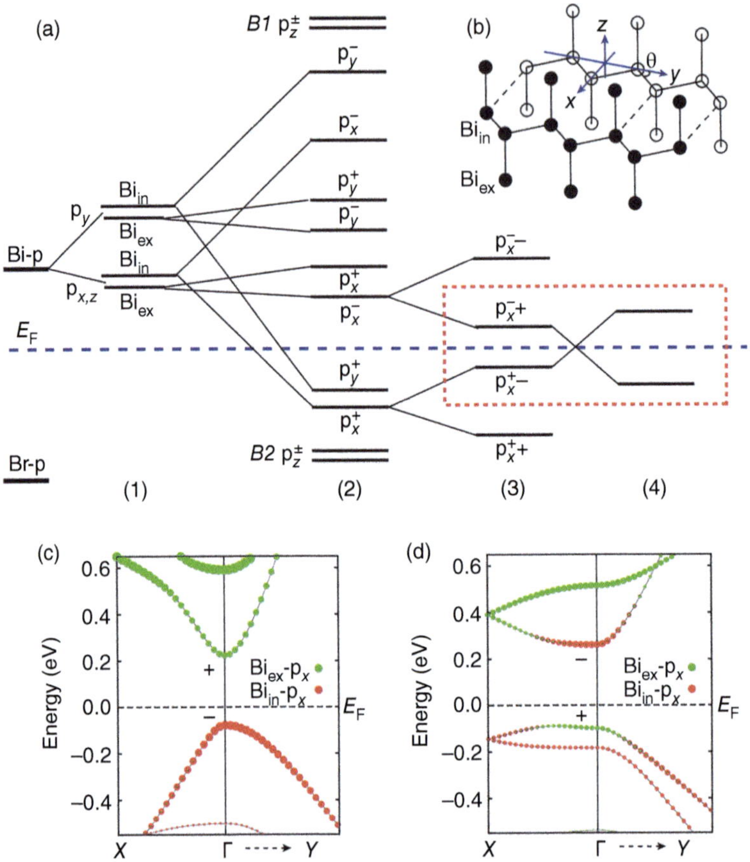

Figure 4.5 (a) Schematic diagram of band evolution in single-layer Bi_4Br_4. (b) Schematic plot of single-layer Bi_4Br_4 structure with the Br atoms neglected. Orbitally resolved band structures (c) without and (d) with SOC. The parity is labeled by "+" and "−" for the conduction band minimum (CBM) and valence band maximum (VBM) at the Γ point. Source: Zhou et al. [89]/with permission of American Chemical Society.

characterized with $Z_2 = 1$, which is further explicitly confirmed by the emergence of the spin-polarized gapless edge states. Remarkably, the angle-resolved photoemission (ARPES) and scanning tunnelling microscopy (STM) measurements establish the 1T′-MX_2 as a new class of QSH insulator with large band gap experimentally [92–94]. The inverted bands are tunable by the vertical electric field that breaks inversion symmetry and introduces a strong Rashba splitting of the doubly degenerate bands. As the electric field increases, band gap becomes zero at a critical strength of 0.142 V Å^{-1} and then reopens. The gap closing and reopening process induces a topological phase transition to a trivial phase with the destruction of helical edge states, revealing the possibility of an all-electrical control of the on/off charge/spin conductance of helical edge states, which is highly desirable for van der Waals (vdW) devices [91].

4.2 Topological Crystalline Insulators

Topological crystalline insulator (TCI) refers to topological quantum state with an insulating bulk energy gap and gapless edge or surface states arising from the crystalline mirror and/or rotation symmetries [3, 95]. Similar to the Z_2 TI, the TCI cannot be adiabatically deformed to an atomic insulator as well while preserving certain crystalline mirror and/or rotation symmetries. The first class of theoretically predicted TCI material is the 3D IV-VI semiconductor SnTe, which crystallizes in rock salt structure [14]. The symmetry responsible for topologically nontrivial character is the crystalline mirror symmetry. In stark contrast, the isostructural compound PbTe in the same rock salt structure is proposed to be topologically trivial. Both SnTe and PbTe have small direct band gaps with the VBM and CBM located at four symmetry-related TRIM points, the L points. The VBM and CBM at a given L form two sets of Kramers doublets with opposite parity eigenvalues. However, their band ordering at L points is inverted, and the first-principles calculations indicate that the SnTe hosts an intrinsically inverted band structure, i.e. the topologically nontrivial one, where the VBM and CBM are derived from the Sn-p and Te-p states, respectively. However, there are two L points on the (110) and each equivalent mirror plane, and SnTe is not a TI owing to an even number of band inversions. Interestingly, the double band inversion gives the nonzero mirror Chern number $C_{\mathrm{M}} = 2$. The C_{M} is given as $C_{\mathrm{M}} = (C_{+i} - C_{-i})/2$, where C_{+i} and C_{-i} are Chern numbers of all occupied bands with opposite mirror eigenvalues $+i$ and $-i$, respectively [96]. The nonzero C_{M} guarantees the existence of topological surface states on the crystal faces that are symmetric with respect to the (110) mirror planes. A key characteristic of TCIs is the presence of topological surface states that are protected by crystalline mirror symmetry, rather than time-reversal symmetry, Dirac points located at non-TRIM points as experimentally observed in SnTe and $Pb_{1-x}Sn_xTe$ alloy [15–18]. Moreover, the TCI state has been theoretically predicted to occur in two dimensions as discussed next.

4.2.1 SnTe Thin Films

The (001) films of 3D TCI SnTe with an odd number of atomic layers are symmetric under the reflection $z \to -z$ about the 2D plane in the middle [97]. In this case, the Bloch states can be chosen to be eigenstates $\pm i$ of the mirror symmetry m_z in all 2D Brillouin zone. For each class of Bloch eigenstates, one can define the corresponding Chern numbers $C_{\pm i}$, resulting in two independent topological invariants – the total Chern number $C = C_{+i} + C_{-i}$ and mirror Chern number $C_{\mathrm{M}} = (C_{+\mathrm{i}} - C_{-\mathrm{i}})/2$. Importantly, even when the C is zero, the C_{M} can be a nonzero integer and thus defines the 2D TCI phase protected by the mirror symmetry $z \to -z$. Because of the bulk topology for 3D TCI SnTe, the topological surface sates (001) surface, which is normal to (110) and (110) mirror planes, can be expected for SnTe. However, when the film thickness is below the penetration length of surface states, there is an energy splitting between the bonding and antibonding states due to the wave function hybridization between the top and bottom surfaces and the band ordering around the X point depends crucially on the competition between the hybridization of the two surfaces. As the film thickness increases, the hybridization becomes weaker, the band gap at X, $E_g(X) = E_{A+}(X) - E_{C-}(X)$, decreases from three to five layers, and then increases monotonically and eventually reaches the value 220 meV above five layers, as schematically shown in Figure 4.6b and c. Clearly, a band inversion occurs from three to five layers with the five-layer film having an inverted band ordering, realizing a 2D TCI with mirror Chern number $C_M = 2$. Around the critical film thickness, the phase transition can be described by the $k \cdot p$ Hamiltonian $H(k) = (\widetilde{v}_x k_x s_x - \widetilde{v}_y k_y s_y)\tau_x + \widetilde{m}\tau_z$, where $\tau_z = \pm 1$ denotes the conduction and valence bands of the TCI film at the X point. The velocities $\widetilde{v}_{x/y}$ and Dirac mass m are derived from the microscopic parameters of surface states in 3D TCIs and their hybridization strengths.

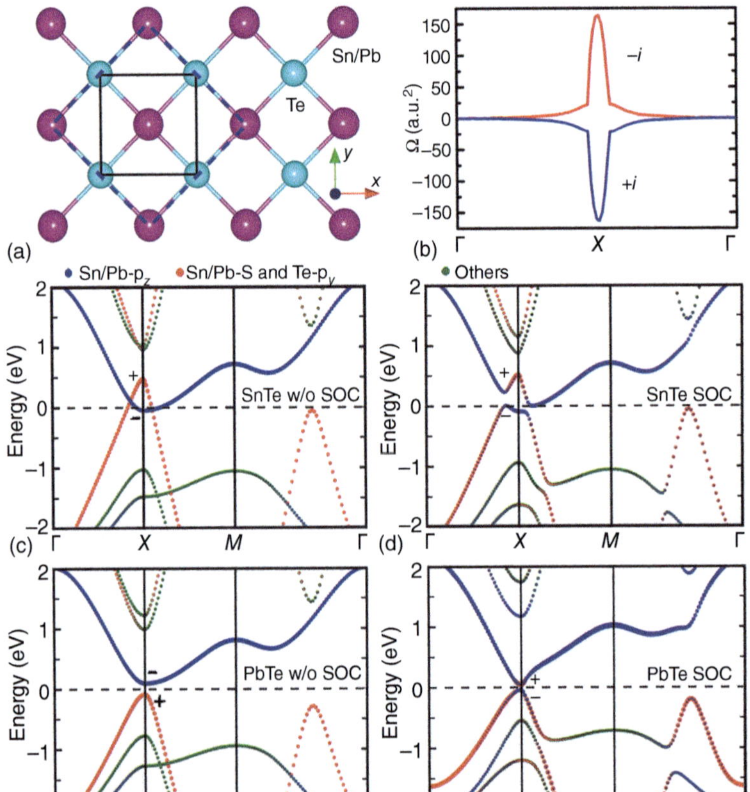

Figure 4.6 (a) Top view of the crystal structure of (001) Sn/PbTe monolayers with the unit cell and the supercell indicated by black solid and blue dashed lines, respectively. (b) Berry curvature distribution associated with $\pm i$ mirror eigenstates of the occupied bands along the Γ-X-Γ path. Orbitally resolved band structures for (001) monolayers of (c), (d) SnTe and (e), (f) PbTe, (c), (e) without, and (d), (f) with SOC, weighted with the contribution of s, p_y, and p_z states. Parities of the CBM and VBM at the X point are labeled by "+" and "−." The Fermi level is indicated with a dashed line. Source: Niu et al. [98]/with permission of American Physical Society.

A hallmark of 2D TCIs is the presence of gapless edge states and the mirror Chern number $C_M = 2$ suggests that there are two pairs of counter-propagating edge states within the nontrivial band gap. This is confirmed in the band structure calculation of a SnTe nanoribbon parallel to [98], using the recursive Green's function method. Within the energy window of the SOC gap, edge states with opposite mirror eigenvalues cross each other and connect the valence and conduction bands. Interestingly, unlike helical edge states in a 2D TI, the band crossings of gapless edge states shown here are located at non-TRIM points, so that they are protected solely by the mirror symmetry $z \to -z$ but not the time-reversal symmetry. A band gap opens up in these edge states when the mirror symmetry is broken under a perpendicular electric field.

4.2.2 IV–VI Monolayers

IV–VI semiconductors SnTe and PbTe share the same face-centered-cubic NaCl-type structure with the band structure of 3D bulk SnTe at the L point inverted relative to PbTe [14], which results in the first realization of the 3D TCI state in bulk SnTe and the 2D TCI state in SnTe multilayers [14, 97]. On the other hand, PbTe both in bulk and in (001) thin films is in the normal insulator state [14]. However, the first-principles calculations indicate that both the SnTe and PbTe as well as the other IV–VI (001) monolayers are 2D TCIs [98–100]. In Figure 4.6a the top view of such a 2D monolayer is shown, with Sn (Pb) and Te atoms forming two square sublattices positioned in the mirror plane $z = 0$. Accordingly, all Bloch states in the system can be labeled with the mirror eigenvalues $\pm i$ with respect to this symmetry plane.

To get preliminary insight into the topological properties of the systems, Figure 4.6 presents the orbitally resolved band structures of SnTe and PbTe monolayers with and without SOC. In the absence of SOC for SnTe, energy bands with the Sn-s and Te-p_y orbital character (positive parity with the inversion center at the Sn atom) overlap around the X point with the Sn-p_z states (negative parity). For PbTe at the X point without SOC a direct band gap appears with the VBM and CBM dominated by the Pb-s and Te-p_y orbitals (positive parity) and the Pb-p_z orbital (negative parity), respectively. Turning on SOC leads to an insulating character in both systems (calculated band gaps are 0.05 eV for SnTe and 0.09 eV for PbTe), and to the band inversion in PbTe, so that with SOC the band structure is inverted in both systems.

Owing to an even number of X points in the Brillouin zone, neither the SnTe monolayer nor the PbTe monolayer is a 2D TI. However, taking the mirror symmetry into account, for both SnTe and PbTe monolayers, band inversion results in the realization of a 2D TCI state. To show this explicitly, the mirror Chern number C_M is calculated. Figure 4.6b presents the distribution of the Berry curvature of all occupied bands with a mirror eigenvalue $\pm i$. The main contribution to the Berry curvature comes from the region around X, with its values having an opposite sign for opposite eigenvalues. The Chern number for each polarization is, respectively, $C_{+i} = -2$ and $C_{-i} = 2$, yielding the total Chern number of all occupied states $C = 0$ and the mirror Chern number $C_M = -2$, proving clearly the TCI nature of (001)-oriented SnTe and PbTe monolayers. The 2D TCI state is further explicitly confirmed by the emergence of the gapless edge states in thin nanoribbons of the monolayers.

4.2.3 Topological Phase Transition Between 2D TCI and TI

Both the TCI and TI can be originated from a topologically trivial system through a topological phase transition by tuning the alloy composition [16, 17, 101, 102] or the crystal lattice [103]. Similarly, a topological phase transition between TCI and TI is possible as well and this phase transition is helpful for both the future use and the fundamental understanding of the gapless edge states that are protected by different symmetries. As discussed earlier, the band inversion can be considered as a heuristic scenario for both the TIs and TCIs with an odd and even number, and thus the number change of the band inversions implies a topological phase transition, for example in TlS and TlSe monolayers with a rectangle lattice [104].

The orbitally resolved band structures without and with SOC are plotted in Figure 4.7. As we can see, the band structures of TlS and TlSe are quite similar. In the absence of SOC, the lowest unoccupied bands are dominated by the Tl-p_z orbital over the whole Brillouin zone, while the highest occupied bands around the X and Y points are dominated by the M-p_x and M-p_y orbitals, respectively. It is, such as TlSe, a direct gap insulator with an energy gap of 0.15 eV (0.21 eV for TlS) at the Y point while the energy gap is 0.21 eV (0.25 eV for TlS) at the X point. When SOC is switched on, the orbital characters at both the X and Y points are inverted and a direct gap of 0.12 eV (0.03 eV for TlS) appears at the X point. The even number of band inversions at the TRIM points means that TlS and TlSe monolayers

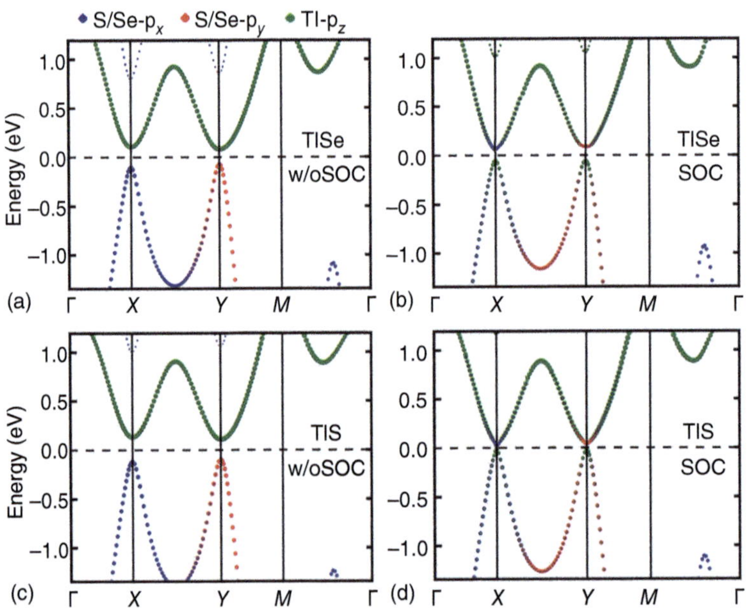

Figure 4.7 Orbitally resolved band structures for the monolayers of (a and b) TlSe and (c and d) TlS (a and c) without and (b and d) with SOC, weighted with the S/Se-p_x, S/Se-p_y, and Tl-p_z characters. The S/Se-p_x and S/Se-p_y orbitals contribute differently around the X and Y points. The Fermi level is shifted into the inverted energy gap. Source: Niu et al. [104]/with permission of American Chemical Society.

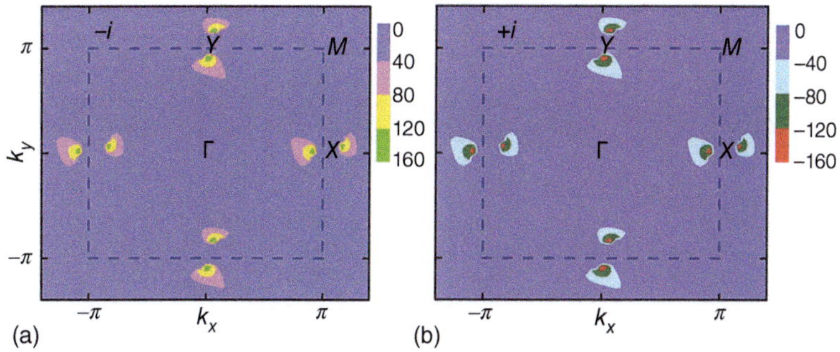

Figure 4.8 The Berry curvature distribution associated with (a) $-i$ and (b) $+i$ mirror eigenstates, respectively, of the occupied bands over the 2D Brillouin zone. Source: Niu et al. [104]/with permission of American Chemical Society.

cannot be 2D Z_2 TIs, while they could be the 2D TCIs owing to the presence of the mirror symmetry in combination with the band inversion taking place at the mirror plane $z = 0$. Indeed, the opposite sign of the Berry curvature for opposite mirror eigenvalues, as shown in Figure 4.8, results in Chern numbers for each polarization $C_{-i} = 2$ and $C_{+i} = -2$. Thus, the total Chern number is zero, but the mirror Chern number is $C_M = -2$, indicating that the TlS and TlSe monolayers are 2D TCIs.

Then, we show the strain-induced phase transition using TlSe monolayer as an example. The magnitude of strain is described by a/a_0 (b/b_0), here a_0 (b_0) and a (b) denote the lattice parameters of the unstrained and strained systems, respectively. The calculated energy gaps at X and Y points without and with SOC versus the uniaxial strain are presented in Figure 4.9. The energy gaps as well as the band topology can be effectively modified by uniaxial strain along both a- and b-axes. Under compressive strain, the band inversion takes place even without SOC, and the 2D TCI phase remains with enhanced gap when SOC is taken into account, indicating that the 2D TCI phase is robust against compressive strain. The normal energy gap increases with the tensile strain increasing at both the X and Y points for the case without SOC. Considering SOC, a band gap closing and reopening occurs individually at the X and Y points under different critical strains (vertical solid lines in Figure 4.9). When the lattice parameter lies between the two critical values, the band inversion occurs only at the X points for a-axis and Y points for b-axis. To determine the band topology of the strain-induced 2D insulators, the spin Chern number C_S is calculated. For both spins, the Chern numbers are respectively $C_\pm = \mp 1$, leading to a spin Chern number $C_S = -1$. Thus, a phase transition from 2D TCI to 2D TI can be effectively tuned by uniaxial strain. With further increasing strain, the band inversion disappears at both the X and Y points, and the system becomes a trivial insulator.

The values of $C_M = -2$ and $C_S = -1$ indicate that there are two pairs and one pair of gapless edge states in the bulk energy gap for 2D TCI and 2D TI, respectively. To illustrate this, the edge state band structures are calculated using maximally localized Wannier functions (MLWFs), which can reproduce the band dispersion of TlSe without and with 2% uniaxial strain (along the b-axis) quite precisely. Figure 4.10

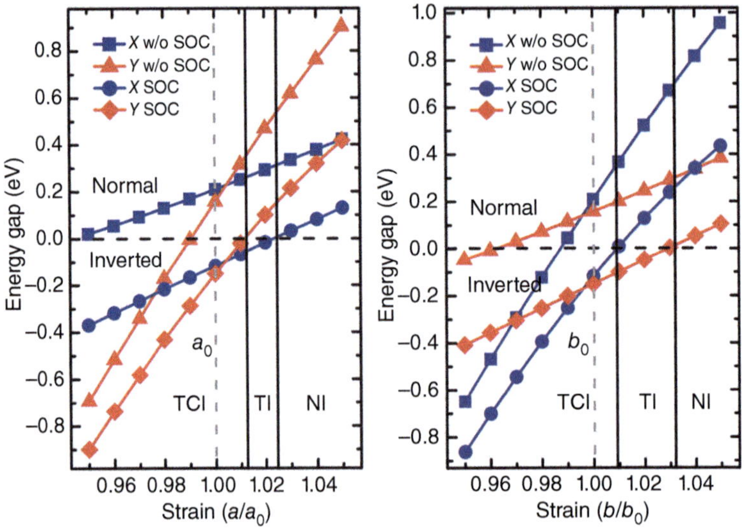

Figure 4.9 Variation of the energy gaps at *X* and *Y* points for TlSe monolayer versus the uniaxial strain along (a) *a*- and (b) *b*-axes. The negative values indicate the inverted energy gaps. For case with SOC, a topological phase transition occurs accompanied by sign changes of the energy gap. Source: Niu et al. [104]/with permission of American Chemical Society.

displays the edge states of an 80-atom wide TlSe nanoribbon edged with Tl atoms. One can clearly see that two pairs of nontrivial edge states cross at Γ and X for the TCI and one pair of that crosses at Γ for the TI phase, in direct agreement with the values of mirror Chern number C_M and spin Chern number C_S. In contrast to 2D TI, whose edge state is determined by time-reversal symmetry, the key property of the 2D TCI relies on crystalline mirror symmetry, resulting in the gapless edge state is not spin degenerate [1–4]. In order to indicate the spin texture, the matrix element of the Pauli matrices σ_α $(\alpha = x, y, z)$ are computed on the basis of the MLWFs and the edge states are colored with the expectation value of the σ_z. The edge states, as shown in Figure 4.10, are spin polarized and change directions of spin when crossing the time-reversal invariant points for the 2D TCI. At the time-reversal invariant points, the expectation value of the σ_z is zero. This is obviously different from the situation of the 2D TI as shown in Figure 4.10b, where the opposite spins are degenerate at the time-reversal invariant points.

4.2.4 Dual Topological Insulator

In the presence of time-reversal and crystalline mirror symmetry, the coexistence of TI and TCI phases has been predicted in three dimensions for $Bi_{1-x}Sb_x$ [96] and Bi chalcogenides [105–108], and thus they exhibit a dual topological character (DTC). Recently, unusual topological surface states for a 3D DTC system have been observed experimentally [106–108]. In the 2D case, graphene may be a prototypical example of a DTC [55, 97]. However, the extremely small band gap of graphene makes it very difficult to verify the DTC in this material experimentally [109]. In many cases of the

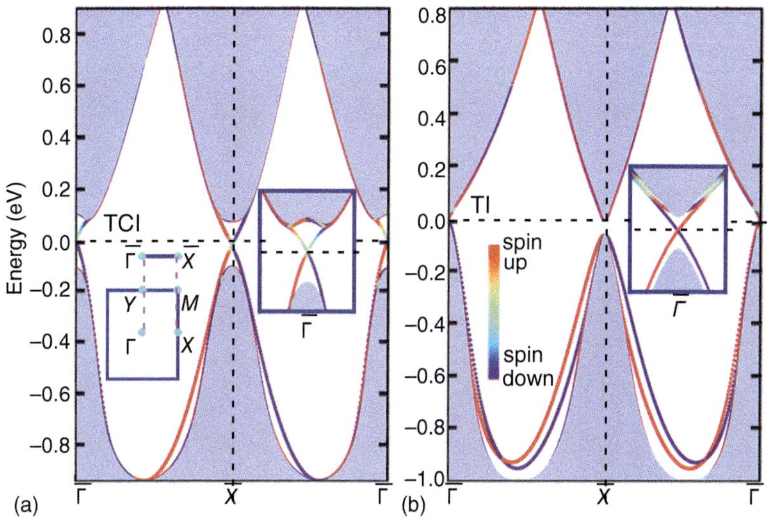

Figure 4.10 Band structures of 1D nanoribbon edged with Tl atoms for TlSe in (a) TCI phase and (b) TI phase. The corresponding projected bulk band structures are represented by the shaded areas. Edge states are colored with the expectation value of σ_z for indicating the spin polarization on one particular side of 1D nanoribbon. Insets in (a) and (b) show the 2D and projected 1D Brillouin zones, and the corresponding zoom-in at the Γ point. Source: Niu et al. [104]/with permission of American Chemical Society.

proposed 2D TIs, the complex structures and the lack of mirror symmetry prevent the formation of the 2D TCI phase. On the other hand, 2D TCIs have been limited to square lattices with mirror Chern number $|C_M| = 2$, for which the even number of band inversions leads to a vanishing Z_2 invariant.

While the bulk Na_3Bi in a hexagonal $P6_3/mmc$ structure is a topological Dirac semimetal [19, 21], i.e. 3D counterpart of grapheme, Na_3Bi monolayer is demonstrated to host both the 2D TI and TCI phases with an odd number of band inversion [110]. In Figure 4.11b and c, the side and top view of the Na_3Bi monolayer are presented, with Bi and Na atoms forming a honeycomb lattice. Unlike in the bulk material, inversion symmetry is broken in a Na_3Bi monolayer, but the mirror symmetry $z \rightarrow -z$ is preserved. Figure 4.12 presents the orbitally resolved band structures of the Na_3Bi monolayer without and with SOC. In the absence of SOC, Bi-p_x and Bi-p_y orbitals contribute to the VBM while the CBM is dominated by Bi-s orbitals with a direct band gap of 0.16 eV. Switching on SOC leads to an s-p band inversion that occurs at the Γ point, and the insulating character is preserved with a band gap of 0.31 eV.

The existence of the mirror symmetry $z \rightarrow -z$ for the Na_3Bi monolayer offers the possibility of realizing a 2D TCI. Indeed, the calculated Chern numbers for mirror eigenvalues $+i$ and $-i$ are, respectively, $C_{\pm i} = \mp 1$, leading to a mirror Chern number $C_M = -1$. Here, the band inversion occurs at the Γ point, i.e. an odd number of band inversions is acquired. Spin Chern number C_S is calculated to identify the relationship between the 2D TI and the odd number of band inversions in the Na_3Bi

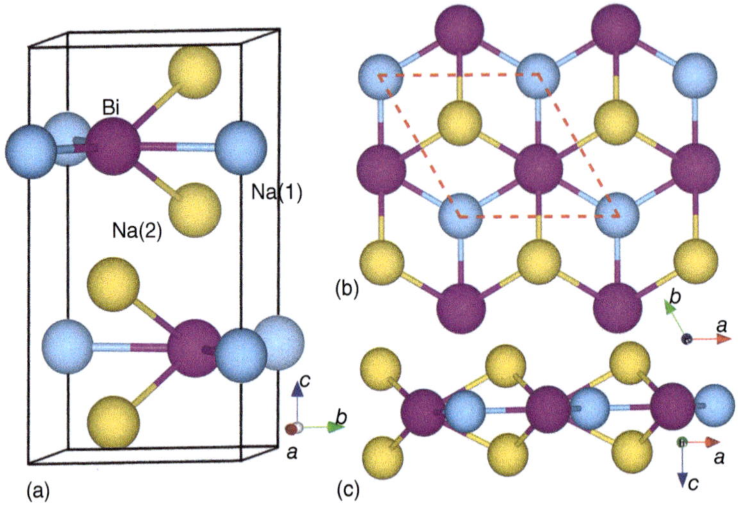

Figure 4.11 (a) Crystal structure of bulk Na_3Bi with *P*6_3*/mmc* symmetry. (b) Top and (c) side view of the honeycomb Na_3Bi monolayer, where the unit cell is indicated by the dashed lines. Source: Niu et al. [110]/with permission of American Physical Society.

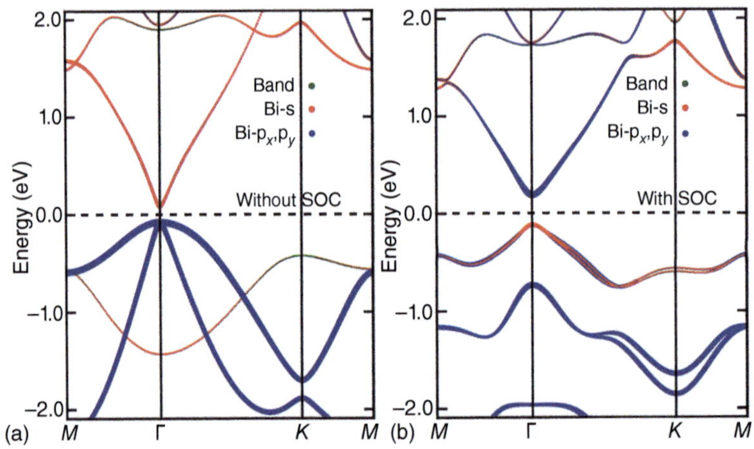

Figure 4.12 (a) Orbitally resolved band structures for Na_3Bi monolayer (a) without and (b) with SOC, weighted with the contribution of Bi-s and Bi-p_x,p_y states. The Fermi level is indicated with a dashed line. Source: Niu et al. [110]/with permission of American Physical Society.

monolayer. The Chern number for each spin manifold is $C_+ = -1$ and $C_- = 1$, yielding the spin Chern number $C_S = -1$. This clearly demonstrates the 2D TI nature of the Na_3Bi monolayer. Therefore, the Na_3Bi monolayer exhibits the DTC with respect to the 2D TI and TCI phases. First-principles calculations indicate that the similar ternary compounds, Na_2MgPb and Na_2CdSn, can possess the properties of both 2D TI and TCI phases as well, realizing dual TIs with nonzero Z_2 invariant $Z_2 = 1$ and mirror Chern number $C_M = -1$ [102]. Remarkably, Na_2MgPb, which gains a large nontrivial band gap of 0.58 eV, turns out to be a potential material for room temperature applications [111].

A pair of gapless edge states in the 2D gap, as shown in Figure 4.13a, further confirms the nontrivial topology of Na_3Bi monolayer. Generally, time-reversal symmetry breaking generates a gap in the surface/edge states of TIs, while mirror symmetry breaking is indispensable for the formation of a band gap in the surface/edge states of TCIs [1–4]. One way to destroy these symmetries is to introduce magnetism in the system. To mimic a magnetic environment, the matrix elements of the Pauli matrices σ_α ($\alpha = x, y, z$) are computed on the basis of MLWFs, which allows to consider the effect of an exchange field applied along different directions. For an exchange field perpendicular to the mirror plane, $H_{mag} = B_\perp \sigma_z$, time-reversal symmetry is broken while mirror symmetry is maintained. In this case, as shown in Figure 4.13b for the

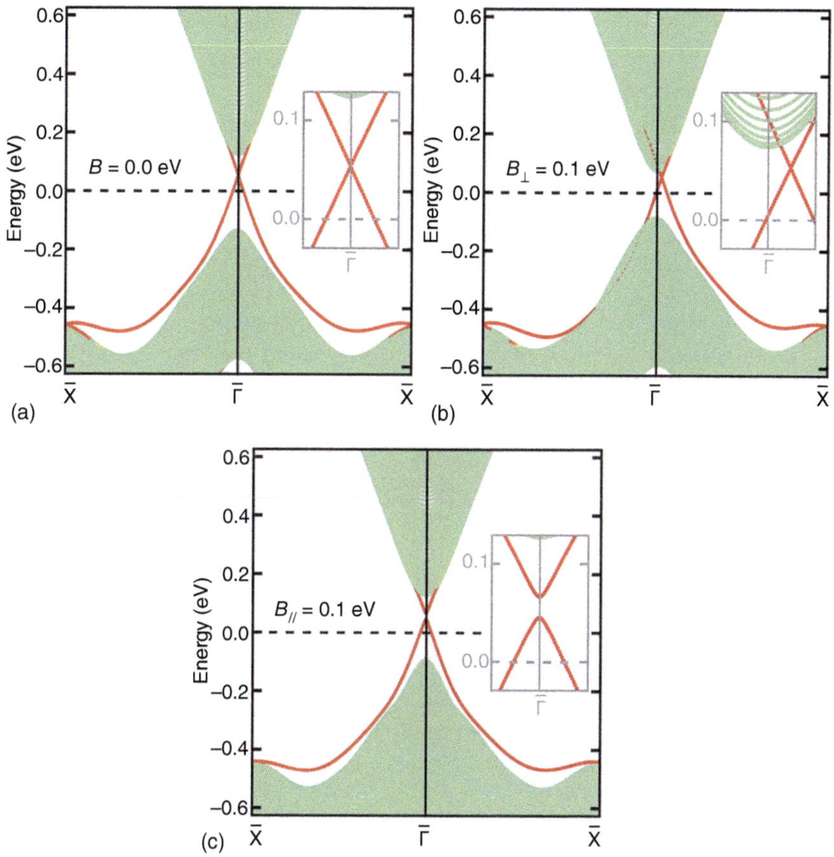

Figure 4.13 Localization-resolved edge states of the Na_3Bi monolayer (a) without a magnetic field, (b) with a magnetic field perpendicular to the mirror plane, and (c) with a magnetic field within the mirror plane. The insets show the corresponding zoom-in at the Γ point. Color from light green to red represents the weight of atoms located from the middle to one edge of the ribbon structures. Source: Niu et al. [110]/with permission of American Physical Society.

Bi-Na(1) termination, the Dirac point moves slightly away from the Γ point, while a band gap does not open as a consequence of the 2D TCI phase's survival. If the exchange field, on the other hand, is in the plane, $H_{\text{mag}} = B_{//}\sigma_x$, both time-reversal and mirror symmetries are broken and the edge states become gapped as shown in Figure 4.13c.

4.2.5 TCI in 2D Ferromagnets

Merging the fields of topology and magnetism expands the scope of fundamental quantum phenomena with novel functionalities for topological spintronics enormously. It is expected that crystalline mirror symmetry can give rise to the TCI phase in 2D magnets; however, the material realization of 2D TCI with intrinsic magnetic order remains elusive [112]. Tight-binding model and first-principles calculations reveal that NpSb monolayer is a long-awaited 2D TCI with intrinsic out-of-plane ferromagnetic (FM) order [113]. Remarkably, when rotating the magnetization into the plane a higher-order TI phase with a parity-based invariant $\nu_{2\text{D}} = 1$ is achieved, and in-gap topological corner states emerge.

In many magnetic 2D (in the xy plane) systems the M_z symmetry of reflection with respect to the xy plane is preserved when the magnetization points out of the plane (along z), which provides a recipe to define the mirror Chern number C_M and achieve a TCI state. When the magnetization direction is varied, M_z symmetry is naturally broken and $(d-1)$-dimensional edge states become gapped. However, for certain directions of the magnetization other mirror symmetries are restored that result in the emergence of topologically protected corner states, i.e. $(d-2)$-dimensional boundary modes, appearing inside the band gap of the 2D bulk and $(d-1)$-dimensional edges, thus manifesting the formation of magnetic higher-order TI state. To show this clearly from simple arguments, as sketched in Figure 4.14a, a four-band tight-binding model for the 2D square lattice is used with the Hamiltonian expressed as $H = H_0 + H_\text{B}$, where

$$H_0 = [m - t(\cos k_x + \cos k_x)]\tau_z - \lambda(\sin k_x\sigma_x + \sin k_y\sigma_y)\tau_x,$$

$$H_\text{B} = B\sigma_x \sin\theta\cos\varphi + B\sigma_y \sin\theta\sin\varphi + B\sigma_z\cos\theta.$$

The onsite energy, the magnitude of the nearest-neighbor hopping, and the strength of SOC correspond to parameters m, t, and λ, respectively; σ and τ are the vectors of spin and suborbital Pauli matrices. The FM order is represented by the Zeeman term H_B with an amplitude of exchange field given by B and the magnetization direction defined by spherical angles θ and φ.

For an out-of-plane FM ordering, $\theta = 0^\circ$, the mirror symmetry M_z remains intact with $M_z^2 = -1$, while the time-reversal symmetry is broken, meaning that the Hamiltonian can be separated into two decoupled mirror subspaces based on the two mirror eigenvalues $\pm i$:

$$H^{\pm i}(k_x, k_y) = \pm[m + B - t(\cos k_x + \cos k_y)]\sigma_z - \sin k_x\sigma_x \mp \sin k_y\sigma_y.$$

The calculated Chern numbers for two opposite mirror subspaces are, respectively, $\text{C}_{+i} = 1$ and $\text{C}_{-i} = -1$. Thus, the mirror Chern number is $C_\text{M} = +1$, which proves

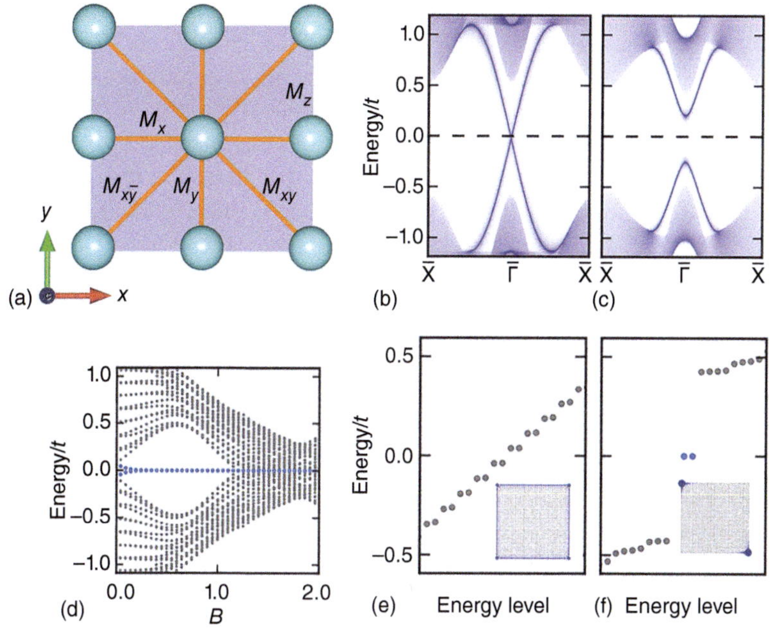

Figure 4.14 (a) Sketch of a 2D square lattice, in which the mirror symmetries depend sensitively on the magnetization direction. Edge spectra of the tight-binding model with an (b) out-of-plane ($\theta = 0°$) and (c) in-plane magnetization ($\theta = 90°$ and $\phi = 45°$) for the exchange field of $B = 0.5t$. (d) Energy spectrum of the 40×40 nanoflake of the model versus the magnitude of B. Corner states are highlighted with blue circles. Energy levels for a nanoflake with an (e) out-of-plane and (f) in-plane magnetization. Insets show the spatial weight of the edge states and corner states. Source: Mao et al. [113]/with permission of American Physical Society.

the TCI nature of the considered 2D FM square lattice. This can be further explicitly confirmed by observing the emergence of gapless edge states at all edges as displayed in Figure 4.14b and e. It is widely known that a gap opens up in the edge states when the protecting M_z symmetry is broken, for example by rotating the magnetic moments into the xy plane, $\theta = 90°$. However, for $\varphi = 0°$ and $\varphi = 90°$, M_x and M_y are preserved, respectively, which keeps the edges, perpendicular to the corresponding mirror planes, gapless. Remarkably, for $\varphi = 45°$, although the edges are gapped out as shown in Figure 4.14c, the in-gap states, which are localized in the corners of a finite sample, arise as visible from the spectrum analysis of a 0D nanoflake in Figure 4.14f. This signals the emergence of higher-order TI phase, characterized by the winding number ν_{2D} [114, 115],

$$(-1)^{\upsilon_{2D}} = \prod_{i=1}^{4} (-1)^{[N_{occ}^{-}(\Gamma_i)/2]},$$

where $N_{occ}^{-}(\Gamma_i)$ is the number of occupied states with an odd parity at TRIM Γ_i. As expected, we find that at the Γ point, the parity eigenvalues of two occupied states are all odd, whereas those for the other three momenta (two X and one M) are all even, yielding $\nu_{2D} = 1$ and thus further explicitly confirming the higher-order TI state.

Similar to SnTe, bulk NpSb crystallizes in the face-centered-cubic NaCl structure with space group $Fm\bar{3}m$. Thus, when constructing a (001)-oriented NpSb monolayer, a 2D square lattice with Np and Sb atoms positioned in the mirror plane $z = 0$ is obtained. In absence of SOC, under the ground state with an FM coupling, the majority bands are metallic with a band touching at the Γ point, while the minority bands exhibit a gap of 195 meV resulting in a pronounced half-metallic behavior near the Fermi energy as illustrated in Figure 4.15a and b. Remarkably, SOC opens a large gap in the majority channel, thereby making the whole system insulating with the corresponding band gaps of 220 meV and 86 meV for the out-of-plane and in-plane directions, respectively. Moreover, as illustrated in Figure 4.15c and d, the fat-band analysis of the orbital contributions to the electronic states suggests that SOC drives a band inversion between Sb-$p_{x/y}$ and Np-d_{x2-y2} states both in the minority and majority channels, implying the formation of a topologically nontrivial insulator.

The magnetic space group of NpSb monolayer with out-of-plane magnetization is $P4/mm'm'$ that possesses the M_z symmetry [116]. For two opposite mirror

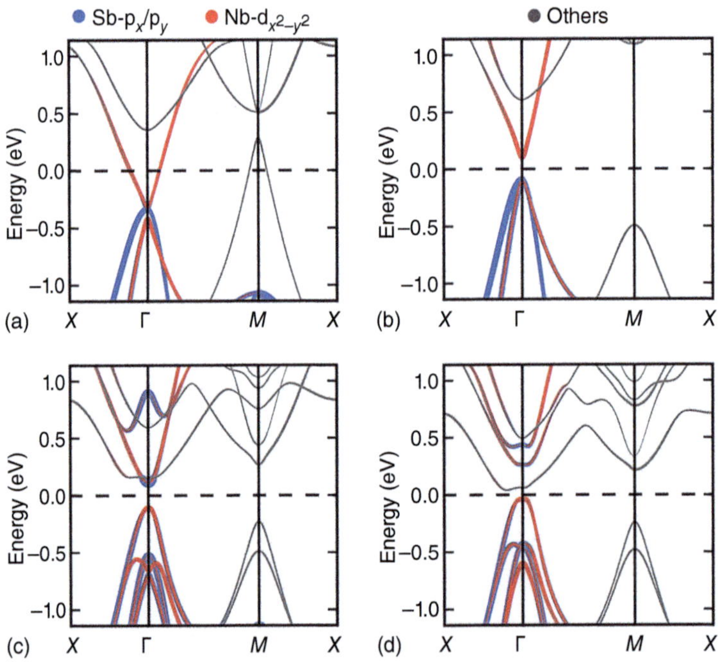

Figure 4.15 Band structure of (a) majority and (b) minority states of NpSb monolayer in the absence of SOC and that of NpSb monolayer with (c) an out-of-plane and (d) an in-plane magnetization along x including SOC. The bands are orbitally weighted with the contribution of Sb-$p_{x/y}$ and Np-d_{x2-y2} states. The Fermi level is indicated with a dashed line. Source: Mao et al. [113]/with permission of American Physical Society.

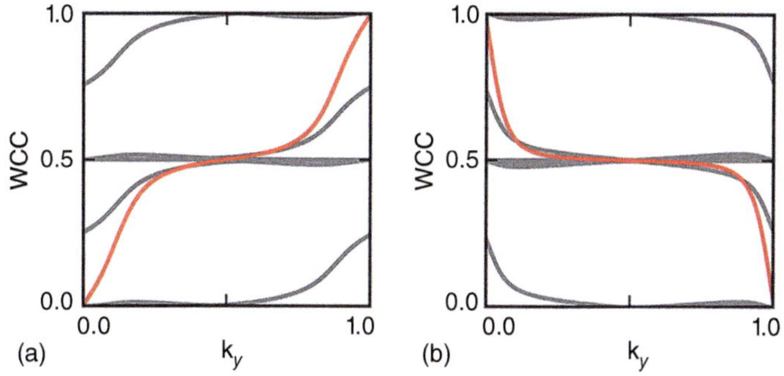

Figure 4.16 Evolution of Wannier charge centers (WCCs) for NpSb monolayer with an out-of-plane magnetization associated with (a) $+i$ and (b) $-i$ mirror eigenstates, indicating a nonzero mirror Chern number $C_M = 1$. Source: Mao et al. [113]/with permission of American Physical Society.

subspaces, the Chern numbers are calculated by the Wilson loop method with the Wilson loop matrix given by:

$$W_{(k_x+2\pi,k_y)\leftarrow(k_x,k_y)} = \lim_{N\to\infty} F_{N-1}F_{N-2}\cdots F_1F_0,$$

where $[F_i]_{mn} = \langle u_m(2\pi(i+1)/N, k_y) \,|\, u_n(2\pi i/N, k_y)\rangle$; $|u_{nk}\rangle$ is the lattice periodic part of the nth Bloch state at the k point (k_x, k_y) and N is the number of k_x points. As shown in Figure 4.16, the Chern numbers of all occupied bands for the opposite mirror eigenvalues $+i$ and $-i$ are $C_{+i} = 1$ and $C_{-i} = -1$, respectively, yielding $C_M = 1$ and confirming the fact that NpSb monolayer is a 2D FM TCI. On the other hand, the magnetic space group turns into $Cmm'm'$ for the in-plane magnetization along (100). The parities at (Γ, X1, X2, M) are ($-$, +, +, +), which leads to $\nu_{2D} = 1$, thus revealing the existence of the FM higher-order TI phase in the NpSb monolayer. The emergence of edge and corner states unambiguously demonstrate further the realization of topologically distinct phases as controlled by the direction of the magnetization.

4.3 Quantum Anomalous Hall Effect

In 1980, Klaus von Klitzing et al. discovered that, as a function of external magnetic field, the Hall conductance shows a sequence of plateaus at very low temperature [43]. The values of these conductance plateaus equal to an integer C multiples of e^2/h, where the integer C originally known as the TKNN number, which was derived by Thouless et al. from the Kubo formula in 1982, and now is characterized as a topological invariant named "Chern number" [68]. The concept of Chern number in mathematics describes the number of holes on a closed manifold. This discovery is of great significance, because it brings a new way to understand the phases of matters beyond the Landau's symmetry-breaking theory and opens up the field of topological electronic states in condensed matter physics. Interestingly, in 1988, Haldane proposed that the quantum Hall effect can be achieved even in the absence of external magnetic field based on the well-known Haldane model given on a honeycomb lattice [44]. It realizes the quantized version of the anomalous Hall effect, namely quantum anomalous Hall (QAH) effect. The Hamiltonian of Haldane model is given by:

$$H = \Delta \sum_{i} (-)^{\tau_i} c_i^{+} c_j + t_1 \sum_{\langle i,j \rangle} \left(c_i^{+} c_j + h.c. \right) + t_2 \sum_{\langle\langle i,j \rangle\rangle} \left(i c_i^{+} c_j + h.c. \right)$$

where i and j run over all sites, $\tau_i = \{1, 2\}$ is the sublattice index of the site, and t_1 is the nearest-neighbor hopping strength for first-neighbor pairs indicated by $<ij>$. This model describes spinless electrons hopping between two sites on a honeycomb lattice. Crucially, Haldane added the last term, which describes a complex hopping amplitude it_2 to each second-nearest-neighbor pair $\ll ij \gg$. The Hermitian conjugate term indicated by "h.c." describes corresponding hopping amplitudes in the reverse direction. This term breaks time-reversal symmetry as the complex conjugation reverses its sign. This model can realize nonzero Chern numbers in a parameter space. Although the Haldane model theoretically proved that the quantum Hall effect (QHE) can be realized without the external exchange field, its material realization has so far remained elusive and is unrealistic in experiment.

In fact, the progress on QAH effect was comparatively slow until the discovery of the QSH effect, protected by the time-reversal symmetry (T) and characterized by Z_2 invariant and helical gapless edge states as discussed earlier [1, 2]. The QSH insulator emerges in 2D systems without magnetization, and its gapless helical edge states can be viewed as two copies of the chiral edge states in the QAH insulator [47]. Breaking the time-reversal symmetry usually by a magnetic impurity or substrate, the QSH insulator can be switched into the QAH insulator, such as in magnetic doped 2D thin films of the Bi_2Se_3 family [49]. The Bi_2Se_3 family, Bi_2Te_3, Bi_2Se_3, and Sb_2Te_3, are 3D TIs [12, 13], and their thin films can cross over to the 2D TIs depending on the thickness of the films [117, 118]. The *3d* transition metal elements Ti, V, Cr, and Fe atoms, which have a stable 3+ chemical state, are used to avoid introducing free carriers into the system. The calculated density of states (DOS) reveal that Cr and Fe doping have FM insulating state, while Ti or V doping are metallic. By substituting the Bi sites, the local environment of dopants forms an octahedral that splits the *3d* shell into t_{2g} and e_g manifolds. For which, the high-spin state is always obtained due to the large Hund's rule, and thus a gap is obtained between the spin-up and spin-down states for Fe^{3+} with a $t_{2g}^{3\uparrow} e_g^{2\uparrow} t_{2g}^{0\downarrow} e_g^{0\downarrow}$ configuration and between the t_{2g} and e_g for Cr^{3+} with a $t_{2g}^{3\uparrow} e_g^{0\uparrow} t_{2g}^{0\downarrow} e_g^{0\downarrow}$ configuration. While for the Ti- and V-doped cases, their t_{2g} states are partially occupied, resulting in the metallic states. Once the insulating FM order is obtained, the QAH effect can be achieved in 2D thin films of Bi_2Se_3 family with different thickness. The anomalous Hall conductance is indeed quantized in units of e^2/h when the Fermi level is located inside the energy gap. Experimentally, based on the prediction, the QAH effect is first confirmed in Cr-doped $(Bi,Sb)_2Te_3$ [50], but extreme requirements, such as the highly precise controlling of the extrinsic impurities and the ultralow temperature due to the small bulk energy gap, are required, which greatly prohibits the potential device applications of this fundamental quantum phenomenon [61, 119, 120].

To realize the high-temperature QAH effect, extensive investigations have been carried out and numerous material candidates have been theoretically proposed, such as the decorated 2D TIs, heterostructures, quantum wells, and layered materials [72, 121–132]. Interestingly, for H-decorated Bi, the band gap of QAH states can be as large as 0.35 eV. The fully H-decorated Bi is demonstrated to be the QSH insulator protected by the time-reversal symmetry [72]. When the hydrogen atoms are removed from one side of fully H-decorated Bi while keeping the other side hydrogenated, the insulating FM ordering is obtained with a magnetic moment of 1.0 μ_B per unit cell, carried by the p_z states of the unhydrogenated Bi atoms. To identify the QAH effect, the anomalous Hall conductivity $\sigma_{xy} = (e^2/h)C$, where C is the Chern number, is calculated as an integral of the Berry curvature of occupied state $\Omega(k)$ over the Brillouin zone. Figure 4.17c presents the σ_{xy} versus the position of the Fermi level. When the Fermi level is located within the insulating band gap, the Chern number C indeed acquires an integer value of +1, revealing clearly the QAH effect in semihydrogenated Bi.

In addition, both the time-reversal and inversion symmetries are broken in half H-decorated Bi, leading to the band structures at K and K' showing different patterns as illustrated in Figure 4.17b. Valleys K and K' are distinguishable. Moreover, Berry curvatures, as shown in Figure 4.17d, which is mainly localized in the vicinity of K and K', have opposite signs around the two valleys, realizing the valley-polarized QAH effect, which exhibits characters of both the QAH and quantum valley Hall (QVH) effects [128, 133]. The underlying physics of the formation of valley-polarized QAH effect is revealed by the evolution of the band gap at valleys K and K', shown in Figure 4.17e, versus the SOC strength. As the SOC is increased, the band gap closes and reopens at K', while that always opens at the K point, and thus a topological phase transition occurs at K' but not at K. The valley-polarized QAH effect is further confirmed by the edge states shown in Figure 4.17f.

The QAH effect is very hard to realize in experiment. To improve the feasibility of experimental realization and possible applications, 2D layered FM insulators are good candidates and deserve to be investigated in experiment and theory. The layered rare-earth pnictides are well known for their extraordinary magnetic and thermoelectric properties and attract growing attentions for the topological properties in recent years, such as the Weyl semimetal $EuCd_2As_2$ [134–136]. Bulk $EuCd_2As_2$ has a trigonal structure with space group $P\bar{3}m1$, and its crystal structure can be visualized as a stacking of quintuple layers (QLs) along the z direction. The exchange interaction between the interlayer Eu atoms are AFM, and that between the intralayer ones are FM. The magnetic moments on each Eu atoms are 6.9 μ_B and the spin polarization occurs on the half-filled 4f orbitals that are far away from the Fermi level. Similar to the bulk form, magnetic moments of $EuCd_2As_2$ QLs at intralayer Eu sites couple ferromagnetically [131]. In the absence of SOC, as shown in Figure 4.18a and b, the spin-up bands are gapless while a gap appears in the spin-down bands. When

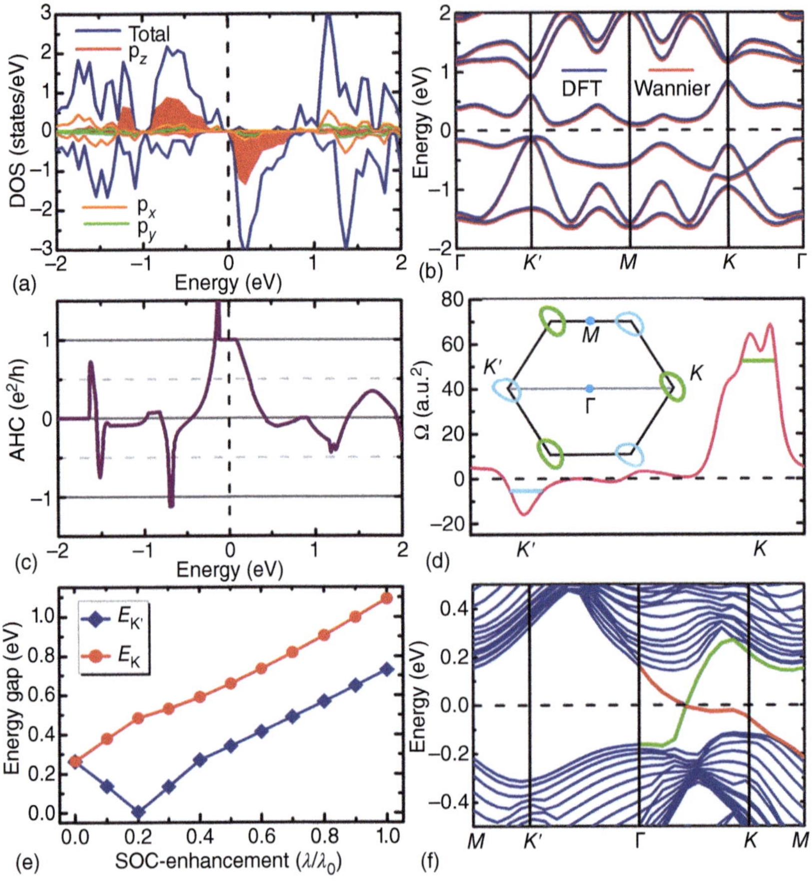

Figure 4.17 (a) Total density of states (DOS) and partial DOS of the unhydrogenated Bi atoms of half H-decorated Bi without SOC. Positive and negative values indicate spin-up and spin-down channels, respectively. (b) Wannier and first-principles band structures with SOC for half H-decorated Bi. The Wannier results have been shifted down by 30 meV for visibility. (c) Anomalous Hall conductivity as a function of the position of the Fermi level E_F. (d) Berry curvature distribution of the occupied bands in the K-Γ-K' direction. The inset shows the contour of the Berry curvature distribution (as marked in main panel) around valleys K and K'. (e) The energy gaps at valleys K (E_K) and K' ($E_{K'}$) as a function of SOC strength. (f) Band structure of half H-decorated Bi exhibiting valley-polarized QAH states. The states located at different edges are indicated by different colors. Source: Niu et al. [72]/with permission of American Physical Society.

taking the SOC effect into account, the system prefers to have an out-of-plane magnetization. Remarkably, as shown in Figure 4.18c, the SOC opens up a gap of 72 meV for the spin-up bands, implying the nontrivial QAH effect with the band inversion in spin-up channel of the 2D FM $EuCd_2As_2$ QLs. To show this explicitly, Figure 4.18d displays the edge states of a $EuCd_2As_2$ nanoribbon terminated by zigzag chains of Cd and As atoms. Within the nontrivial gap, it is clear that one chiral edge state bridges the conduction and valence bands, which is consistent with the calculated Chern number $C = 1$.

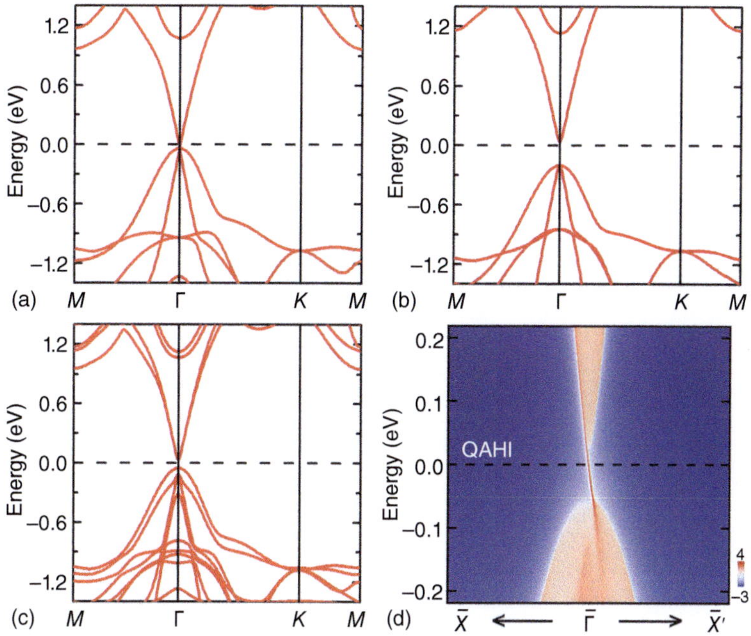

Figure 4.18 Band structures of $EuCd_2As_2$ QLs (a), (b) without SOC and (c) with SOC. The (a) spin-up bands are gapless while a gap appears in the (b) spin-down bands. (d) Band structures of a $EuCd_2As_2$ nanoribbon terminated by zigzag chains of Cd and As atoms in the QAH phase. Source: Niu et al. [131]/with permission of American Physical Society.

4.4 Antiferromagnetic Topological Insulators

Another notable example of the topological states is the AFM TI, which was worked out in 2010 by Mong et al. [137] and recently proposed theoretically [30, 31] and observed experimentally in 3D $MnBi_2Te_4$ [32, 33]. Although the time-reversal symmetry T is broken, a combination of time-reversal and primitive-lattice translational symmetry $T_{1/2}$, $S = T\,T_{1/2}$, protects the topologically nontrivial phase and gives rise to the Z_2 classification of insulating AFM phases [137]. The research on AFM topological state is also reached out to two dimensions [138, 139]. For example, the QSH effect is reported to coexist with the superconductivity in AFM FeSe monolayers, but the inverted gap lies below the Fermi level [51]. As discussed earlier, it has long been known that a TI phase can be obtained through a gap opening induced in a topological semimetal. The most famous example is graphene, where the Dirac point is protected by symmetry, but only when the SOC is neglected, and a 2D TI is obtained when switching the SOC on [55]. Similarly, the band inversion (band gap opening) can be considered as a heuristic scenario for obtaining the 2D magnetic topological states. If the band inversion occurs only for one spin channel, it will carry a nonzero Chern number, i.e. for example $C_+ = 0$ but $C_- = \pm 1$, and results in the QAH effect with an integer total Chern number $C = C_+ + C_- = \pm 1$, shown in Figure 4.19. While the band inversion (band gap opening) emerges in two spin channels, two

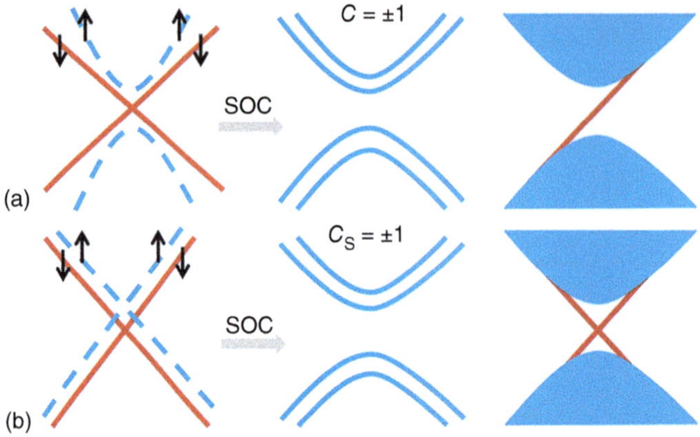

Figure 4.19 Schematics of the band inversion (band gap opening) for (a) QAH effect with Chern number $C = \pm 1$ and (b) 2D magnetic TIs with spin Chern number $C_S = \pm 1$. Source: Wang et al. [140]. Reproduced with permission of Royal Society of Chemistry.

opposite spin channels carry nonzero Chern numbers with opposite signs, i.e. +1 and/or −1, yielding the total Chern number $C = C_+ + C_- = 0$ and the spin Chern number $C_S = (C_+ - C_-)/2 = \pm 1$.

Following this line of thought, a magnetic topologically insulating phase can indeed be obtained as manifested in the 2D antiferromagnet [52]. Distinct from graphene, symmetry-protected Dirac semimetals can be obtained even in the presence of SOC, in which the appearance of Dirac points strongly depends on the combination of time-reversal symmetry and nonsymmorphic symmetry [141]. Starting with such a 2D Dirac semimetal, a gap can be opened by introducing the AFM ordering to break the time-reversal symmetry, as shown in a four-band tight-binding model with an intralayer out-of-plane AFM ordering [141, 142]:

$$\begin{aligned} H =& [Re(M)\tau_x - \mathrm{Im}(M)\tau_y]\sigma_0 - 2t_{\mathrm{in}}(\cos k_x + \cos k_y)\tau_z\sigma_z \\ &+ t_{\mathrm{SOC}}\tau_z(\sigma_y \sin k_x - \sigma_x \sin k_y) + \lambda_{\mathrm{mag}}\tau_z\sigma_z \end{aligned}$$

which is sketched in Figure 4.20a. Here, $M = \left(t_1 + t_2 e^{ik_y}\right) \times \left(1 + e^{-ik_x}\right)$. τ_α and σ_α ($\alpha = x, y, z$) are the Pauli matrices of the sublattices and spin degrees of freedom, respectively. Clearly, the combined symmetry $S = T\,T_{1/2}$ is broken when the hopping energy $t_1 \neq t_2$ in the first term. The second and third terms indicate the intrinsic and Rashba SOC, and the fourth term represents the AFM ordering with an out-of-plane easy axis.

As shown in Figure 4.20b, neglecting the AFM ordering, the tight-binding model hosts the Dirac semimetal phase with one Dirac point at the X point, where the Dirac point relies on the nonsymmorphic symmetry $\{C_{2x} \mid 1/2\ 0\}$ where C_{2x} is the twofold rotation symmetry and (1/2 0) is half of the lattice translation along the x-axis [52, 141]. Switching on the AFM term, as illustrated in Figure 4.20c, the conduction and valence bands are no longer degenerate at the X point. As expected, the breaking of time-reversal symmetry is the effect to lift the fourfold Dirac band crossing that results in the presence of an insulating state, i.e. band gap opening for both the spin-up and spin-down channels. The topologically nontrivial nature of the gap can be explicitly confirmed via calculations of the Wannier charge centers (WCCs),

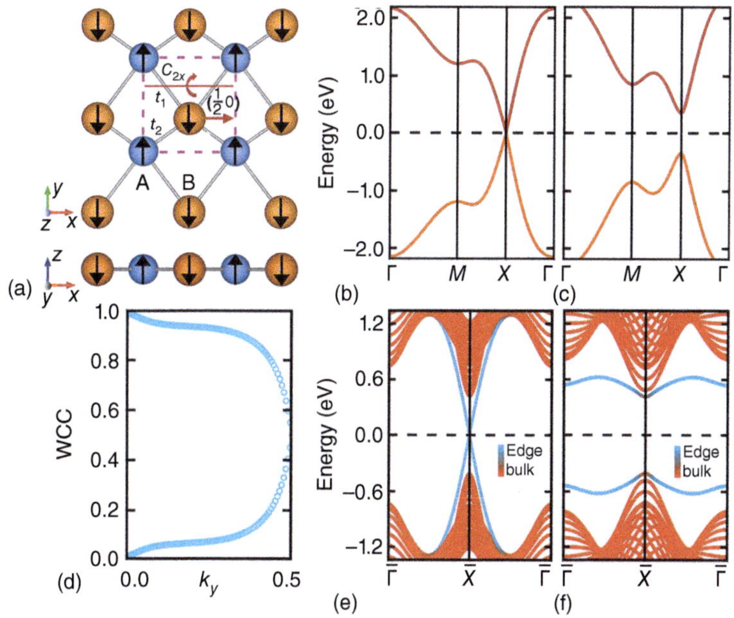

Figure 4.20 (a) Sketch of the tight-binding model for a 2D antiferromagnet on a square lattice. C_{2x} represents a twofold rotation symmetry and (1/2 0) denotes a half of the lattice translation along the *x*-axis. Band structures (b) without and (c) with out-of-plane antiferromagnetism; (d) and (e) display the Wannier charge centers (WCCs) and edge states with the nonsymmorphic symmetry $\{C_{2x} \mid 1/2\ 0\}$. (f) A gap opens up in the spectrum of a one-dimensional nanoribbon without $\{C_{2x} \mid 1/2\ 0\}$. The color transition from red to blue represents the weight of atoms located from the middle to one edge of the ribbon structures. Source: Niu et al. [52]/with permission of American Physical Society.

which result in $Z_2 = 1$, and the emergence of the exotic edge states in the nanoribbons with $\{C_{2x} \mid 1/2\,0\}$ symmetry at the edges, shown in Figure 4.20d and e. Moreover, a gap opens up in the edge states without the $\{C_{2x} \mid 1/2\,0\}$.

First-principles calculations identified that the intrinsic AFM *X*Mn*Y* (*X* = Sr and Ba, *Y* = Sn and Pb) QLs are experimentally feasible examples of the predicted topological states with a stable crystal structure and giant magnitude of the nontrivial band gaps [52]. Bulk *X*Mn*Y* crystallizes in the tetragonal crystal structure with space group *P4/nmm* and exhibits a layered structure with stacked QLs along the *z*-axis. The QLs consist of six atoms with two Mn atoms in the middle. The calculations of magnetic properties of *X*Mn*Y* QLs show that the magnetic moments on each Mn are about 5 μ_B, and therefore the Mn are in a half-filled $3d^5$ configuration, leading to intrinsic antiferromagnetism with large values of the energy difference between FM and AFM orderings. Under AFM ordering without net magnetic moments, the minority and majority spin bands of *X*Mn*Y* are degenerate in the absence of SOC. As shown in Figure 4.21a, the Mn-d_{yz} and Mn-d_{zx} orbitals are doubly degenerate as a result of the D_{2d} symmetry of the crystal group and contribute to the VBM, while the CBM is dominated by a single Mn-d_{z2} orbital with a small direct band gap of 12 meV. Switching on SOC leads to the inversion of the orbital characters around the Γ point, and, remarkably, the bands remain doubly degenerate with an insulating band gap of 147 meV. The 2D AFM TI phase is obtained with nonzero spin Chern number

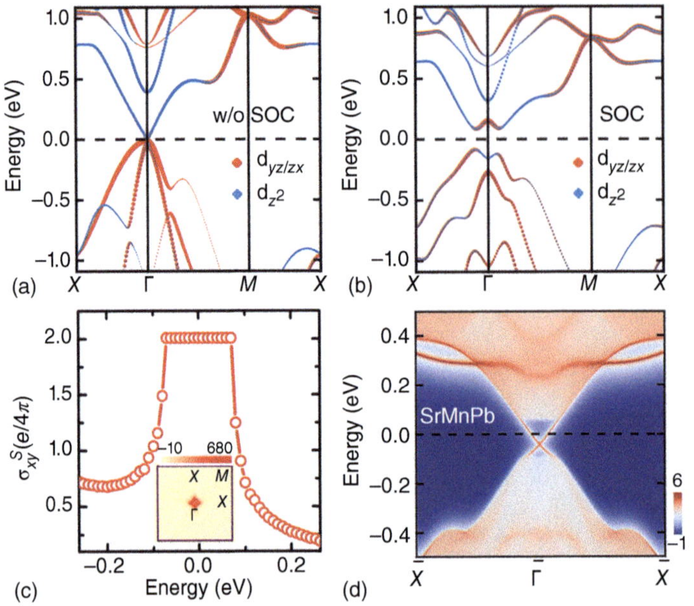

Figure 4.21 Orbitally resolved band structures for SrMnPb QLs (a) without and (b) with SOC, weighted with the contribution of Mn-$d_{yz/zx}$ and Mn-d_{z2} states. The Fermi level is indicated with a dashed line. (c) Energy dependence of the spin Hall conductivity σ^S_{xy}. (Inset) K-space distribution of spin Berry curvature within the SOC gap. (d) Calculated localized density of states (LDOS) of edge states with the color range from blue to red represents the higher LDOS. Source: Niu et al. [52]/with permission of American Physical Society.

C_S, as displayed in Figure 4.21c, the quantization of σ^S_{xy}, $\sigma^S_{xy} = C_S e/(2\pi)$, within the insulating region, which arises mainly from the spin Berry curvature near the Γ point, is clearly visible. The nonzero C_S is further confirmed by the gapless edge states shown in Figure 4.21d. In addition, similar to the helical edge states in 2D TIs, the edge states of 2D AFM TIs are spin polarized, with the spin polarization direction locked with their momentum. Moreover, after investigations of 57 tetragonal antiferromagnets in the crystal structure of outstanding AFM CuMnAs, nine experimentally feasible candidates are predicted to be intrinsic 2D AFM TIs characterized by $Z_2 = 1$ and gapless edge states, although the 2D AFM insulator is scarce [143].

In fact, predicted 2D magnetic TIs can be obtained in both the FM and AFM configurations, as demonstrated in the well-known Kane–Mele model [45, 55, 140, 144]:

$$H = -t\sum_{\langle i,j\rangle} c_i^+ c_j + i\lambda_{\mathrm{SOC}} \sum_{\langle\langle ij\rangle\rangle} v_{ij} c_i^+ s^z c_j + \lambda_{\mathrm{mag}} \sum_i \mu_i c_i^+ B{\cdot}s c_i,$$

where the first and second terms represent the nearest-neighbor hopping and the next nearest-neighbor intrinsic SOC, respectively. $\mu_i = \pm 1$ has opposite values on the two sublattices for the AFM configuration, while $\mu_i = 1$ for the FM configuration in the third term. As shown in Figure 4.22, the electronic structures of the tight-binding model can indeed be insulating regardless of the magnetic orderings. Interestingly, for both the AFM and FM configurations, SOC-induced band inversions (band

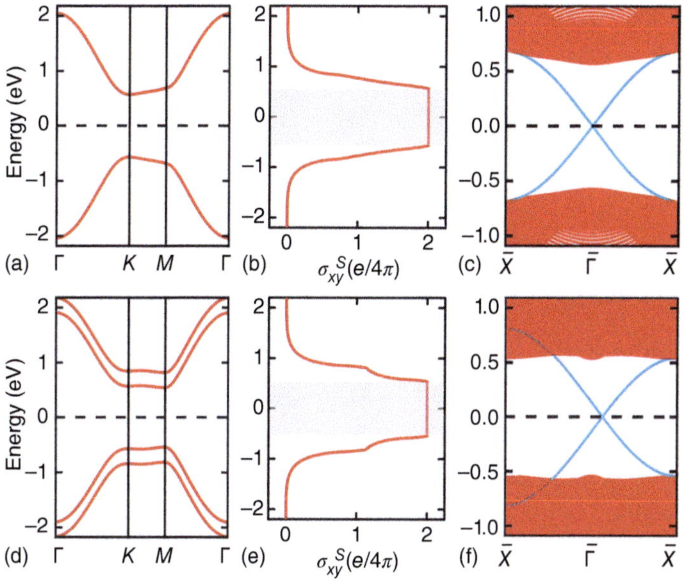

Figure 4.22 Band structures of (a) AFM and (d) FM configurations in honeycomb magnets. (b and e) Spin Hall conductivity σ_{xy}^S and (c and f) energy bands of the armchair nanoribbons for (b and c) AFM and (e and f) FM configurations, confirming the 2D magnetic TI phase. Source: Wang et al. [140]. Reproduced with permission of Royal Society of Chemistry.

gap opening) emerge simultaneously at the spin-up and spin-down channels. The magnetic TI phase is explicitly confirmed by the quantized spin Hall conductivity σ_{xy}^S and emergence of gapless edge states in the 1D nanoribbons with the armchair termination.

For material realization, the magnetic TI phase appears in a layered material of $EuCd_2Bi_2$ QLs under both FM and AFM configurations [140]. Bulk $EuCd_2Bi_2$ crystallizes in the trigonal structure with space group $P\bar{3}m1$, which has a layered crystal structure that can be visualized as a stacking of QLs, with the Eu layer sandwiched by two Cd–Bi layers, along the z direction. Owing to the $4f^{14}$ orbitals of Eu atoms being half-filled, the calculated magnetic moment is about 7 μ_B per Eu^{2+} with the high-spin configuration. In the absence of SOC, as illustrated in Figure 4.23a and e for the AFM and FM $EuCd_2Bi_2$ QLs, both the spin-up and spin-down bands are gapless with the band crossings exactly at the Γ point. The spin-up and spin-down bands are degenerate, as the combined time-reversal and inversion symmetry is preserved, for the AFM ordering, while they are spin polarized for the FM ordering with a splitting of 72.4 meV at the Γ point. Turning on SOC leads to an insulating character for AFM and FM $EuCd_2Bi_2$ QLs with different magnetization directions, and clearly, both the electronic and topological properties are robust against the magnetic phases, accompanied by an integer spin Chern number $C_S = 1$ and a pair of gapless edge states. Moreover, the giant band gap, which can reach as much as 750 meV, is much larger than those in known magnetic TIs [30, 31].

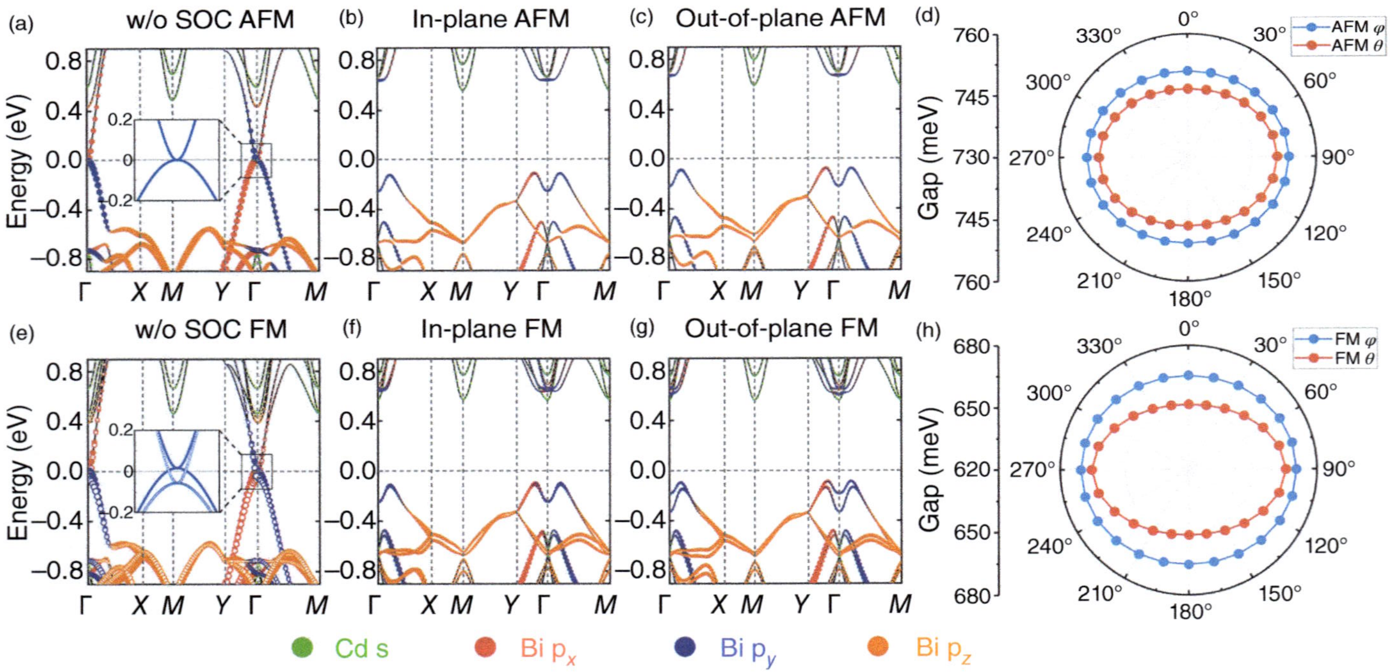

Figure 4.23 The orbitally resolved band structures of $EuCd_2Bi_2$ QLs under (a) AFM and (e) FM configurations without SOC. The solid and empty circles present the spin-up and spin-down bands, respectively. After including SOC, gaps open up in the band structures of (b) in-plane AFM, (c) out-of-plane AFM, (f) in-plane FM, and (g) out-of-plane FM configurations. Band gaps of (d) AFM and (h) FM $EuCd_2Bi_2$ QLs with respect to the magnetic directions θ and φ. Source: Wang et al. [140]. Reproduced with permission of Royal Society of Chemistry.

4.5 Mixed Topological Semimetals

Similar to the nontrivial surface states of TIs and TCIs, metallic points with band crossings also exist in the bulk states called topological semimetals, including mainly the Dirac point and Weyl point refer to the Dirac semimetal and Weyl semimetal, respectively [4–6]. If the metallic points form lines, they are referred to as nodal lines and the system is called as topological nodal-line semimetal. These topological semimetals have been theoretically proposed and experimentally confirmed in three dimensions [19–29, 145–147], revealing a plethora of fascinating properties such as the ultrahigh mobility, nonlinear optical response, and anomalous magnetoresistance [148–151]. However, in two dimensions, the material realization of topological semimetals has been elusive so far and a gap is usually introduced once the SOC comes into account.

While magnets have been successfully fabricated in two dimensions [152, 153], combining the band topology and magnetism holds great opportunities for exploring fundamental topological quantum physics. The concept of a mixed Weyl semimetal is introduced in the mixed phase space of the crystal momentum $k = (k_x, k_y)$ and the magnetization direction $\hat{m}$ as illustrated in Figure 4.24 [154, 155]. Remarkably, in such topologically nontrivial magnetic states, the strength of spin–orbit torques and the Dzyaloshinskii–Moriya interaction can exceed by far that of the conventional metals [155]. The emergence of metallic points, i.e. mixed Weyl points, in mixed Weyl semimetal correlates with discrete jumps of the Chern number C with respect to the magnetization direction and of the mixed Chern number $\mathbb{Z} = 1/(2\pi)\int \Omega_{yx}^{\hat{m}k} dk_x d\theta$ with respect to the crystal momentum, illustrated in Figure 4.24b. Here, the mixed Berry curvature of all occupied states $| u_{kn}^{\theta}\rangle$ is given by $\Omega_{yi}^{\hat{m}k} = 2\text{Im}\sum_n^{\text{occ}} \langle \partial_\theta u_{kn}^{\theta} \mid \partial_{ki} u_{kn}^{\theta} \rangle$ and θ is the angle that the magnetization $\hat{m} = (\sin\theta, 0, \cos\theta)$ makes with the z-axis. There are two different types of such mixed Weyl points: First, the combined time-reversal and mirror symmetry can enforce topological phase transitions accompanied by a band gap closing, which is robust against perturbations that preserve the protective symmetry, as the magnetization direction θ is fixed. Second, generic band crossings may arise due to the complex interplay of exchange interaction and SOC in systems of low symmetry. In addition, nodal points can form closed lines in the mixed phase space (k_x, k_y, θ) and is characterized by a nontrivial Berry phase $\gamma = \oint_c A \cdot d\ell$, where A is the Berry connection defined by $A = i\sum_n^{\text{occ}} \langle u_{kn}^{\theta} \mid \nabla u_{kn}^{\theta} \rangle$ where ∇ stands for $(\partial_{k_x}, \partial_{k_y}, \partial_\theta)$, in analogy to the 3D topological nodal-line semimetals [136–138]. While crystalline mirror symmetry underlies the emergence of nodal line in the mirror plane illustrated in Figure 4.24c, mixed topological nodal-line semimetals host additionally a distinct type of nodal lines that can be thought of as series of nodal points evolve with the magnetization direction θ as depicted in Figure 4.24d.

The existence of mixed topological semimetals can be established by using an additional exchange field term $B \cdot \sigma$ on top of the nonmagnetic 2D TIs and/or TCIs [146]. TlSe [104], Na_3Bi [110], and GaBi [156] are taken as examples. The TlSe is a 2D TCI if no exchange field is applied [104]. Introducing an exchange field leads to an exchange splitting between spin-up and spin-down states and brings conduction

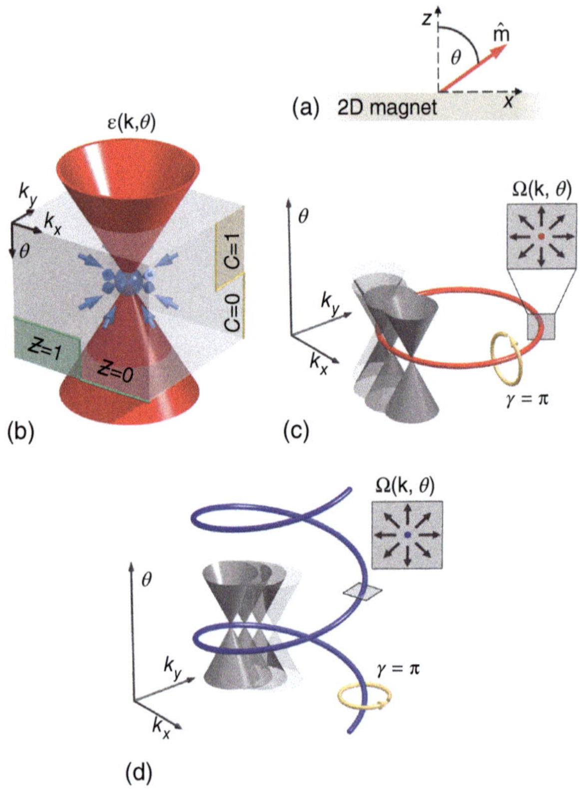

Figure 4.24 (a) The magnetization direction $\hat{m} = (\sin\theta, 0, \cos\theta)$ of a 2D magnet encloses the angle θ with the z-axis perpendicular to the film plane. (b) Acting as sources or sinks of the Berry curvature, emergent band crossings in the mixed phase space of crystal momentum $k = (k_x, k_y)$ and θ can be identified with jumps of the momentum Chern number C and the mixed Chern number Z upon passing through the nodal points. (c) If the magnetic system is symmetric with respect to reflections at $z = 0$, nodal lines with the Berry phase $\gamma = \pi$ may manifest in the corresponding (k_x, k_y)-plane of the mixed phase space. (d) Mixed topological semimetals can host additionally a very distinct type of nodal lines that are one-dimensional manifolds evolving in θ as well as in k. Source: Niu et al. [154]/Springer Nature/CC BY 4.0.

and valence bands closer together. Under small exchange fields, as illustrated in Figure 4.25a, the TCI character survives even if the mirror symmetry M is broken, referred to as the M-broken TCI in analogy to the T-broken QSH insulator. Increasing the magnitude B leads to a band gap closure at the critical value B_c. If the exchange field exceeds this critical value, reopening of the energy gap is accompanied by the realization of QAH phase for any magnetization direction with finite out-of-plane component. Interestingly, for the in-plane magnetized system, $\theta = 90°$, that exhibits the combined time-reversal and mirror symmetry, the gap closes over a wide range of fields $B > B_c$. Figure 4.26a indicates that the gap closing is mediated by four isolated mixed Weyl points with bands of opposite spin crossing slightly off the X and Y points. It corresponds to a change of the total Chern number

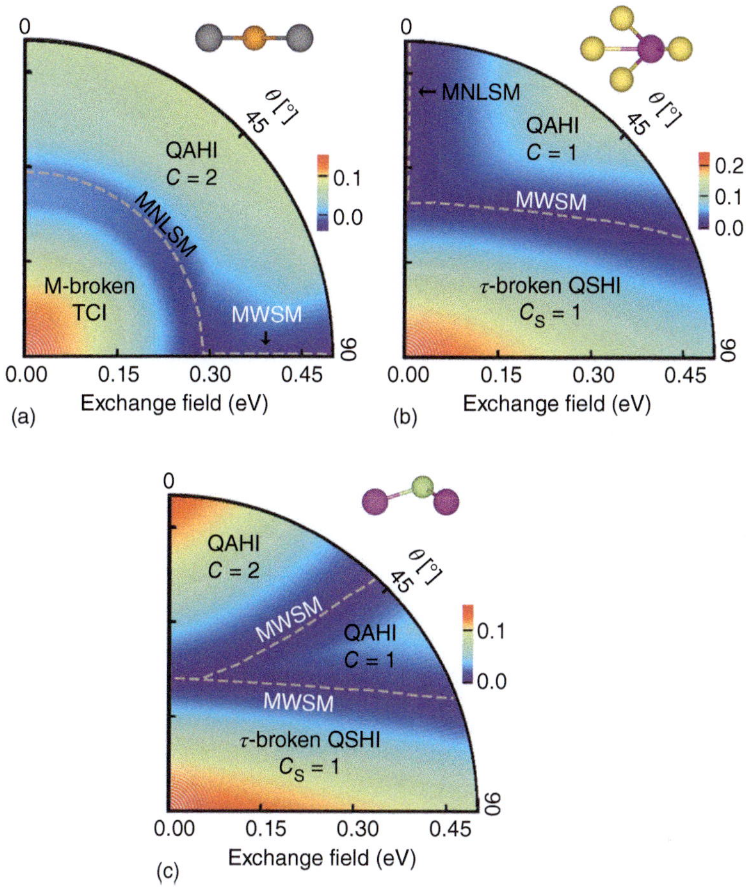

Figure 4.25 Phase diagrams of (a) TlSe, (b) Na_3Bi, and (c) GaBi with respect to the magnitude B and the direction $\hat{m} = (\sin\theta, 0, \cos\theta)$ of the applied exchange field. Side views of the unit cells highlight differences in the crystalline symmetries, and colors represent the value of the global band gap in eV. The emergent metallic states are labeled as either mixed Weyl semimetal (MWSM) or mixed nodal-line semimetal (MNLSM). Source: Niu et al. [154]/Springer Nature/CC BY 4.0.

C, such as from +2 to −2 for $\theta = 90°$ with positive unit charge and from −2 to +2 for $\theta = 270°$ with negative unit charge. In total, the topological charge over the full phase space vanishes. The topologically nontrivial mixed topology further leads to exotic edge states in finite ribbons of TlSe as shown in Figure 4.26b. For Na_3Bi and GaBi, where the combined time-reversal and mirror symmetry is absent, the mixed Weyl points emerge only at the boundaries between the T-broken QSH phase and QAH phases with different Chern numbers, Figure 4.26b and c.

On the other hand, for TlSe, the mixed Weyl point is realized for a range of θ but at a fixed value of exchange field about $B_c = 0.29$ eV. In the spirit of Figure 4.24d, this presents a truly mixed nodal line, a 1D manifold of states, which evolves not only in k-space but also in θ. The occurrence of such mixed nodal lines is purely accidental

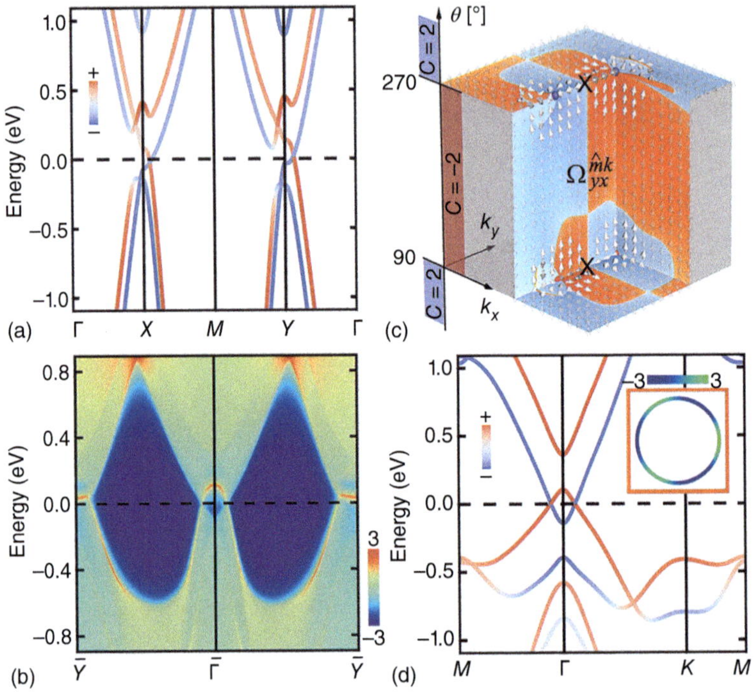

Figure 4.26 (a) Spin-resolved band structure and (b) energy dispersion of a finite ribbon for TlSe with an in-plane exchange field of magnitude $B = 0.5$ eV. (c) Around the X point, for example, the emergence of nodal points with opposite topological charge for reversed in-plane directions θ of the magnetization. (d) Spin-resolved band structure of Na_3Bi with an out-of-plane exchange field of magnitude $B = 0.5$eV. Source: Niu et al. [154]/Springer Nature/CC BY 4.0.

and thus its material realization sets an exciting challenge. As another distinct type, mixed nodal line can evolve in the k-space for a fixed magnetization direction θ, see Figure 4.24c, corresponding to the mirror symmetry. The mirror symmetry survives when exchange field is perpendicular to the film. As shown in Figures 4.25b and 4.26d, the band gap remains closed for Na_3Bi with $\theta = 0°$ above the critical magnitude B_c, and remarkably, a nodal line is formed with the highest occupied and lowest unoccupied bands cross each other around the Γ point. The nodal line remains intact, though with different strengths of exchange field, under various distortions of the lattice that preserves the mirror symmetry. However, it disappears when the mirror symmetry is broken upon turning the exchange field away from the z-axis.

The mixed Weyl points can exist in several 2D magnets such as doped graphene and/or semihydrogenated bismuth [155]. Interestingly, first-principles calculations demonstrate further the possibility of their realization in vdW crystal of VOI_2 monolayer with intrinsic FM ordering [154]. Bulk VOI_2 has a layered structure characterized by the orthorhombic space group *Immm*, and its monolayer contains two I, one O, and one V atoms as shown in the inset of Figure 4.27a. The ground state of VOI_2 monolayer is FM with a spin magnetization of about 1 μ_B per unit cell. As

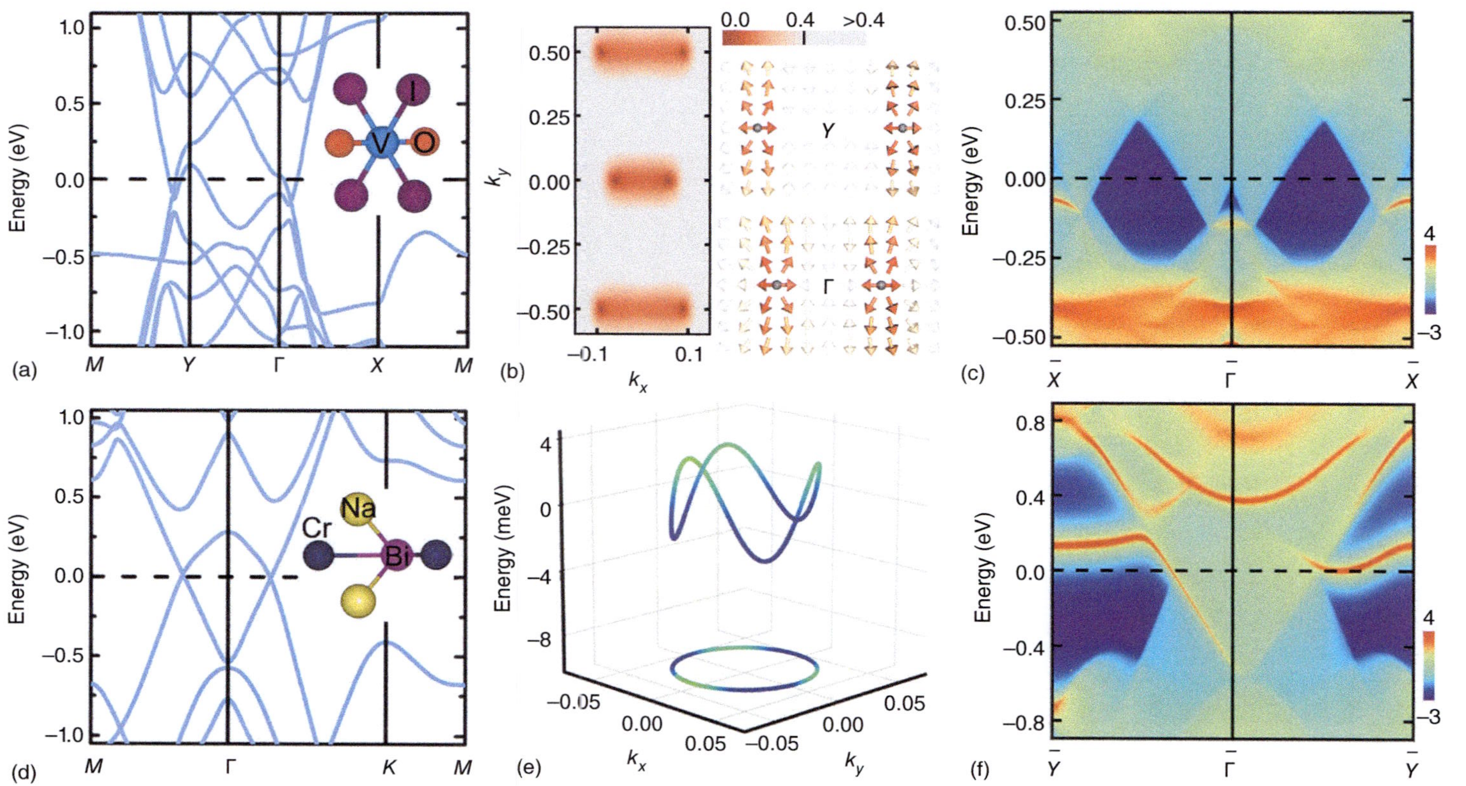

Figure 4.27 Realization of mixed topological semimetals in two-dimensional ferromagnets. Including SOC, the electronic band structures of the stable single-layer compounds (a) VOI_2 and (d) Na_2CrBi display band crossings with nontrivial topological properties in the mixed phase space (k_x, k_y, θ). Side views of the corresponding unit cells are shown as insets. (b) The signatures of the mixed Weyl points for in-plane magnetized VOI_2 manifest in the momentum-resolved direct band gap (color scale in eV) and in the distribution of the Berry curvature field shown as arrows throughout the complex phase space. (e) In the mirror-symmetric plane, the band crossings in perpendicularly magnetized Na_2CrBi form a mixed nodal line that disperses in energy. Band structures of semi-infinite (c) VOI_2 and (f) Na_2CrBi ribbons reveal characteristic edge states due to the nontrivial mixed topology. Source: Niu et al. [154]/Springer Nature/CC BY 4.0.

displayed in Figure 4.27a, the band structures with SOC reveal band crossings along the Γ–X and M–Y paths around the Fermi level. There are four mixed Weyl points in the 2D Brillouin zone as can be seen from the *k*-resolved energy difference between VBM and CBM shown in Figure 4.27b. Further Berry curvature analysis reveals that all four mixed Weyl points host the same topological charge of +1, similar to the magnetized TlSe as discussed earlier. Moreover, when constructing a 1D VOI_2 ribbon along the Γ–Y direction, as shown in Figure 4.27c, four metallic points project onto two pairs of distinct points that are connected by emergent edge states around the *X* and Γ point. In addition, the emergence of mixed nodal lines is theoretically proposed in Na_2CrBi monolayer, replacing one Na atom of Na_3Bi with Cr atom, as illustrated in Figure 4.27d. Including SOC, band crossings emerge in the perpendicularly magnetized Na_2CrBi around the Γ point, and, in the full 2D Brillouin zone, these band crossings forge a nodal loop. As in the previous case of magnetized Na_3Bi monolayer, the nodal line is gapped out as soon as the crystalline mirror symmetry is broken.

References

1 Hasan, M.Z. and Kane, C.L. (2010). *Rev. Mod. Phys.* 82: 3045.
2 Qi, X.-L. and Zhang, S.-C. (2011). *Rev. Mod. Phys.* 83: 1057.
3 Ando, Y. and Fu, L. (2015). *Annu. Rev. Condens. Matter Phys.* 6: 361.
4 Bansil, A., Lin, H., and Das, T. (2016). *Rev. Mod. Phys.* 88: 021004.
5 Yan, B. and Felser, C. (2017). *Annu. Rev. Condens. Matter Phys.* 8: 337.
6 Armitage, N.P., Mele, E.J., and Vishwanath, A. (2018). *Rev. Mod. Phys.* 90: 015001.
7 Po, H.C., Vishwanath, A., and Watanabe, H. (2017). *Nat. Commun.* 8: 50.
8 Bradlyn, B., Elcoro, L., Cano, J. et al. (2017). *Nature (London)* 547: 298.
9 Watanabe, H., Po, H.C., and Vishwanath, A. (2018). *Sci. Adv.* 4: eaat8685.
10 Song, Z., Zhang, T., Fang, Z., and Fang, C. (2018). *Nat. Commun.* 9: 3530.
11 Tang, F., Po, H.C., Vishwanath, A., and Wan, X. (2019). *Nat. Phys.* 15: 470.
12 Zhang, H., Liu, C.-X., Qi, X.-L. et al. (2009). *Nat. Phys.* 5: 438.
13 Xia, Y., Qian, D., Hsieh, D. et al. (2009). *Nat. Phys.* 5: 398.
14 Hsieh, T.H., Lin, H., Liu, J. et al. (2012). *Nat. Commun.* 3: 982.
15 Tanaka, Y., Ren, Z., Sato, T. et al. (2012). *Nat. Phys.* 8: 800.
16 Dziawa, P., Kowalski, B.J., Dybko, K. et al. (2012). *Nat. Mater.* 11: 1023.
17 Xu, S.-Y., Liu, C., Alidoust, N. et al. (2012). *Nat. Commun.* 3: 1192.
18 Yan, C., Liu, J., Zang, Y. et al. (2014). *Phys. Rev. Lett.* 112: 186801.
19 Wang, Z., Sun, Y., Chen, X.-Q. et al. (2012). *Phys. Rev. B* 85: 195320.
20 Wang, Z., Weng, H., Wu, Q. et al. (2013). *Phys. Rev. B* 88: 125427.
21 Liu, Z.K., Zhou, B., Zhang, Y. et al. (2014). *Science* 343: 864.
22 Borisenko, S., Gibson, Q., Evtushinsky, D. et al. (2014). *Phys. Rev. Lett.* 113: 027603.
23 Neupane, M., Xu, S.-Y., Sankar, R. et al. (2014). *Nat. Commun.* 5: 3786.
24 Xu, S.-Y., Liu, C., Kushwaha, S.K. et al. (2015). *Science* 347: 294.

25 Huang, S.-M., Xu, S.-Y., Belopolski, I. et al. (2015). *Nat. Commun.* 6: 7373.
26 Weng, H., Fang, C., Fang, Z. et al. (2015). *Phys. Rev. X* 5: 011029.
27 Lv, B.Q., Xu, N., Weng, H.M. et al. (2015). *Nat. Phys.* 11: 724.
28 Xu, S.-Y., Belopolski, I., Alidoust, N. et al. (2015). *Science* 349: 613.
29 Yang, L.X., Liu, Z.K., Sun, Y. et al. (2015). *Nat. Phys.* 11: 728.
30 Zhang, D., Shi, M., Zhu, T. et al. (2019). *Phys. Rev. Lett.* 122: 206401.
31 Li, J., Li, Y., Du, S. et al. (2019). *Sci. Adv.* 5: eaaw5685.
32 Gong, Y., Guo, J., Li, J. et al. (2019). *Chin. Phys. Lett.* 36: 076801.
33 Otrokov, M.M., Klimovskikh, I.I., Bentmann, H. et al. (2019). *Nature (London)* 576: 416.
34 Wu, J., Liu, F., Sasase, M. et al. (2019). *Sci. Adv.* 5: eaax9989.
35 Li, H., Gao, S.-Y., Duan, S.-F. et al. (2019). *Phys. Rev. X* 9: 041039.
36 Hao, Y.-J., Liu, P., Feng, Y. et al. (2019). *Phys. Rev. X* 9: 041038.
37 Chen, Y.J., Xu, L.X., Li, J.H. et al. (2019). *Phys. Rev. X* 9: 041040.
38 Vidal, R.C., Zeugner, A., Facio, J.I. et al. (2019). *Phys. Rev. X* 9: 041065.
39 Hu, C., Gordon, K.N., Liu, P. et al. (2020). *Nat. Commun.* 11: 97.
40 Zhang, T., Jiang, Y., Song, Z. et al. (2019). *Nature* 566: 475.
41 Vergniory, M.G., Elcoro, L., Felser, C. et al. (2019). *Nature* 566: 480.
42 Tang, F., Po, H.C., Vishwanath, A., and Wan, X. (2019). *Nature* 566: 486.
43 v. Klitzing, K., Dorda, G., and Pepper, M. (1980). *Phys. Rev. Lett.* 45: 494.
44 Haldane, F.D. (1988). *Phys. Rev. Lett.* 61: 2015.
45 Kane, C.L. and Mele, E.J. (2005). *Phys. Rev. Lett.* 95: 146802.
46 Bernevig, B.A., Hughes, T.L., and Zhang, S.-C. (2006). *Science* 314: 1757.
47 Qi, X.-L., Wu, Y.-S., and Zhang, S.-C. (2006). *Phys. Rev. B* 74: 085308.
48 Liu, C.-X., Qi, X.-L., Dai, X. et al. (2008). *Phys. Rev. Lett.* 101: 146802.
49 Yu, R., Zhang, W., Zhang, H.-J. et al. (2010). *Science* 329: 61.
50 Chang, C.-Z., Zhang, J., Feng, X. et al. (2013). *Science* 340: 167.
51 Wang, Z.F., Zhang, H., Liu, D. et al. (2016). *Nat. Mater.* 15: 968.
52 Niu, C., Wang, H., Mao, N. et al. (2020). *Phys. Rev. Lett.* 124: 066401.
53 Shamim, S., Beugeling, W., Shekhar, P. et al. (2021). *Nat. Commun.* 12: 3193.
54 Shockley, W. (1939). *Phys. Rev.* 56: 317.
55 Kane, C.L. and Mele, E.J. (2005). *Phys. Rev. Lett.* 95: 226801.
56 Bernevig, B.A., Hughes, T.L., and Zhang, S.-C. (2006). *Science* 314: 1757.
57 Novoselov, K.S., Geim, A.K., Morozov, S.V. et al. (2004). *Science* 306: 666.
58 Peres, N.M.R. (2010). *Rev. Mod. Phys.* 82: 2673.
59 Bradlyn, B., Elcoro, L., Cano, J. et al. (2017). *Nature* 547: 298.
60 König, M., Wiedmann, S., Brüne, C. et al. (2007). *Science* 318: 766.
61 Ren, Y., Qiao, Z., and Niu, Q. (2016). *Rep. Prog. Phys.* 79: 066501.
62 Knez, I., Du, R.R., and Sullivan, G. (2011). *Phys. Rev. Lett.* 107: 136603.
63 Fu, L. and Kane, C.L. (2006). *Phys. Rev. B* 74: 195312.
64 Xiao, D., Chang, M.-C., and Niu, Q. (2010). *Rev. Mod. Phys.* 82: 1959.
65 Fu, L. and Kane, C.L. (2007). *Phys. Rev. B* 76: 045302.
66 Yang, Y., Xu, Z., Sheng, L. et al. (2011). *Phys. Rev. Lett.* 107: 066602.
67 Prodan, E. (2011). *Phys. Rev. B* 83: 195119.

68 Thouless, D.J., Kohmoto, M., Nightingale, M.P., and den Nijs, M. (1982). *Phys. Rev. Lett.* 49: 405.
69 Yao, Y., Kleinman, L., MacDonald, A.H. et al. (2004). *Phys. Rev. Lett.* 92: 037204.
70 Liu, C.-C., Feng, W., and Yao, Y. (2011). *Phys. Rev. Lett.* 107: 076802.
71 Xu, Y., Yan, B.H., Zhang, H.J. et al. (2013). *Phys. Rev. Lett.* 111: 136804.
72 Niu, C., Bihlmayer, G., Zhang, H. et al. (2015). *Phys. Rev. B* 91: 041303(R).
73 Chuang, F.-C., Yao, L.-Z., Huang, Z.-Q. et al. (2014). *Nano Lett.* 14: 2505.
74 Murakami, S. (2006). *Phys. Rev. Lett.* 97: 236805.
75 Wada, M., Murakami, S., Freimuth, F., and Bihlmayer, G. (2011). *Phys. Rev. B* 83: 121310.
76 Liu, Z., Liu, C.-X., Wu, Y.-S. et al. (2011). *Phys. Rev. Lett.* 107: 136805.
77 Drozdov, I.K., Alexandradinata, A., Jeon, S. et al. (2014). *Nat. Phys.* 10: 664.
78 Nagao, T., Sadowski, J.T., Saito, M. et al. (2004). *Phys. Rev. Lett.* 93: 105501.
79 Koroteev, Y.M., Bihlmayer, G., Chulkov, E.V., and Blügel, S. (2008). *Phys. Rev. B* 77: 045428.
80 Wada, M., Murakami, S., Freimuth, F., and Bihlmayer, G. (2011). *Phys. Rev. B* 83: 121310(R).
81 Zhou, M., Ming, W., Liu, Z. et al. (2014). *Proc. Natl. Acad. Sci. U. S. A.* 111: 14378.
82 Li, L., Zhang, X., Chen, X., and Zhao, M. (2015). *Nano Lett.* 15: 1296.
83 Nemanick, E.J., Hurley, P.T., Brunschwig, B.S., and Lewis, N.S. (2006). *J. Phys. Chem. B* 110: 14800.
84 Knapp, D., Brunschwig, B.S., and Lewis, N.S. (2010). *J. Phys. Chem. C* 114: 12300.
85 Ma, Y., Dai, Y., Kou, L. et al. (2015). *Nano Lett.* 15: 1083.
86 Reis, F., Li, G., Dudy, L. et al. (2017). *Science* 357: 287.
87 Weng, H., Dai, X., and Fang, Z. (2014). *Phys. Rev. X* 4: 011002.
88 Fjellvåg, H. and Kjekshus, A. (1986). *Solid State Commun.* 60: 91.
89 Zhou, J.-J., Feng, W., Liu, C.-C. et al. (2014). *Nano Lett.* 14: 4767.
90 von Benda, H., Simon, A., and Bauhofer, W. (1978). *Z. Anorg. Allg. Chem.* 438: 53.
91 Qian, X., Liu, J., Fu, L., and Li, J. (2014). *Science* 346: 1344.
92 Tang, S., Zhang, C., Wong, D. et al. (2017). *Nat. Phys.* 13: 683.
93 Chen, P., Pai, W.W., Chan, Y.-H. et al. (2018). *Nat. Commun.* 9: 2003.
94 Song, Y.-H., Jia, Z.-Y., Zhang, D. et al. (2018). *Nat. Commun.* 9: 4071.
95 Fu, L. (2011). *Phys. Rev. Lett.* 106: 106802.
96 Teo, J.C.Y., Fu, L., and Kane, C.L. (2008). *Phys. Rev. B* 78: 045426.
97 Liu, J., Hsieh, T.H., Wei, P. et al. (2014). *Nat. Mater.* 13: 178.
98 Niu, C., Buhl, P.M., Bihlmayer, G. et al. (2015). *Phys. Rev. B* 91: 201401(R).
99 Wrasse, E.O. and Schmidt, T.M. (2014). *Nano Lett.* 14: 5717.
100 Liu, J., Qian, X., and Fu, L. (2015). *Nano Lett.* 15: 2657.
101 Xu, S.-Y., Xia, Y., Wray, L.A. et al. (2011). *Science* 332: 560.
102 Sato, T., Segawa, K., Kosaka, K. et al. (2011). *Nat. Phys.* 7: 840.
103 Sa, B., Zhou, J., Sun, Z. et al. (2012). *Phys. Rev. Lett.* 109: 096802.
104 Niu, C., Buhl, P.M., Bihlmayer, G. et al. (2015). *Nano Lett.* 15: 6071.

105 Rauch, T., Flieger, M., Henk, J. et al. (2014). *Phys. Rev. Lett.* 112: 016802.
106 Weber, A.P., Gibson, Q.D., Ji, H. et al. (2015). *Phys. Rev. Lett.* 114: 256401.
107 Eschbach, M., Lanius, M., Niu, C. et al. (2017). *Nat. Commun.* 8: 14976.
108 Avraham, N., Nayak, A.K., Steinbok, A. et al. (2020). *Nat. Mater.* 19: 610.
109 Yao, Y., Ye, F., Qi, X.-L. et al. (2007). *Phys. Rev. B* 75: 041401.
110 Niu, C., Buhl, P.M., Bihlmayer, G. et al. (2017). *Phys. Rev. B* 95: 075404.
111 Mao, N., Hu, X., Niu, C. et al. (2019). *Phys. Rev. B* 100: 205116.
112 Xu, Y., Song, Z., Wang, Z. et al. (2019). *Phys. Rev. Lett.* 122: 256402.
113 Mao, N., Hu, X., Wang, H. et al. (2021). *Phys. Rev. B* 103: 195152.
114 Park, M.J., Kim, Y., Cho, G.Y., and Lee, S.B. (2019). *Phys. Rev. Lett.* 123: 216803.
115 Ahn, J., Kim, D., Kim, Y., and Yang, B.-J. (2018). *Phys. Rev. Lett.* 121: 106403.
116 Perez-Mato, J., Gallego, S., Tasci, E. et al. (2015). *Annu. Rev. Mater. Sci.* 45: 217.
117 Liu, C.X., Zhang, H., Yan, B. et al. (2010). *Phys. Rev. B* 81: 041307(R).
118 Lu, H.Z., Shan, W.Y., Yao, W. et al. (2010). *Phys. Rev. B* 81: 115407.
119 Weng, H., Yu, R., Hu, X. et al. (2015). *Adv. Phys.* 64: 227.
120 Liu, C.-X., Zhang, S.-C., and Qi, X.-L. (2016). *Annu. Rev. Condens. Matter Phys.* 7: 301.
121 Qiao, Z., Yang, S.A., Feng, W. et al. (2010). *Phys. Rev. B* 82: 161414(R).
122 Zhang, H., Lazo, C., Blügel, S. et al. (2012). *Phys. Rev. Lett.* 108: 056802.
123 Wang, J., Lian, B., Zhang, H. et al. (2013). *Phys. Rev. Lett.* 111: 136801.
124 Zhang, H., Xu, Y., Wang, J. et al. (2014). *Phys. Rev. Lett.* 112: 216803.
125 Wu, S.-C., Shan, G., and Yan, B. (2014). *Phys. Rev. Lett.* 113: 256401.
126 Qiao, Z., Ren, W., Chen, H. et al. (2014). *Phys. Rev. Lett.* 112: 116404.
127 Xu, G., Wang, J., Felser, C. et al. (2015). *Nano Lett.* 15: 2019.
128 Zhou, J., Sun, Q., and Jena, P. (2017). *Phys. Rev. Lett.* 119: 046403.
129 Wang, Z.F., Liu, Z., Yang, J., and Liu, F. (2018). *Phys. Rev. Lett.* 120: 156406.
130 Sun, H., Xia, B., Chen, Z. et al. (2019). *Phys. Rev. Lett.* 123: 096401.
131 Niu, C., Mao, N., Hu, X. et al. (2019). *Phys. Rev. B* 99: 235119.
132 Li, Y., Li, J., Li, Y. et al. (2020). *Phys. Rev. Lett.* 125: 086401.
133 Pan, H., Li, Z., Liu, C.-C. et al. (2014). *Phys. Rev. Lett.* 112: 106802.
134 Rahn, M.C., Soh, J.-R., Francoual, S. et al. (2018). *Phys. Rev. B* 97: 214422.
135 Hua, G., Nie, S., Song, Z. et al. (2018). *Phys. Rev. B* 98: 201116(R).
136 Xu, Y., Song, Z., Wang, Z. et al. (2019). *Phys. Rev. Lett.* 122: 256402.
137 Mong, R.S.K., Essin, A.M., and Moore, J.E. (2010). *Phys. Rev. B* 81: 245209.
138 Fang, C., Gilbert, M.J., and Bernevig, B.A. (2013). *Phys. Rev. B* 88: 085406.
139 Liu, C.-X., Zhang, R.-X., and VanLeeuwen, B.K. (2014). *Phys. Rev. B* 90: 085304.
140 Wang, H., Mao, N., Hu, X. et al. (2021). *Mater. Horiz.* 8: 956.
141 Young, S.M. and Kane, C.L. (2015). *Phys. Rev. Lett.* 115: 126803.
142 Liu, C.-C., Jiang, H., and Yao, Y. (2011). *Phys. Rev. B* 84: 195430.
143 Mao, N., Wang, H., Hu, X. et al. (2020). *Phys. Rev. B* 102: 115412.
144 Wang, H., Mao, N., Niu, C. et al. (2020). *Mater. Horiz.* 7: 2431.
145 Kim, Y., Wieder, B.J., Kane, C.L., and Rappe, A.M. (2015). *Phys. Rev. Lett.* 115: 036806.
146 Ru, Y., Weng, H., Fang, Z. et al. (2015). *Phys. Rev. Lett.* 115: 036807.
147 Bian, G., Chang, T.-R., Sankar, R. et al. (2016). *Nat. Commun.* 7: 10556.

148 Liang, T., Gibson, Q., Ali, M.N. et al. (2015). *Nat. Mater.* 14: 280.
149 Wu, L., Patankar, S., Morimoto, T. et al. (2017). *Nat. Phys.* 13: 350.
150 Huang, X., Zhao, L., Long, Y. et al. (2015). *Phys. Rev. X* 5: 031023.
151 Moll, P.J.W., Nair, N.L., Helm, T. et al. (2016). *Nature* 535: 266.
152 Gong, C., Li, L., Li, Z. et al. (2017). *Nature* 546: 265.
153 Huang, B., Clark, G., Navarro-Moratalla, E. et al. (2017). *Nature* 546: 270.
154 Niu, C., Hanke, J.-P., Buhl, P.M. et al. (2019). *Nat. Commun.* 10: 3179.
155 Hanke, J.-P., Freimuth, F., Niu, C. et al. (2017). *Nat. Commun.* 8: 1479.
156 Crisostomo, C.P., Yao, L.-Z., Huang, Z.-Q. et al. (2015). *Nano. Lett.* 15: 6568.

5

Calculation of Excited-State Properties

5.1 Green's Function Many-Body Perturbation Theory

It has been already known that electronic excitations constitute the center concept of most of the commonly measured spectra. In low-dimensional materials, the many-body effects, i.e. electron–electron correlations and electron–hole interactions, are of paramount importance for understanding the excited properties such as the absorption and emission. It has been identified that strong excitonic effects play a crucially important role in the optical response of low-dimensional materials (such as nanosheets, nanotubes, nanoribbons, and nanowires) due to the strong electron–hole Coulomb interactions, with tightly bound excitons featured the absorption spectrum and large binding energy assigned to the bound excitons in low-dimensional materials [1–19]. In this sense, the excitonic effects are the center concept to the novel many-particles phenomena in low-dimensional materials such as exciton–exciton coupling, trions, spatially indirect interlayer excitons, and the valley excitons [20–22]. In applications such as optoelectronics, photovoltaics, and photocatalysis [23–25], correctly understanding the excitonic effects is primarily important, especially for the two-dimensional materials. In Figure 5.1, concept of the exciton is indicated, and in Figure 5.2 the excitonic effects dominating the optical absorption in low-dimensional materials are shown.

In this section, only the most important ideals and equations are summarized. More details for the background of many-body perturbation theory are available in previous literatures [27–44].

On the basis of a set of Green's function equations, the many-body perturbation theory starts with a one-particle propagator. In accordance to the following equation, one-particle Green's function can be defined as:

$$G(\mathbf{r}_1t_1, \mathbf{r}_2t_2) = -i\langle N,0|T[\widehat{\Psi}(\mathbf{r}_1t_1)\widehat{\Psi}^{\dagger}(\mathbf{r}_2t_2)]|N,0\rangle$$

where $|N,0\rangle$ stands for the ground state of the system with N electrons, $\widehat{\Psi}^{\dagger}(\mathbf{r}t)$ and $\widehat{\Psi}(\mathbf{r}t)$ are the fermion creation and annihilation operators in the Heisenberg representation, respectively; T is the Wick's time-ordering operator. If $t_1 > t_2$,

$$G(\mathbf{r}_1t_1, \mathbf{r}_2t_2) = -i\langle N,0|\widehat{\Psi}(\mathbf{r}_1t_1)\widehat{\Psi}^{\dagger}(\mathbf{r}_2t_2)|N,0\rangle$$

Calculations and Simulations of Low-Dimensional Materials: Tailoring Properties for Applications, First Edition. Ying Dai, Wei Wei, Yandong Ma, and Chengwang Niu.

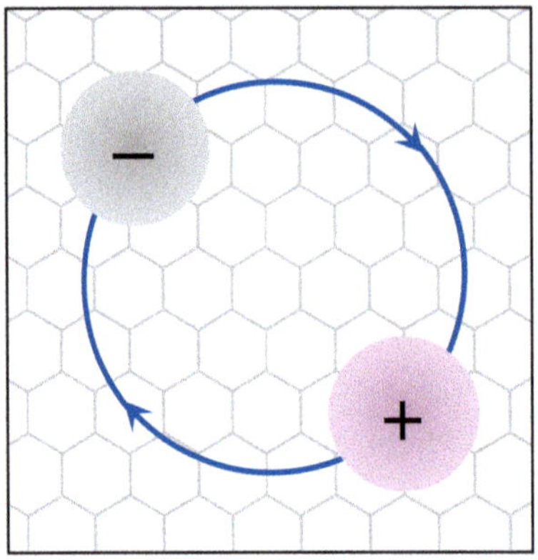

Figure 5.1 Exciton, i.e. the Coulomb bounded electron–hole pair, which can be free to move through the material. Source: Wei Wei.

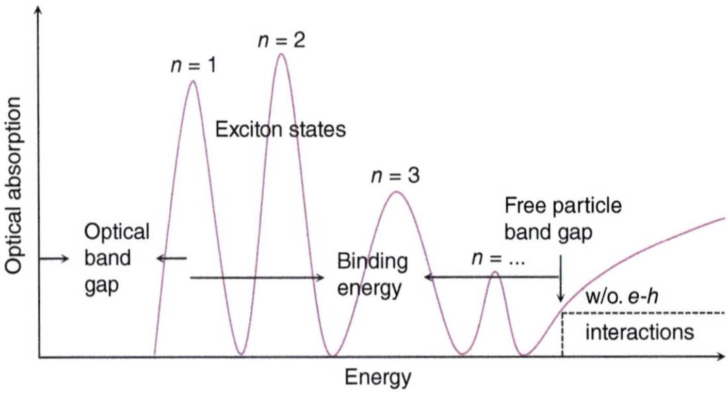

Figure 5.2 Typical excitonic absorption in low-dimensional materials; absorption peaks in low energy correspond to the excitonic states. Source: Wei et al. [26]/with permission of John Wiley & Sons.

it gives the probability amplitude finding the electron in $\mathbf{r}_1$ at time t_1 after an electron is added in $\mathbf{r}_2$ at time t_2. In case of $t_1 < t_2$,

$$G(\mathbf{r}_1 t_1, \mathbf{r}_2 t_2) = i\langle N, 0|\widehat{\Psi}^{\dagger}(\mathbf{r}_2 t_2)\widehat{\Psi}(\mathbf{r}_1 t_1)|N, 0\rangle$$

it describes the situation for hole, and the removal of an electron is equivalent to the creation of a hole. In light of the Fourier transformation, one-particle Green's function of Lehmann representation in the energy space can be written as:

$$G(\mathbf{r}_1, \mathbf{r}_2; \omega) = \sum_i \frac{f_i(\mathbf{r}_1) f_i^*(\mathbf{r}_2)}{\omega - E_i + i\eta \operatorname{sgn}(E_i - \mu)}$$

where the energies E_i and Lehmann amplitudes $f_i(\mathbf{r})$ can be defined as:

$$E_i = \begin{cases} E_{N+1,i} - E_{N,0} & \text{if } E_i > \mu \\ E_{N,0} - E_{N-1,i} & \text{if } E_i < \mu \end{cases}$$

$$f_i(\mathbf{r}) = \begin{cases} \langle N, 0|\widehat{\Psi}(\mathbf{r})|N+1, i\rangle & \text{if } E_i > \mu \\ \langle N-1, i|\widehat{\Psi}(\mathbf{r})|N, 0\rangle & \text{if } E_i < \mu \end{cases}$$

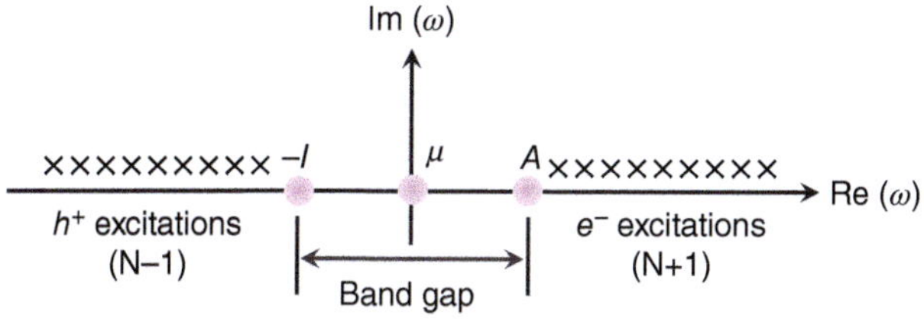

Figure 5.3 Poles of the Green's function, with I being the ionization energy, A the electron affinity, and chemical potential μ is located in the middle of the band gap. Source: Wei et al. [26]/with permission of John Wiley & Sons.

here μ is the chemical potential, and η is a positive real infinitesimal. In these equations, $E_{N,0}$ is the ground-state total energy, and $E_{N\mp1,i}$ and $|N\mp1, i\rangle$ are the total energy and wave function, respectively. It should be emphasized that E_i is actually also the pole of $G(\mathbf{r}_1, \mathbf{r}_2; \omega)$, see Figure 5.3. In the quasi-particle approximation, one-particle Green's function can be expressed in terms of quasi-particle energies and wave functions as:

$$G(\mathbf{r}_1, \mathbf{r}_2; \omega) = \sum_i \frac{\Psi_i^{\mathrm{QP}}(\mathbf{r}_1)\Psi_i^{\mathrm{QP}*}(\mathbf{r}_2)}{\omega - E_i^{\mathrm{QP}}}$$

In this case, the quasi-particle equation can be obtained:

$$\left[-\frac{1}{2}\nabla^2 + V_H(\mathbf{r}) + V_{\mathrm{ext}}(\mathbf{r})\right]\Psi_i^{\mathrm{QP}}(\mathbf{r}) + \int \Sigma\left(\mathbf{r}, \mathbf{r}'; E_i^{\mathrm{QP}}\right)\Psi_i^{\mathrm{QP}}(\mathbf{r}')dr' = E_i^{\mathrm{QP}}\Psi_i^{\mathrm{QP}}(\mathbf{r})$$

which finds the similarity with the single-particle equation in density functional theory:

$$\left[-\frac{1}{2}\nabla^2 + V_H(\mathbf{r}) + V_{\mathrm{ext}}(\mathbf{r})\right]\Psi_i^{\mathrm{DFT}}(\mathbf{r}) + V_{\mathrm{xc}}(\rho(\mathbf{r}))\Psi_i^{\mathrm{DFT}}(\mathbf{r}) = E_i^{\mathrm{DFT}}\Psi_i^{\mathrm{DFT}}(\mathbf{r})$$

In comparison to the equation in density functional theory, physically, quasi-particle energies and wave functions are more meaningful and, therefore, can be directly linked to the experimentally measured electronic structures.

In the framework of Green's function, self-energy reads:

$$\Sigma(1,2) = i\int d34v(1^+, 3)[-G_2(1,3; 4, 3^+) + G(1,4)G(3, 3^+)]G^{-1}(4,2)$$

where 1, 2, 3, and 4 are for combined space and time coordinates, e.g. $(1) = (\mathbf{r}_1, t_1)$, v is the bare Coulomb potential, G_2 is the two-particle Green's function, and $(1^+) = (\mathbf{r}_1, t_1 + \eta)$. In particular, Σ can be expressed by W, the effective electron–electron interaction, which is defined as:

$$W(1,2) = \int v(1,3)\varepsilon^{-1}(3,2)d3$$

here ε is the dielectric function, and W is weaker than v. Once the exact one-particle Green's function is known, according to Hedin's equations, exact self-energy can in principle be calculated. If the contribution from $\delta\Sigma/\delta G$ in the vertex function Γ is ignored, the self-energy operator reduces to:

$$\Sigma(1,2) = iG(1,2)W(1^+, 2)$$

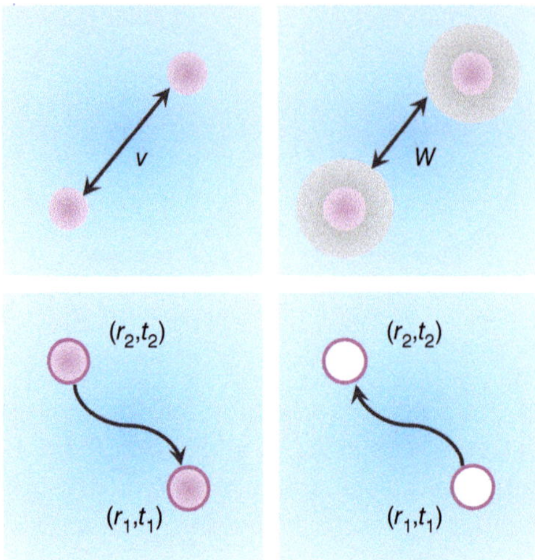

Figure 5.4 (Top) Concept of the quasi-particle; *v* is the bare Coulomb interaction, and *W* is the dynamically screened interaction. (Bottom) One-particle Green's function. In the context of second quantized formulation, one-particle Green's function (or propagator) can be defined. Source: Wei et al. [26]/with permission of John Wiley & Sons.

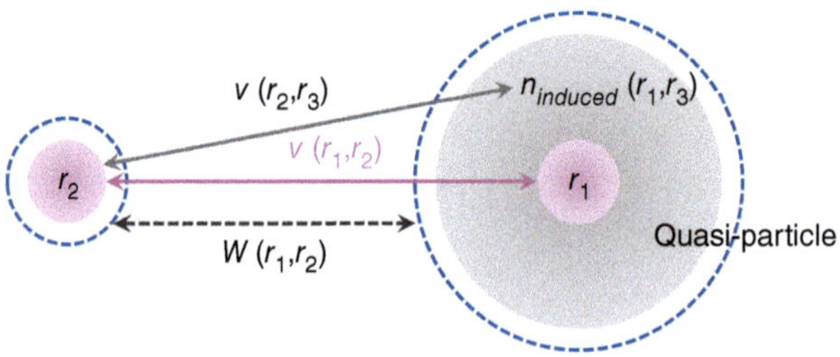

Figure 5.5 Effective interaction between quasi-particles. $W(\mathbf{r}_1, \mathbf{r}_2)$ means the dynamically screened interaction, which is the "true" interaction between quasi-particles; $v(\mathbf{r}_1, \mathbf{r}_2)$ is the bare Coulomb interaction, while $v(\mathbf{r}_2, \mathbf{r}_3)$ is the interaction between $\mathbf{r}_2$- and $\mathbf{r}_1$-induced polarized charge (quasi-hole) $\mathbf{r}_3$. Source: Wei et al. [26]/with permission of John Wiley & Sons.

i.e. the well-known *GW* approximation. In Figure 5.4, the concepts of quasi-particle and one-electron Green's function are shown, and effective interactions between two quasi-particles (i.e. *W*) are schematically shown in Figure 5.5.

If Σ is expressed by *W*, self-energy can be deduced from Hedin's equations. As shown in Figure 5.6, on the basis of Hedin's equations and Dyson's equation, self-energy and interacting Green's function can be obtained by the iteration process. In the first step, first-principles calculations based on density functional theory should be performed, which is chosen as the starting point to get the ground-state single-particle eigenvalues and eigenfunctions. In the following, one-particle Green's function and dielectric function can be constructed on top of the results obtained from last step. In the third step, the screened Coulomb potential and the self-energy are addressed. In closing, quasi-particle energies and wave functions are obtained. In principle, the simplest *GW* calculation is considered as the one-shot G_0W_0, namely, there is no iteration with the density functional theory

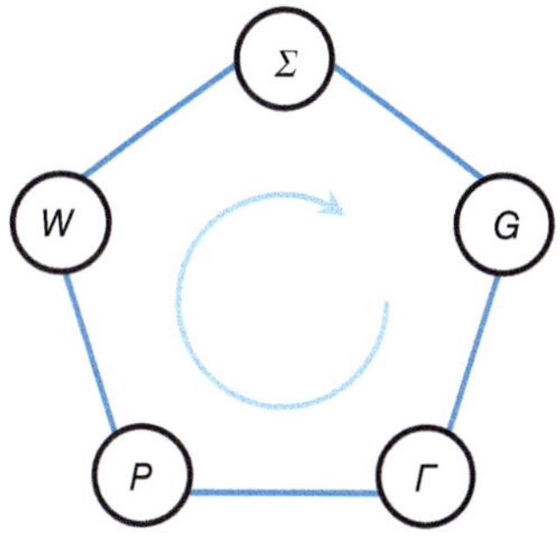

Figure 5.6 The self-energy Σ and one-particle Green's function *G* can be determined using Hedin's equations and Dyson's equation iteratively. Source: Wei et al. [26]/with permission of John Wiley & Sons.

results being a zero approximation to the quasi-particle eigenvalues and wave functions.

In general, the two-particle Green's function takes

$$G_2(1,2;1',2') = (-i)^2\langle N,0|T[\hat{\Psi}(1)\hat{\Psi}(2)\hat{\Psi}^{\dagger}(2')\hat{\Psi}^{\dagger}(1')]|N,0\rangle$$

It is natural that the two-particle Green's function describes the propagation of coupled electron–electron, electron–hole (exciton), and hole–hole pairs. As for the electron–hole pair of interest, only the orderings $t_1, t_{1'} > t_2, t_{2'}$ and $t_1, t_{1'} < t_2, t_{2'}$ are taken into account, then the two-particle Green's function can be expressed as:

$$\begin{aligned} G_2^{\text{exciton}}&(1,2;1',2') \\ &= -\Theta\left(\tau - \frac{1}{2}|\tau_1| - \frac{1}{2}|\tau_2|\right) \\ &\quad \times \sum_S \exp[-i(E_{N,S} - E_{N,0})\tau]\chi_S\left(r_1, r_{1'}; \tau_1\right)\widetilde{\chi}_S\left(r_2, r_{2'}; \tau_2\right) \\ &\quad - \Theta\left(-\tau - \frac{1}{2}|\tau_1| - \frac{1}{2}|\tau_2|\right) \\ &\quad \times \sum_S \exp[i(E_{N,S} - E_{N,0})\tau]\widetilde{\chi}_S\left(r_1, r_{1'}; \tau_1\right)\chi_S\left(r_2, r_{2'}; \tau_2\right) \end{aligned}$$

where $E_{N,S}$ is the total energy of the excited state S, $E_{N,S} - E_{N,0}$ is the excitation energy, and χ_S is the exciton wave function. In the framework of Bethe–Salpeter equation (BSE), the motion of two-particle Green's function obeys:

$$\begin{aligned} L(1,2;1',2') &= G(1,2')G(2,1') \\ &\quad + \int G(1,3)G(3',1')K(3,4';3',4)L(4,2;4',2')d(3,3';4',4) \end{aligned}$$

where $L(1, 2; 1', 2')$ is the two-particle correlation function, and $K(3, 4'; 3', 4)$ is the two-particle interaction kernel, which is composed of two contributions, i.e. the exchange and direct terms. In the case of $t_1, t_{1'} > t_2, t_{2'}$, the BSE turns out to be an eigenvalue problem:

$$(E_c - E_v)A_{vc}^S + \sum_{v'c'} K_{vc,v'c'}^{AA}(\Omega_S)A_{v'c'}^S = \Omega_S A_{vc}^S$$

here the first term describes the uncorrelated electron–hole pair, and the second term accounts for the bound electron–hole interaction. In the equation, $K^{AA}_{vc,v'c'}$ takes

$$K^{AA}_{vc,v'c'}(\Omega_S) = \langle c, v'|v|v, c'\rangle + \left(-\frac{i}{2\pi}\right) \times \int_{-\infty}^{+\infty} d\omega \ \exp(-i\omega\gamma)\langle c, v'|W(\omega)|c', v\rangle$$
$$\times \left[\frac{1}{\Omega_S - \omega - (E_{c'} - E_v) + i\eta} + \frac{1}{\Omega_S + \omega - (E_c - E_{v'}) + i\eta}\right]$$

where Ω_S is the exciton energy, c (v) refers to the conduction (valence) band, and γ is a positive real infinitesimal.

In the practice to obtain quasi-particle energies and optical response within many-body perturbation theory, GW+BSE scheme is usually chosen. In Figure 5.7, a flow chart is shown for the GW+BSE calculations.

It has been well known that the many-body perturbation theory is a method rooting in the Green's function for accurately calculating the excited-state phenomena. In general, the noninteracting Green's function G^0 takes the form:

$$G^0_{n\mathbf{k}}(\omega) = \frac{f_{n\mathbf{k}}}{\omega - \varepsilon_{n\mathbf{k}} - i0^+} + \frac{1 - f_{n\mathbf{k}}}{\omega - \varepsilon_{n\mathbf{k}} + i0^+}$$

here $|n\mathbf{k}\rangle$ is used to label the single-particle levels, and n and $\mathbf{k}$ are the band index and the grid generic vector for sampling the Brillouin zone, respectively. In the equation, $f_{n\mathbf{k}}$ is the occupation factor, and $\varepsilon_{n\mathbf{k}}$ is the Kohn–Sham energies. In terms of the Dyson equation, relation between G^0 and the exact Green's function can be expressed as:

$$G_{n\mathbf{k}}(\omega) = \left[\left(G^0_{n\mathbf{k}}(\omega)\right)^{-1} - \Sigma_{n\mathbf{k}}(\omega) + V^{\mathrm{XC}}_{n\mathbf{k}}\right]^{-1}$$

where $\Sigma_{n\mathbf{k}}(\omega) = \Sigma^{\mathrm{X}}_{n\mathbf{k}} + \Sigma^{\mathrm{C}}_{n\mathbf{k}}(\omega)$, with $\Sigma^{\mathrm{X}}_{n\mathbf{k}}$ and $\Sigma^{\mathrm{C}}_{n\mathbf{k}}(\omega)$ being the exchange (X) and correlation (C) part, respectively. It should be pointed out that, additionally, the

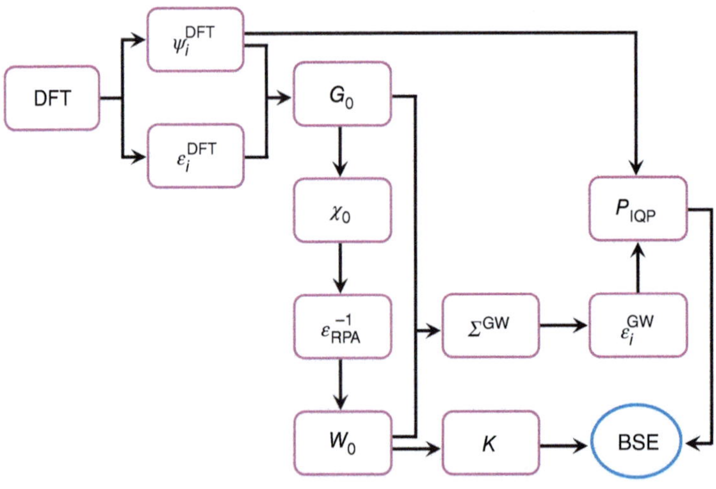

Figure 5.7 Flow chart for the GW+BSE calculations starting from the density functional theory. Source: Wei et al. [26]/with permission of John Wiley & Sons.

self-energy is a function of G^0 and of $\epsilon^{-1}(\mathbf{r}_1, \mathbf{r}_2; \omega)$ (inverse dynamical dielectric function). In case of knowing the ϵ^{-1}, the correlation part of the self-energy can be immediately written out. In particular, the equation of motion of ϵ^{-1} reads from the reducible response function χ, and the GW approximation could be written down as χ is given by the random phase approximation (RPA). In accordance to the plasmon–pole approximation (PPA) for GW self-energy, ϵ^{-1} can be approximated by the single-pole function:

$$\epsilon^{-1}_{\mathbf{GG}'}(\mathbf{q}, \omega) \approx \delta_{\mathbf{GG}'} + R_{\mathbf{GG}'}(\mathbf{q}) \left[\left(\omega - \Omega_{\mathbf{GG}'}(\mathbf{q}) + i0^+\right)^{-1} - \left(\omega + \Omega_{\mathbf{GG}'}(\mathbf{q}) - i0^+\right)^{-1} \right]$$

where $R_{\mathbf{GG}'}$ and $\Omega_{\mathbf{GG}'}$ can be addressed by using the PPA to get the exact ϵ^{-1} at $\omega = 0$ and $\omega = iE_{\mathrm{PPA}}$, with E_{PPA} as a suitable artificially defined parameter. In principle, the macroscopic dielectric function is defined with the microscopic inverse dielectric function:

$$\epsilon_{\mathrm{M}}(\omega) \equiv \lim_{\mathbf{q}\to 0} \frac{1}{[\epsilon(\mathbf{q}, \omega)^{-1}]_{\mathbf{G}=0,\mathbf{G}'=0}}$$

where ϵ is the matrix in the space of reciprocal vectors $\mathbf{G}$. In general, the RPA to the dielectric function and polarizability is inadequate in describing the electronic correlations for the response function χ. In order to overcome the drawbacks of the RPA, a more elaborate equation of motion for χ with the consideration of electron–electron correlations will be employed, that is, the BSE introduced by using the electron–hole Green's function. After defining the noninteracting electron–hole Green's function and the corresponding interacting electron–hole Green's function, the BSE reduces to an eigenvalue problem of the Hamiltonian:

$$H_{\substack{nn'\mathbf{k} \\ mm'\mathbf{k}'}} = \left(\varepsilon_{n\mathbf{k}} - \varepsilon_{n\mathbf{k}'}\right) \delta_{nm}\delta_{n'm'}\delta_{\mathbf{kk}'} + \left(f_{n'\mathbf{k}} - f_{n\mathbf{k}}\right) \left[2\overline{V}_{\substack{nn'\mathbf{k} \\ mm'\mathbf{k}'}} - W_{\substack{nn'\mathbf{k} \\ mm'\mathbf{k}'}} \right]$$

which is in general non-Hermitian. In practice, however, Tamm–Dancoff approximation is usually adopted, in which only the electron–hole pairs at positive energy are considered. As a result, the Hamiltonian turns out to be Hermitian. In the last step, the dielectric function can be calculated from

$$\epsilon_{\mathrm{M}}(\omega) \equiv \lim_{\mathbf{q}\to 0} \frac{8\pi}{|\mathbf{q}|^2 \Omega N_q} \sum_{nn'\mathbf{k}} \sum_{mm'\mathbf{k}'} \rho^*_{n'n\mathbf{k}}(\mathbf{q}, \mathbf{G}) \rho_{m'm\mathbf{k}'}(\mathbf{q}, \mathbf{G}') \times \sum_{\lambda} \frac{A^{\lambda}_{n'n\mathbf{k}} \left(A^{\lambda}_{m'm\mathbf{k}'}\right)^*}{\omega - E_\lambda}$$

where $A^{\lambda}_{n'n\mathbf{k}} = \langle n'n\mathbf{k}|\lambda\rangle$ is the eigenvectors of H. In respect to insulators with large band gap, nevertheless, E_λ may go into the single-particle gap. It can be deduced that strong band gap renormalization and excitonic effects are characterized in two-dimensional materials.

It should be pointed out that although real-time time-dependent density functional theory (TDDFT) has also been demonstrated to be useful to calculate the optical properties of finite systems, exciton interactions due to the periodic boundary conditions cannot be screened in TDDFT, thus not suitable for calculating the solid-state systems. In periodic boundary conditions, as in most of theoretical calculations, a vital question arises, that is, the charge–charge interactions between two nearby cells will play if charge transfer exciton is created in a unit cell. It is sensible, therefore, to consider such corrections by using a large enough cell to make the charge–charge interaction effects negligible.

5.2 Excitonic Effects and Band Gap Renormalization in Two-Dimensional Materials

It has been intensively demonstrated that the *GW*+BSE strategy shows great success in calculating the optical absorption of a large number of two-dimensional materials. It has been already illustrated that the strongly bound excitons dominate the optical response of MoS_2 and the same for other group-VI semiconducting single-layer transition metal dichalcogenides (TMDCs) (such as $MoSe_2$, WS_2, and WSe_2) [45–57]. As an example, MoS_2 shows strongly bound excitonic states with novel *k*-space characteristics. In other words, excitonic effects are responsible for the absorption with a binding energy of 0.96 eV assigned to the first bound exciton [57]; see Figure 5.8 for the definition of exciton binding energy. In particular, characteristic A and B excitons due to the spin–orbit coupling induced Zeeman-like valence band splitting are observed, which are found to be located at 1.88 and 2.02 eV, respectively. In this work, the number of *k* points, the significantly high energy cutoff for the dielectric matrix, and the number of Bloch bands are emphasized in importance, which are necessary to reproduce experimental results. In general, for other two-dimensional TMDCs, the optical spectra are also characterized by the low-energy A and B excitons, with these bound

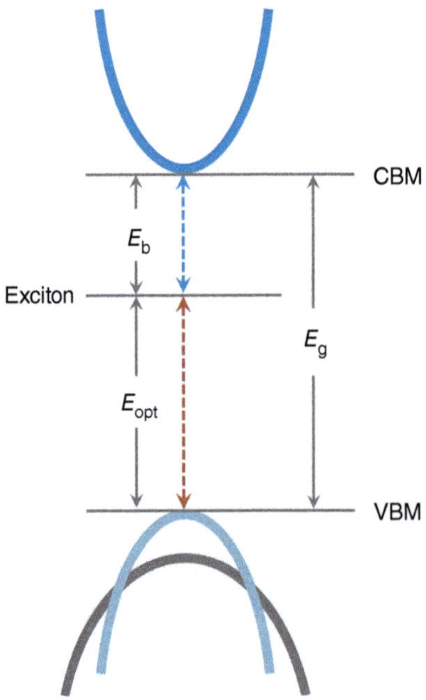

Figure 5.8 Definition of the exciton binding energy (E_b), E_g is the quasi-particle band gap, E_{opt} corresponds to the optical gap. CBM, conduction band minimum; VBM, valence band maximum. Source: Wei Wei.

excitons (binding energy >0.5 eV) being confined to a (near) two-dimensional geometry because of the reduced screening in the structures of reduced dimensionality. In the optical absorption spectra of different two-dimensional TMDCs, differences lie in the position (excitation energy) and binding energy of these bound excitons. At different levels of many-body perturbation theory, such as self-consistent *GW*, results show otherwise obvious different exciton excitation and binding energies. In addition, more accurate parameters give rise to apparently different optical absorption. In recent studies, two-dimensional TMDCs materials are often used in photocatalytic and photovoltaic experiments; the large exciton binding energy should be reduced to harness high light–energy conversion efficiency.

It should be emphasized that even the same simulation technology can give dramatically different results with respect to the band structure and optical spectrum. On the basis of *GW* approximation, for example, band gaps for monolayer MoS_2 available from literatures are 2.41 [46], 2.97 [47], 2.82 [51], and 2.84 eV [57]. In addition, the excitation energy and binding energy for the characteristic A exciton of MoS_2 show significant discrepancy. In particular, MoS_2 A exciton could be positioned at 1.80 [47], 1.78 [51], and 1.88 eV [57]. In my opinion, the differences can be attributed to the exact *GW*+BSE parameters used in practice calculations, such as the *k*-point mesh, the potential for subsequent charge density and wave function, the empty bands for dielectric function, etc. In addition, the *GW* scheme plays an important role in affecting the band gap and thus the absorption profile. In particular, such as non-self-consistent G_0W_0 gives rise to different band gap from other *GW* schemes like full *GW* and GW_0. In some cases, even direct–indirect band gap transition can be found when using dissimilar *GW* methods.

In respect to two-dimensional photocatalysts, phosphorene has also drawn extensive attention for its unique optical and physical properties. In particular, phosphorene shows highly anisotropic optical response, that is, light polarization along armchair direction shows strong absorption, while it indicates transparency along the zigzag direction. It is chiefly absorbent across the infrared-light range and part of the visible-light range, making it also an ideal candidate as an optical linear polarizer with a wide energy window. In case of phosphorene, the puckered geometry is response for the highly anisotropic electronic structures dictating unique excitonic effects, and the exotic correlations between photoinduced carriers indicate unique photoexcitation processes that are closely associated with the involved photocatalytic behavior. As for monolayer phosphorene, the self-energy correction enlarges the band gap from 0.8 to 2 eV, and the lowest-energy optical absorption peak is reduced to 1.2 eV because of a huge exciton binding energy (0.8 eV) [58]. In addition, the band gap, exciton binding energies, absorption, and linear polarization energy window of phosphorene can all be broadly tuned by changing the number of stacked layers. It therefore serves as a convenient and efficient method for engineering the excited-state properties of materials. In particular, the interlayer interaction and the corresponding coupling reduce the perpendicular quantum confinement, resulting in smaller band gaps and weaker excitonic effects for few-layer black phosphorous.

In analogy to black phosphorus, Group-IV monochalcogenides assume anisotropic optical properties. In practice, G_0W_0 approximation gives rise to fundamental band gaps of 2.85 and 1.70 eV for GeS and GeSe monolayers, respectively [59]. As can be expected, excitons are strongly bound, especially for GeS. As the light polarization is along the zigzag direction, the lowest-energy excitons in GeS and GeSe show binding energies of 1.05 and 0.4 eV, respectively. In case of GaSe, strong excitonic effects are also confirmed with the binding energies of 0.66 and 0.34 eV assigned to the 1*s* and 2*s* excitons, respectively [60].

In a recent work, using first-principles *GW*+BSE scheme in combination with the $\mathbf{k}\cdot\mathbf{p}$ theory, a robust linear scaling law between the quasi-particle band gap and the exciton binding energy, $E_b = E_g/4$, was found for two-dimensional semiconductors [53]. It is conclusive that this relationship is independent of the lattice configuration, bonding feature, and the topological property of the semiconductors, a similar conclusion has also been drawn previously [61]. In view of physics, the larger the energy gap is, the weaker the screening becomes, which naturally corresponds to a smaller exciton radius and hence the higher binding energy. In this sense, a proportional relationship between the band gap and exciton binding energy is taken for granted. It is nevertheless clear that the universal parameter-free slope as well as the formulas of the band gap and exciton binding energy are not valid for the bulk semiconductors. It is conclusive that such a difference essentially arises from the nonlocal and local screening effect for two-dimensional and bulk materials. In the two-dimensional case, the electron–hole interaction is long ranged and thus notably affects the exciton radius. In contrast, for the bulk case, the strong screening weakens the electron–hole interaction rapidly, generally leading to a larger exciton radius, and hence a smaller exciton binding energy. In these ways, excitonic effects thus are closely related to the different photocatalytic mechanisms. In comparison to the standard charge transfer processes, effective light–energy conversion could be reached through exciton exchange and/or dipole–dipole interactions in exciton-based energy transfer processes [62]. In such processes, net charge carrier transfer from donor to acceptor is not involved, thus suggesting interesting photocatalytic behaviors. In Figure 5.9, the relationship between band gap and exciton binding energy is shown.

In comparison to standard density functional theory, with many-body perturbation theory a more correct optical absorption of two-dimensional materials can be obtained. In spite of this, open questions are still there. In general, dark excitons are generated under light irradiation, which plays a determinative role in determining the light–matter interactions and the light–energy conversion efficiency. In general, bright–dark exciton interconversion suggests an exotic approach to tune the quantum yield efficiency [63–67]. In addition, charged exciton (the trion) and exciton–exciton (the so-called biexciton) interactions are of importance to understand the photon absorption and the photoexcited charge carrier behavior in especially low-dimensional materials. It is of paramount importance that, as discussed earlier, the formation of interlayer exciton (or charge transfer exciton) in van der Waals (vdW) homostructures and heterostructures of two-dimensional TMDCs means an exciting phenomenon, opening up a new avenue in photocatalytic

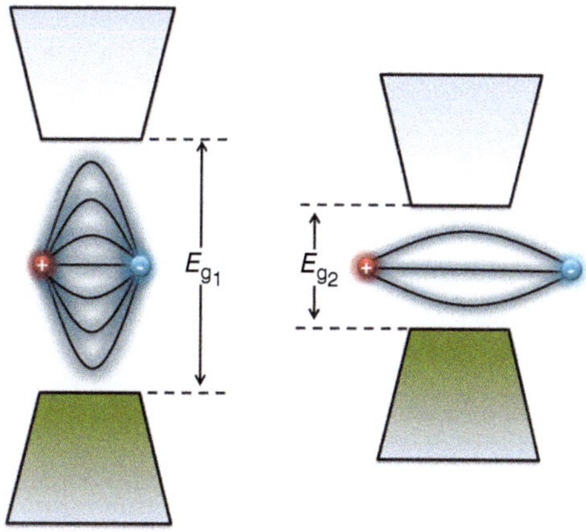

Figure 5.9 Relationship between the energy gap and exciton: for a system with large band gap, the screening is weak and the exciton binds strongly, giving rise to a relatively narrow spatial extension and large exciton binding energy. In contrast, for a system with a small band gap, the screening is strong and the exciton binds loosely, leading to a relatively wide spatial extension and small exciton binding energy. Source: Jiang et al. [53]/with permission of American Physical Society.

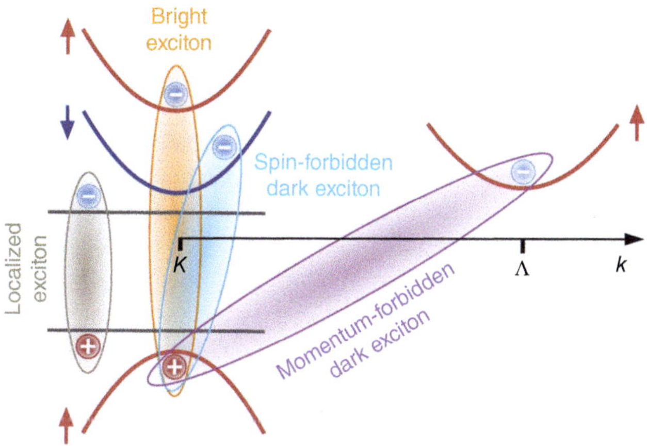

Figure 5.10 Momentum-forbidden dark excitons consist of electrons and holes located at different valleys in the momentum space. Spin-forbidden dark excitons consist of electrons and holes with opposite spin. These states cannot be accessed by light due to the lack of required momentum transfer and spin-flip, respectively. Localized excitons are those electrons and holes trapped into an impurity-induced potential. Source: Mueller and Malic [66]/Springer Nature/CC BY 4.0.

and photovoltaic applications. In this context, the light–matter interaction becomes more complex, and more advanced theory and simulation methods should be developed. In Figure 5.10, different kinds of dark excitons are shown.

5.3 Electron–Phonon Effects on the Excited-state Properties

In this section, one can refer to the pioneering works by Marini [68, 69] and Giustino, Louie, and Cohen [70, 71] and references therein. In general, optical features are characterized at room temperature, and of course the temperature will affect the light–energy conversion efficiency. In density functional theory-based studies without considering the temperature effects, important features will be overlooked, for example, the intensity of the excitonic absorption and the broadening, which is directly related to the temperature and to the photoexcited charge carrier nonradiative relaxation time. In other words, usually the ground- and even the excited-state calculations from the first-principles are done at 0 K and, thus, ignore the effects of thermal lattice vibrations on the optical properties. It should be pointed out that the thermal lattice vibrations show non-negligible effects on the electronic structures, since they change the spectral function. As a consequence, sharp discrepancy between theoretical and experimental observations emerges. In this respect, including the electron–phonon interaction to take the temperature into account in theoretical simulations is thus of significant importance.

In general, a first evaluation of the electronic energies could be obtained from density functional theory, and the phonon modes and electron–phonon coupling matrix elements are then gained by using density functional perturbation theory. On the basis of the obtained matrix, change in electronic energies due to lattice vibrations can be captured. In the third step, temperature-dependent BSE can be solved to determine the change in the optical response, exciton energies, and linewidths as temperature changes. In the framework of many-body perturbation theory, two self-energy diagrams corresponding to the lowest nonvanishing terms of a perturbative treatment should be estimated. In particular, the Fan self-energy, as the first-order term, can be written as

$$\Sigma^{\mathrm{Fan}}_{n,\mathbf{k}}(\omega, T) = \sum_{n'\mathbf{q}\lambda} \frac{\left|g^{\mathbf{q}\lambda}_{nn'\mathbf{k}}\right|^2}{N_q} \left[\frac{N_{\mathbf{q}}(T) + 1 - f_{n'\mathbf{k}-\mathbf{q}}}{\omega - \varepsilon_{n'\mathbf{k}-\mathbf{q}} - \omega_{\mathbf{q}\lambda} - i0^+}\right] \times \left[\frac{N_{\mathbf{q}}(T) + f_{n'\mathbf{k}-\mathbf{q}}}{\omega - \varepsilon_{n'\mathbf{k}-\mathbf{q}} + \omega_{\mathbf{q}\lambda} - i0^+}\right]$$

where $\varepsilon_{n,\,\mathbf{k}}$ are the eigenvalues from density functional theory, $\omega_{\mathbf{q},\,\lambda}$ are the phonon frequencies, and $f_{n,\,\mathbf{k}}$ and $N_{\mathbf{q}}(T)$ are the Fermi and Bose distributions of electrons and phonons, respectively. In the case of the Debye–Waller (DW) self-energy, corresponding to the second-order term, it reads:

$$\Sigma^{DW}_{n,\mathbf{k}}(T) = \frac{1}{N_q} \sum_{\mathbf{q}\lambda} \Lambda^{\mathbf{q}\lambda,-\mathbf{q}\lambda}_{nn\mathbf{k}} [2N_{\mathbf{q}\lambda}(T) + 1]$$

In this equation, $\Lambda^{\mathbf{q}\lambda,\mathbf{q}'\lambda'}_{nn'\mathbf{k}}$ stands for the amplitude of the second-order scattering $|n\mathbf{k}\rangle \rightarrow |n'\mathbf{k} - \mathbf{q} - \mathbf{q}'\rangle \otimes |\mathbf{q}\lambda\rangle \otimes |\mathbf{q}'\lambda'\rangle$. In the expression of the two self-energy terms, phonon population introduces the temperature. In the following, the electron–phonon interaction can be related to a fully interacting electron propagator as:

$$G_{n\mathbf{k}}(\omega, T) = \left[\omega - \varepsilon_{n\mathbf{k}} - \Sigma^{\mathrm{Fan}}_{n\mathbf{k}}(\omega, T) - \Sigma^{DW}_{n\mathbf{k}}(T)\right]^{-1}$$

where the complex poles correspond to the electronic excitations. It is helpful to get the Eliashberg functions to find the phonon modes contributing to the electron–phonon coupling:

$$g^2F_{n\mathbf{k}}(\omega) = \sum_{\lambda\mathbf{q}} \left[\frac{\sum_{n'} \left| g_{nn'\mathbf{k}}^{\mathbf{q}\lambda} \right|^2 N_{\mathbf{q}}^{-1}}{\varepsilon_{n\mathbf{k}} - \varepsilon_{n'\mathbf{k}+\mathbf{q}}} \right] \delta(\omega - \omega_{\mathbf{q}\lambda}) - \sum_{\lambda\mathbf{q}} \left[2\frac{\sum_{n'} \Lambda_{nn'\mathbf{k}}^{\mathbf{q}\lambda} N_{\mathbf{q}}^{-1}}{\varepsilon_{n\mathbf{k}} - \varepsilon_{n'\mathbf{k}}} \right] \delta(\omega - \omega_{\mathbf{q}\lambda})$$

In this situation, the temperature-dependent BSE can be solved, where the corresponding excitonic Hamiltonian is:

$$H_{ee'hh'}^{FA} = [E_e + \Delta E_e(T) - E_h - \Delta E_h(T)]\delta_{eh,e'h'} + (f_e - f_h)\Xi_{ee'hh'}$$

in which E_e and E_h are electron and hole energies, respectively; f_e and f_h are the occupations, and $\Xi_{ee'hh'}$ is the BS kernel. In the equation, $\Delta E_e(T)$ and $\Delta E_h(T)$ represent the renormalization to the electron and hole energies induced by the electron–phonon interaction, respectively. In closing, the dielectric function depends explicitly on the temperature:

$$\varepsilon(\omega, T) \propto \sum_X [S_X(T)]^2 \text{Im} \left[\frac{1}{\omega - E_X(T)} \right]$$

where $S_X(T)$ is the exciton oscillator strength.

In a recent work, band gap renormalization and electron–phonon coupling in two-dimensional WSe_2 are comprehensively discussed [72]. In consideration of the polaronic energies, solving a coupled electron–hole BSE results in the in-plane torsional acoustic phonon branch mainly accounting for the characteristic A and B excitons buildup. In particular, the A, B, and C excitons in MoS_2 (from the vertical optical transitions at the Γ point of the Brillouin zone of two-dimensional TMDCs) behave differently with respect to temperature, i.e. with different nonradiative linewidths [73]. In detail, the longitudinal acoustic (LA) phonons dominantly contribute to A and B excitons, while LA and the optical modes near 225 cm^{-1} couple to the C exciton. It should be pointed out that a zero-point energy renormalization of 31 meV mainly due to the polaronic interaction is found, with negligible contributions from the lattice anharmonicities. It is obvious that results taking the electron–phonon effects into account are very helpful for studying the light-energy/current conversion behavior of two-dimensional TMDCs at finite temperatures, where the exciton physics will govern the conversion efficiency. In case of two-dimensional MoS_2, similar results with that of two-dimensional WSe_2 are obtained and discussed in a similar way.

It is already known that even at $T \to 0$ K, zero-point motion effect of atoms will play a crucial role in affecting the excited-state properties such as the excitation position and spectrum width. It has been revealed that this is also true for monolayer GeS [74]. In particular, at the band edge the longitudinal mode B_{2u} couples efficiently with the electronic states. In addition, the electronic states that give rise to the exciton E_1 couple mostly with the vibrational modes A_{1g} and B_{3g}^2. It therefore suggested that one should include the electron–phonon interaction effects in order to properly describe group-VI monochalcogenides. It is of interest that previous work indicates an anomalous temperature dependence of the band gap of two-dimensional black phosphorus (BP) [75]. As temperature increases, particularly, the fundamental band gap increases instead of decreases. It is a result of electron–phonon coupling with distinct optical modes. It is also evidenced that in semiconductors the thermal anomalies can be identified from the correlation between harmonic and anharmonic effects.

5.4 Nonlinear Optical Response

It has been illustrated that many-body perturbation theory cannot be used to simulate the light emission by laser pulse excitation, a nonequilibrium phenomenon. In 1994, Kadanoff and Baym developed the generalization of many-body perturbation theory to nonequilibrium conditions, where the Kadanoff–Baym equations (KBE) make nonequilibrium Green's function theory accessible [76]. In combination with density functional theory, Marini et al. simplified the KBE, enabling a parameter-free theory to predict and reproduce ultrafast and nonlinear phenomena: the time-dependent BSE [77]. In case of weak perturbations, it reduces to the standard BSE. At the same time, however, it can naturally describe the optical excitations beyond the linear regime. In Figure 5.11, perturbation (by a strong laser field) induced elemental processes are schematically illustrated. In the year of 2016, progress has been made by Marini et al. [78] In particular, electron–electron, electron–phonon, and electron–photon interactions could be considered in the complete BKE on the basis of the nonequilibrium Green's function theory.

In this context, light emission thus can be simulated, in principle, on the basis of Coulomb hole and screened exchange (COHSEX) self-energy and the collision integral S^{dyn}. In practice, S^{dyn} is split into three terms according to the different carrier interactions:

$$\frac{d}{dT}\rho(T) + i[h_{\text{ext}}(T), \rho(T)] = -S^{\text{coh}}[\rho](T) - S^{\text{dyn}}_{e-e}[\rho](T) - S^{\text{dyn}}_{e-p}[\rho](T) - S^{\text{dyn}}_{e-\gamma}[\rho](T)$$

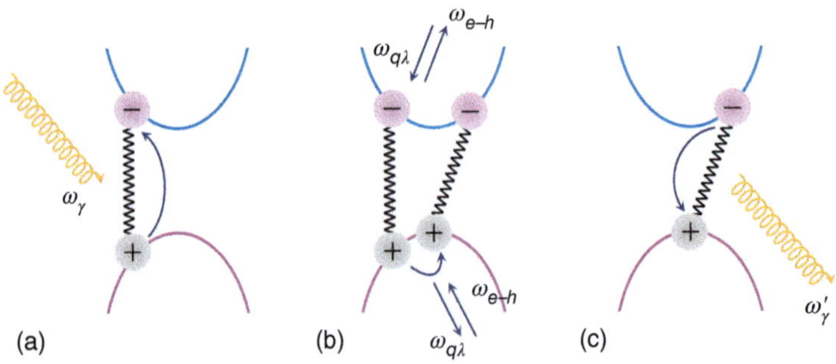

Figure 5.11 Short and intense laser pulse induced different processes in a semiconductor. (a) Electron–hole pairs generation. (b) Carrier collisions. (c) Carrier relaxation to photon (c). $\omega_\gamma/\omega'_\gamma$, $\omega_{q\lambda}$, and ω_{e-h} are photon energy, phonon energy, and electron–hole pair energy, respectively. Source: Wei et al. [26]/with permission of John Wiley & Sons.

In the publications by de Melo and Marini, one can find more about the basic equations and calculation details for light emission. In addition to the temperature, from the aspects of theoretical simulation, the photoluminescence signal varies with the feature of the laser and the pulse duration. In fact, however, there is no comparison between the light emission based on the aforementioned method and that of experimental result. It should be emphasized here that lattice relaxation occurs in excited states and then photon emits. In general, therefore, emission energy should not be the same as the absorption energy only if one can relax the excited state.

5.5 Optical Properties of van der Waals Heterostructures of Two-Dimensional Materials

In two-dimensional TMDCs, the exciton binding energies are usually hundreds of millielectronvolts, two orders of magnitude larger than those in bulk semiconductors such as Si and GaAs [79]. On one hand, large exciton binding energy guarantees that the exciton states can survive at room temperature. On the other hand, however, large exciton binding energy is not in favor of electron–hole pair dissociation into free carrier to participate in the photochemical or photoelectrochemical interactions. In vdW heterostructures of two-dimensional TMDCs, the excited-state dynamics will be profoundly affected, thus providing a useful method to modulate and optimize the dynamic response of the individual monolayers.

In recent years, advances have been made in vdW heterostructures of two-dimensional TMDCs, many of them have been realized and characterized. In particular, two-dimensional TMDCs are semiconductors with direct band gap with two degenerate copies of energy gaps located at the K and K' points (or valleys) of the Brillouin zone. It is conclusive that extremely strong excitonic effects and contrasting properties for different valleys are the two hallmarks of this class of materials, thus providing a fruitful material platform to study interlayer exciton phenomena. It can be envisaged that rich exciton physics and unique optoelectronic, photocatalytic, and photovoltaic applications in these materials are waiting for investigation. It should be pointed out that different kinds of stacking patterns for bilayers have been observed both experimentally and theoretically, and the bilayers are strain-free due to the weak vdW interactions.

In general, stacking two distinct two-dimensional monolayers gives rise to the type-II band alignment, with the valence band maximum (VBM) and conduction band maximum (CBM) located on opposite layers. In particular, when two monolayers of TMDCs are coupled vertically by the vdW interaction, the band alignment is determined by the work function of the individual layers, the interlayer interaction, and the alignment of the crystal axes of the two layers. It is natural that the electronic structure of TMDCs heterobilayers will show a dependence on the relative angle of rotation between the two layers, which presents a huge opportunity for rotational control of interlayer excitons and their response to external electric and magnetic fields [80–86]. In this context, therefore, manipulating interlayer excitons by structural tuning could become another powerful tool. In fact, photocatalytic and photovoltaic devices based on the vdW heterostructures of two-dimensional TMDCs benefit the typical type-II band alignment and large band offset, which could be the driving force for exciton dissociation, significantly affecting the band-to-band interlayer charge transfer, the intralayer exciton formation, and the photogenerated charge carrier ultrafast dynamics. In vdW heterostructures of two-dimensional materials, interlayer coupling causes vertical dipole (or the build-in electric field) and thus affects the exciton binding energy through dipole–dipole interaction. In general, if such a vertical field promotes the separation of exciton, the binding energy thus will be reduced and vice versa. In turn, small binding energy is in favor of the exciton dissociation into free charge carriers to participate in photocatalytic and photovoltaic reactions.

In experiments, ultrafast optical measurements by using pump–probe spectroscopy make it possible to access the charge and energy transfer processes with femtosecond time resolution [87–89]. It has been experimentally demonstrated that interlayer exciton forms following the formation of intralayer exciton. Under light radiation with photon energy larger than the band gap of one or both monolayers, intralayer transitions occur. In analogues to the optical absorption of isolated monolayer TMDCs, characteristic peaks will be observed, while the interlayer excitonic absorption is characterized by the oscillation in lower energy than intralayer excitons. However, the oscillation strength is very small due to the momentum mismatching. In case of a type-II band alignment, hole charge transfer from one monolayer to the other is responsible for the transient response [87]. In the charge transfer process, however, the effects of electron–hole attractive interactions are not fully considered. In addition, error arises due to the current resolution of the instrument to probe the photoexcited charge carrier transfer in stacked monolayers forming Moiré patterns, while the optical measurements make the results vague.

It should be pointed out that band-to-band interlayer excitons, which are different in nature from the experimentally observed charge transfer interlayer excitons, could be obtained from simulation point of view such as *GW*+BSE calculations. As a result of the spatially indirect geometry, however, their near-zero intensities make no contributions to the absorption profile. In case of MoS_2/WS_2 heterobilayer, the lowest A/B transitions in constituent monolayers are 1.93/2.12 eV for MoS_2 and 1.96/2.44 eV for WS_2 [90]. It is of interest that interlayer direct transitions can be observed from simulation results, for example, in MoS_2/WS_2 such transitions are about 50 meV below the low-energy A exciton. It is possible that the interlayer direct transitions undergo bright–dark exciton crossover. In another work, however, such interlayer transitions cannot be found [91]. On the basis of *GW*+BSE scheme, an interlayer transition was found to be 0.14 eV lower than the first intralayer exciton in MoS_2/WS_2 heterostructures [92]. It is therefore indicative that much more attention should be paid to the technical details when dealing with such systems by *GW*+BSE method, and the k-point mesh, cutoff energy, and empty bands should be carefully tested for convergence. It is of importance that electron–phonon effects on the band gap renormalization and absorption profile should also be considered.

References

1 Mak, K.F., Shan, J., and Heinz, T.F. (2011). Seeing many-body effects in single- and few-layer graphene: observation of two-dimensional saddle-point excitons. *Phys. Rev. Lett.* 106: 046401.

2 Yang, L. (2011). Excitonic effects on optical absorption spectra of doped graphene. *Nano Lett.* 11: 3844–3847.

3 Chen, Z. and Wang, X.Q. (2011). Stacking-dependent optical spectra and many-electron effects in bilayer graphene. *Phys. Rev. B* 83: 081405.

4 Trevisanutto, P.E., Holzmann, M., Côté, M., and Olevano, V. (2010). Ab initio high-energy excitonic effects in graphite and graphene. *Phys. Rev. B* 81: 121405.

5 Yang, L. (2011). Excitons in intrinsic and bilayer graphene. *Phys. Rev. B* 83: 085405.

6 Wei, W. and Jacob, T. (2012). Strong charge-transfer excitonic effects in C_4H-type hydrogenated graphene. *Phys. Rev. B* 86: 165444.

7 Wei, W. and Jacob, T. (2013). Strong excitonic effects in the optical properties of graphitic carbon nitride g-C_3N_4 from first principles. *Phys. Rev. B* 87: 085202.

8 Wei, W. and Jacob, T. (2013). Strong many-body effects in silicene-based structures. *Phys. Rev. B* 88: 045203.

9 Wei, W. and Jacob, T. (2013). Electronic and optical properties of fluorinated graphene: a many-body perturbation theory study. *Phys. Rev. B* 87: 115431.

10 Wei, W., Dai, Y., Huang, B., and Jacob, T. (2013). Many-body effects in silicene, silicane, germanene and germanene. *Phys. Chem. Chem. Phys.* 15: 8789–8794.

11 Wei, W., Dai, Y., Huang, B., and Jacob, T. (2013). Enhanced many-body effects in 2- and 1-dimensional ZnO structures: a Green's function perturbation theory study. *J. Chem. Phys.* 139: 144703.

12 Wei, W. and Jacob, T. (2014). Many-body effects in semiconducting single-wall silicon nanotubes. *Beilstein J Nanotechnol.* 5: 19–25.

13 Cudazzo, P., Attaccalite, C., Tokatly, I.V., and Rubio, A. (2010). Strong charge-transfer excitonic effects and the Bose-Einstein exciton condensate in graphane. *Phys. Rev. Lett.* 104: 226804.

14 Bockstedte, M., Marini, A., Pankratov, O., and Rubio, A. (2010). Many-body effects in the excitation spectrum of a defect in SiC. *Phys. Rev. Lett.* 105: 026401.

15 Wirtz, L., Marini, A., and Rubio, A. (2006). Excitons in boron nitride nanotubes: dimensionality effects. *Phys. Rev. Lett.* 96: 126104.

16 Bruno, M., Palummo, M., Marini, A. et al. (2007). From Si nanowires to porous silicon: the role of excitonic effects. *Phys. Rev. Lett.* 98: 036807.

17 Marini, A. and Sole, R.D. (2003). Dynamical excitonic effects in metals and semiconductors. *Phys. Rev. Lett.* 91: 176402.

18 Bernardi, M., Palummo, M., and Grossman, J.C. (2012). Optoelectronic properties in monolayers of hybridized graphene and hexagonal boron nitride. *Phys. Rev. Lett.* 108: 226805.

19 Marinopoulos, A.G., Reining, L., Rubio, A., and Vast, N. (2003). Optical and loss spectra of carbon nanotubes: depolarization effects and intertube interactions. *Phys. Rev. Lett.* 91: 046402.

20 Sun, D., Rao, Y., Reider, G.A. et al. (2014). Observation of rapid exciton–exciton annihilation in monolayer molybdenum disulfide. *Nano Lett.* 14: 5625–5629.

21 Yang, J., Lü, T., Myint, Y.W. et al. (2015). Robust excitons and trions in monolayer $MoTe_2$. *ACS Nano* 9: 6603–6609.

22 Courtade, E., Semina, M., Manca, M. et al. (2017). Charged excitons in monolayer WSe_2: experiment and theory. *Phys. Rev. B* 96: 085302.

23 Lopez-Sanchez, O., Lembke, D., Kayci, M. et al. (2013). Ultrasensitive photodetectors based on monolayer MoS_2. *Nat. Nanotechnol.* 8: 497–501.

24 Radisavljevic, B., Radenovic, A., Brivio, J. et al. (2014). Single-layer MoS_2 transistors. *Nat. Nanotechnol.* 6: 147–150.

25 Yin, Z., Li, H., Li, H. et al. (2012). Single-layer MoS_2 phototransistors. *ACS Nano* 6: 74–80.

26 Wei, W., Huang, B., and Dai, Y. (2020). Photoexcited charge carrier behaviors in solar energy conversion systems from theoretical simulations. *WIREs Comput. Mol. Sci.* 10: e1441.

27 Onida, G., Reining, L., and Rubio, A. (2002). Electronic excitations: density-functional versus many-body Green's-function approaches. *Rev. Mod. Phys.* 74: 601–659.

28 Hedin, L. (1965). New method for calculating the one-particle Green's function with application to electron-gas problem. *Phys. Rev.* 139: A796–A823.

29 Hybertsen, M.S. and Louie, S.G. (1985). First-principles theory of quasiparticles: calculation of band gaps in semiconductors and insulators. *Phys. Rev. Lett.* 55: 1418–1421.

30 Hanke, W. and Sham, L.J. (1979). Many-particle effects in the optical excitations of a semiconductor. *Phys. Rev. Lett.* 43: 387–390.

31 Hedin, L. (1999). On correlation effects in electron spectroscopies and the GW approximation. *J. Phys. Condens. Matter* 11: R489–R528.

32 Strinati, G. (1988). Application of the Green's functions method to the study of the optical properties of semiconductors. *Riv. Nuovo. Cimento.* 11: 1–86.

33 Aryasetiawan, F. and Gunnarsson, O. (1998). The GW method. *Rep. Prog. Phys.* 61: 237–312.

34 Ping, Y., Rocca, D., and Galli, G. (2013). Electronic excitations in light absorbers for photoelectrochemical energy conversion: first principles calculations based on many body perturbation theory. *Chem. Soc. Rev.* 42: 2437–2469.

35 Leng, X., Jin, F., Wei, M., and Ma, Y. (2016). GW method and Bethe–Salpeter equation for calculating electronic excitations. *WIREs Comput. Mol. Sci.* 6: 532–550.

36 Danovich, D. (2011). Green's function methods for calculating ionization potentials, electron affinities, and excitation energies. *WIREs Comput. Mol. Sci.* 1: 377–387.

37 Reining, L. (2018). The GW approximation: content, successes and limitations. *WIREs Comput. Mol. Sci.* 8: e1344.

38 Marini, A., Hogan, C., Grüning, M., and Varsano, D. (2009). Yambo: an ab initio tool for excited state calculations. *Comput. Phys. Commun.* 180: 1392–1403.

39 Deslippe, J., Samsonidze, G., Strubbe, D.A. et al. (2012). BerkeleyGW: a massively parallel computer package for the calculation of the quasiparticle and optical properties of materials and nanostructures. *Comput. Phys. Commun.* 183: 1269–1289.

40 Giannozzi, P., Baroni, S., Bonini, N. et al. (2009). QUANTUM ESPRESSO: a modular and open-source software project for quantum simulations of materials. *J. Phys. Condens. Matter* 21: 395502.

41 Schlipf, M. and Gygi, F. (2015). Optimization algorithm for the generation of ONCV pseudopotentials. *Comput. Phys. Commun.* 196: 36–44.

42 Gonze, X., Beuken, J.M., Caracas, R. et al. (2002). First-principles computation of material properties: the ABINIT software project. *Comput. Mater. Sci.* 25: 478–492.

43 Kresse, G. and Furthmüller, J. (1996). Efficient iterative schemes for ab initio total-energy calculations using a plane-wave basis set. *Phys. Rev. B* 54: 11169–11186.

44 Kresse, G. and Joubert, D. (1999). From ultrasoft pseudopotentials to the projector augmented-wave method. *Phys. Rev. B* 59: 1758–1775.

45 Liu, X. and Guo, W. (2019). Shear strain tunable exciton dynamics in two-dimensional semiconductors. *Phys. Rev. B* 99: 035401.

46 Molina-Sánchez, A., Sangalli, D., Hummer, K. et al. (2013). Effect of spin-orbit interaction on the optical spectra of single-layer, double-layer, and bulk MoS_2. *Phys. Rev. B* 88: 045412.

47 Komsa, H.P. and Krasheninnikov, A.V. (2012). Effects of confinement and environment on the electronic structure and exciton binding energy of MoS_2 from first principles. *Phys. Rev. B* 86: 241201.

48 Raja, A., Selig, M., Berghä, G. et al. (2018). Enhancement of exciton-phonon scattering from monolayer to bilayer WS_2. *Nano Lett.* 18: 6135–6143.

49 Godde, T., Schmidt, D., Schmutzler, J. et al. (2016). Exciton and trion dynamics in atomically thin $MoSe_2$ andWSe_2: effect of localization. *Phys. Rev. B* 94: 165301.

50 Wang, G., Marie, X., Gerber, I. et al. (2015). Giant enhancement of the optical second-harmonic emission of WSe_2 monolayers by laser excitation at exciton resonances. *Phys. Rev. Lett.* 114: 097403.

51 Ramasubramaniam, A. (2012). Large excitonic effects in monolayers of molybdenum and tungsten dichalcogenides. *Phys. Rev. B* 86: 115409.

52 Mai, C., Barrette, A., Yu, Y. et al. (2014). Many-body effects in valleytronics: direct measurement of valley lifetimes in single-layer MoS_2. *Nano Lett.* 14: 202–206.

53 Jiang, Z., Liu, Z., Li, Y., and Duan, W. (2017). Scaling Universality between band gap and exciton binding energy of two-dimensional semiconductors. *Phys. Rev. Lett.* 118: 266401.

54 Kogar, A., Rak, M.S., Vig, S. et al. (2017). Signatures of exciton condensation in a transition metal dichalcogenides. *Science* 358: 1314–1317.

55 He, K., Kumar, N., Zhao, L. et al. (2014). Tightly bound excitons in monolayer WSe_2. *Phys. Rev. Lett.* 113: 026803.

56 Cheiwchanchamnangij, T. and Lambrecht, W.R. (2012). Quasiparticle band structure calculation of monolayer, bilayer, and bulk MoS_2. *Phys. Rev. B* 85: 205302.

57 Qiu, D.Y., Jornada, F.H., and Louie, S.G. (2013). Optical spectrum of MoS_2: many-body effects and diversity of exciton states. *Phys. Rev. Lett.* 111: 216805.

58 Tran, V., Soklaski, R., Liang, Y., and Yang, L. (2014). Layer-controlled band gap and anisotropic excitons in few-layer black phosphorus. *Phys. Rev. B* 89: 235319.

59 Gomes, L.C., Trevisanutto, P.E., Carvalho, A. et al. (2016). Strongly bound Mott-Wannier excitons in GeS and GeSe monolayers. *Phys. Rev. B* 94: 55428.

60 Antonius, G., Qiu, D.Y., and Louie, S.G. (2018). Orbital symmetry and the optical response of single-layer MX monochalcogenides. *Nano Lett.* 18: 1925–1929.

61 Choi, J.H., Cui, P., Lan, H., and Zhang, Z. (2015). Linear scaling of the exciton binding energy versus the band gap of two-dimensional materials. *Phys. Rev. Lett.* 115: 066403.

62 Wang, H., Zhang, X., and Xie, Y. (2018). Photocatalysis in two-dimensional black phosphorus: the roles of many-body effects. *ACS Nano* 12: 9648–9653.

63 Poem, E., Kodriano, Y., Tradonsky, C. et al. (2010). Accessing the dark exciton with light. *Nat. Phys.* 6: 993–997.

64 Malic, E., Selig, M., Feierabend, M. et al. (2018). Dark excitons in transition metal dichalcogenides. *Phys. Rev. Mater.* 2: 014002.

65 Zhang, X.X., You, Y., Yang, S. et al. (2015). Experimental evidence for dark excitons in monolayer WSe_2. *Phys. Rev. Lett.* 115: 257403.

66 Mueller, T. and Malic, E. (2018). Exciton physics and device application of two-dimensional transition metal dichalcogenide semiconductors. *npj 2D Mater. Appl.* 2: 29.

67 You, Y., Zhang, X.X., Berkelbach, T.C. et al. (2015). Observation of biexcitons in monolayer WSe_2. *Nat. Phys.* 11: 377–481.

68 Cannuccia, E. and Marini, A. (2011). Effect of the quantum zero-point atomic motion on the optical and electronic properties of diamond and trans-polyacetylene. *Phys. Rev. Lett.* 107: 255501.

69 Marini, A. (2008). Ab Initio finite-temperature excitons. *Phys. Rev. Lett.* 101: 106405.

70 Giustino, F., Louie, S.G., and Cohen, M.L. (2010). Electron-phonon renormalization of the direct band gap of diamond. *Phys. Rev. Lett.* 105: 265501.

71 Giustino, F., Cohen, M.L., and Louie, S.G. (2007). Electron-phonon interaction using Wannier functions. *Phys. Rev. B* 76: 165108.

72 Mishra, H., Bose, A., and Dhar, A. (2018). Bhattacharya S Exciton-phonon coupling and band-gap renormalization in monolayer WSe_2. *Phys. Rev. B* 98: 045143.

73 Molina-Sánchez, A., Palummo, M., Marini, A., and Wirtz, L. (2016). Temperature-dependent excitonic effects in the optical properties of single-layer MoS_2. *Phys. Rev. B* 93: 15543.

74 Villegas, C.E., Rocha, A.R., and Marini, A. (2016). Electron-phonon scattering effects on electronic and optical properties of orthorhombic GeS. *Phys. Rev. B* 94: 134306.

75 Villegas, C.E., Rocha, A.R., and Marini, A. (2016). Anomalous temperature dependence of the band gap in black phosphorus. *Nano Lett.* 16: 5095–5101.

76 Pines, D., Kadanoff, L.P., and Baym, G. (1994). *Quantum Statistical Mechanics.* New York: Perseus Books.

77 Attaccalite, C., Grüning, M., and Marini, A. (2011). Real-time approach to the optical properties of solids and nanostructures: time-dependent Bethe-Salpeter equation. *Phys. Rev. B* 84: 245110.

78 de Melo, P.M.M.C. and Marini, A. (2016). Unified theory of quantized electrons, phonons, and photons out of equilibrium: a simplified ab initio approach based on the generalized Baym-Kadanoff ansatz. *Phys. Rev. B* 93: 155102.

79 Jin, C., Ma, E.Y., Karni, O. et al. (2018). Ultrafast dynamics in van der Waals heterostructures. *Nat. Nanotechnol.* 13: 994–1003.

80 Wang, M., Krasnok, A., Zhang, T. et al. (2018). Tunable Fano resonance and plasmon–exciton coupling in single Au nanotriangles on monolayer WS_2 at room temperature. *Adv. Mater.* 30: 1705779.

81 Pak, S., Lee, J., Lee, Y.W. et al. (2017). Strain-mediated interlayer coupling effects on the excitonic behaviors in an epitaxially grown MoS_2/WS_2 van der Waals heterobilayer. *Nano Lett.* 17: 5634–5640.

82 Heo, H., Sung, J.H., Cha, S. et al. (2015). Interlayer orientation-dependent light absorption and emission in monolayer semiconductor stacks. *Nat. Commun.* 6: 7372.

83 Wilson, N.R., Nguyen, P.V., Seyler, K. et al. (2017). Determination of band offsets, hybridization, and exciton binding in 2D semiconductor heterostructures. *Sci. Adv.* 3: e1601832.

84 Yang, L., Sinitsyn, N.A., Chen, W. et al. (2015). Long-lived nanosecond spin relaxation and spin coherence of electrons in monolayer MoS_2 andWS_2. *Nat. Phys.* 11: 830–834.

85 Xu, X., Yao, W., Xiao, D., and Heinz, T.F. (2014). Spin and pseudospins in layered transition metal dichalcogenides. *Nat. Phys.* 10: 343–350.

86 Wu, S., Ross, J.S., Liu, G.B. et al. (2013). Electrical tuning of valley magnetic moment through symmetry control in bilayer MoS_2. *Nat. Phys.* 9: 149–153.

87 Hong, X., Kim, J., Shi, S.F. et al. (2014). Ultrafast charge transfer in atomically thin MoS_2/WS_2 heterostructures. *Nat. Nanotechnol.* 9: 682–686.

88 Rivera, P., Seyler, K.L., Yu, H. et al. (2016). Valley-polarized exciton dynamics in a 2D semiconductor heterostructure. *Science* 351: 688–691.

89 Rivera, P., Schaibley, J.R., Jones, A.M. et al. (2015). Observation of long-lived interlayer excitons in monolayer $MoSe_2$–WSe_2 heterostructures. *Nano Lett.* 15: 2992–2997.

90 Komsa, H.P. and Krasheninnikov, A.V. (2013). Electronic structures and optical properties of realistic transition metal dichalcogenides heterostructures from first principles. *Phys. Rev. B* 88: 085318.

91 Mlinar, V. (2017). Influence of band offset, nanostructuring, and applied electric field on the optoelectronic properties of vertically stacked MoS_2/WS_2 materials. *Phys. Rev. B* 96: 235437.

92 Torun, E., Miranda, H.P.C., Molina-Sánchez, A., and Wirtz, L. (2018). Interlayer and intralayer excitons in MoS_2/WS_2 and $MoSe_2/WSe_2$ heterobilayers. *Phys. Rev. B* 97: 245427.

6

Charge Carrier Dynamics from Simulations

6.1 Time-Dependent Density Functional Theory and Nonadiabatic Molecular Dynamics

In nanoscale materials, charge separation and recombination lay the foundation in photovoltaic and photocatalytic devices, Figure 6.1. In the light–energy conversion process, excitons (Coulomb bound electron–hole pairs) have to dissociate into free carriers to realize high conversion efficiency. It has been discussed in Chapter 5, however, significantly large binding energies are always assigned to the excitons in low-dimensional materials including the emerging two-dimensional candidates for applications in photovoltaics and photocatalysts. It will lead with large possibility to inefficient charge separation and, hence, inefficient light–energy conversion. In two-dimensional materials-based van der Waals (vdW) heterostructures, type-II band alignment as the driving force for charge transfer and the formation of interlayer excitons implies efficient charge separation. In case the large electron–hole Coulomb interaction exceeds the driving force, consequently, the charge separation will be impeded. However, sometimes, this is not the reality. In order to unravel the photoexcitation charge carrier dynamics in pristine and even defective materials, nonadiabatic molecular dynamics (NAMD) in combination with ab initio real-time time-dependent density functional theory (TDDFT) provides a possibility to mimic the time-resolved laser experiments at atomic scale [1–13].

In recent years, NAMD has been implemented with real-time TDDFT and used in simulations of photovoltaic and photocatalytic processes, i.e. carrier separation/recombination, energy relaxation/transfer, and electron–hole elastic/inelastic scattering. In the framework of Kohn–Sham method, the dynamics of interfacial charge carrier transfer including nonadiabatic effects can be described by real-time TDDFT. In detail, the electron density, $\rho(\boldsymbol{r}, t)$, is repressed in the way of DFT as a summation of the densities of the single-particle Kohn–Sham orbitals, $\varphi_i(\boldsymbol{r}, t)$, occupied by N_e electrons:

$$\rho(\boldsymbol{r}, t) = \sum_{i=1}^{N_\mathrm{e}} |\varphi_i(\boldsymbol{r}, t)|^2$$

Calculations and Simulations of Low-Dimensional Materials: Tailoring Properties for Applications, First Edition. Ying Dai, Wei Wei, Yandong Ma, and Chengwang Niu.

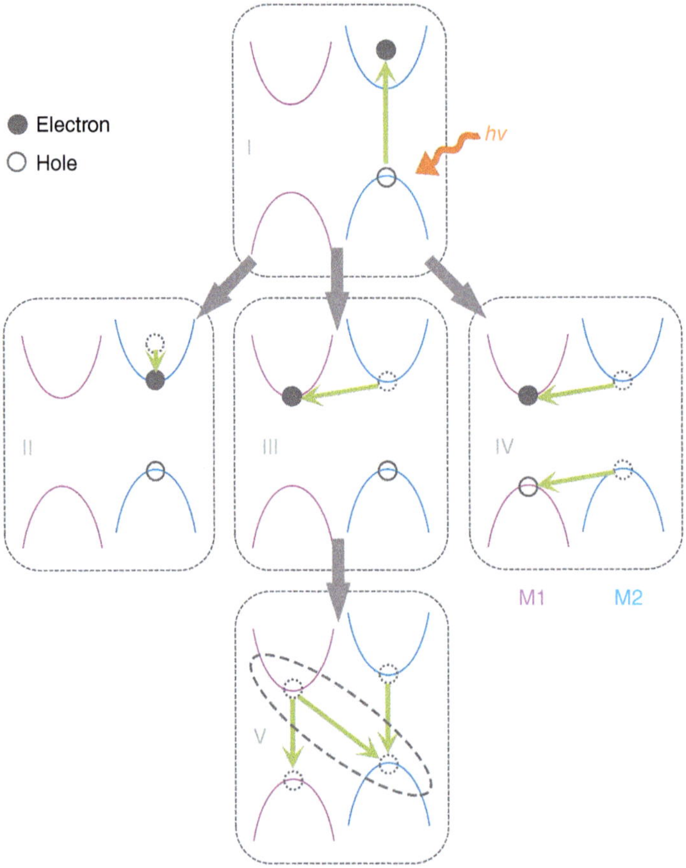

Figure 6.1 Photoinduced processes competing with electron transfer between a donor and an acceptor, including thermal relaxation, energy transfer, and charge recombination. The energy levels of the donor (acceptor) are shown as solid horizontal lines on the left (right). The electron and hole are represented by solid and open circles, respectively. Source: Wei et al. [1]/with permission of John Wiley & Sons.

As used in the Kohn–Sham energy, time-dependent variational principle determines the electron density evolution:

$$E\{\varphi_i\} = \sum_{p=1}^{N_e} \langle \varphi_i | K(\boldsymbol{r}) | \varphi_i \rangle + \sum_{p=1}^{N_e} \langle \varphi_i | V(\boldsymbol{r}; \boldsymbol{R}) | \varphi_i \rangle + \frac{e^2}{2} \iint \frac{\rho(\boldsymbol{r}', t)\rho(\boldsymbol{r}, t)}{|\boldsymbol{r} - \boldsymbol{r}'|} d^3\boldsymbol{r} d^3\boldsymbol{r}' + E_{XC}\{\rho\}$$

In this case, the evolution of Kohn–Sham orbitals takes

$$i\hbar \frac{\partial \varphi_i(\boldsymbol{r}, t)}{\partial t} = H(\boldsymbol{r}, \boldsymbol{R}, t)\varphi_i(\boldsymbol{r}, t)$$

In consideration that the Hamiltonian $H(\boldsymbol{r}, \boldsymbol{R}, t)$ is determined by the overall electron density, the equations are correlated. In the manner of atom-induced external potential, the electron–vibrational coupling can be included in the Hamiltonian $H(\boldsymbol{r}, \boldsymbol{R}, t)$, and expansion of the time-dependent Kohn–Sham orbitals is in adiabatic Kohn–Sham basis $\varphi_i(\boldsymbol{r}, t)$, which can be calculated by the atomic position $\boldsymbol{R}$:

$$\varphi_i(\boldsymbol{r}, t) = \sum_{i=1}^{N_e} c_{ij}(t) \mid \tilde{\varphi}_j(\boldsymbol{r}, \boldsymbol{R}(t)\rangle$$

In the following, the evolution of the expansion coefficients can be obtained from

$$i\hbar \frac{\partial}{\partial t} c_{ij}(t) = \sum_{k}^{N_e} c_{ij}(t)(\tilde{\varepsilon}_k \delta_{jk} + \boldsymbol{d}_{jk} \cdot \dot{\boldsymbol{R}})$$

where $\tilde{\varepsilon}_k$ is the energy of the adiabatic state k, and $\boldsymbol{d}_{jk} \cdot \dot{\boldsymbol{R}}$ is the nonadiabatic coupling between orbital j and k, which is created by atomic motions.

In respect to the classical path approximation (CPA), the fewest-switches surface hopping (FSSH) and the decoherence-corrected surface hopping are used in the real-time TDDFT method [10, 12]. In the method Long et al. used, the most important approximation is the CPA, which indicates that the nuclear motion is not affected by the electron excitation. In this case, only the ground-state molecular dynamics is run to obtain the energies, gradients, and nonadiabatic couplings. In addition, CPA also means that the dynamics is not really performed in the real excited states of the solid state. In this sense, the modified trajectory surface hopping (TSH) is really different from the original version of TSH by Tully [14]. It should be clearly declared that CPA cannot give the truly correct description of the excited states for solid state and cannot describe the chemical bond formation and breaking.

It has been identified that the real-time TDDFT method could be an exotic ab initio quantum strategy to explore the strong field physics beyond linear response theory such as ultrafast photoelectron emission. In these approaches, electron density is treated as propagator in real time through numerical integration for the time-dependent Kohn–Sham equations, giving rise to the evolution of electronic wave functions in time domain together with the movement of ion. As a result, the real-time TDDFT represents in perturbative or non-perturbative regimes a way for real-time tracking of ultrafast dynamics. In this field, Meng et al. developed the real-time ab initio approach (time-dependent *ab initio* package, TDAP) for simulating electron–nuclear dynamics under laser excitations, which has been proved successful in some systems [15–19]. In a model of MoS_2/WS_2 vdW heterostructure, for example, laser-induced ultrafast carrier dynamics can be explained by using this approach [15]. In accordance to the results, the interlayer geometry and therefore the charge transfer quantum dynamics could be modulated. In addition, another possible alternative way for simulating the photoinduced charge carrier dynamics of the extended two-dimensional monolayers and the quasi-two-dimensional vdW heterostructures is the multilayer multiconfiguration time-dependent Hartree (ML-MCTDH) method, in which the description of correlation effects could be more effective [20–23].

6.2 Applications of TDDFT and NAMD in Two-Dimensional Materials

In the light of TDDFT and NAMD strategy, simulation works focusing on the photoexcitation dynamic show great success, particularly in two-dimensional TMDCs materials. In Figure 6.2, electronic energy levels involved in the photoinduced nonequilibrium processes in a type-II donor–acceptor heterostructure is shown.

In general, the performance of materials depends strongly on the morphology and quality of the samples. In MoS_2, S adatom and S vacancy are the most energetically favorable point defects, introducing trap states and destroying the stability and leading to charge carrier loss [2, 24–28]. It is therefore conclusive that defects will accelerate the nonradiative electron–hole recombination. It should be pointed out that, nevertheless, the acceleration mechanisms are different with regard to the adatom and vacancy in MoS_2. As a result of the adatom, the hole trap states are strongly localized and couple weakly with charge carriers and are scarcely populated, and the acceleration mechanism for adatom is also related to the new phonon modes that couple to the electronic subsystem due to the distorted symmetry of MoS_2, as well as the increased nonadiabatic charge–phonon coupling. In contrast, chalcogen vacancy gives rise to deep and relatively localized electron trap, while shallow and less localized hole trapping center. In general, the impurity states within the band gap are assumed to be populated, and that the relaxation is faster as it is related to transitions over smaller energy gaps. In this situation, carrier recombination across the hole trap dominates in MoS_2 with S vacancy. In consideration that it will substantially reduce the lifetimes of charge carriers in MoS_2 than S vacancy, the S adatom should be avoided to guarantee the high performance of TMDCs-based devices. In light–energy conversion processes, nonradiative electron–hole recombination suggests the dominating channel for carrier loss and energy dissipation. In pristine MoS_2, TDDFT and NAMD calculations indicate a timescale of 388 ps for direct recombination of a conduction band electron and

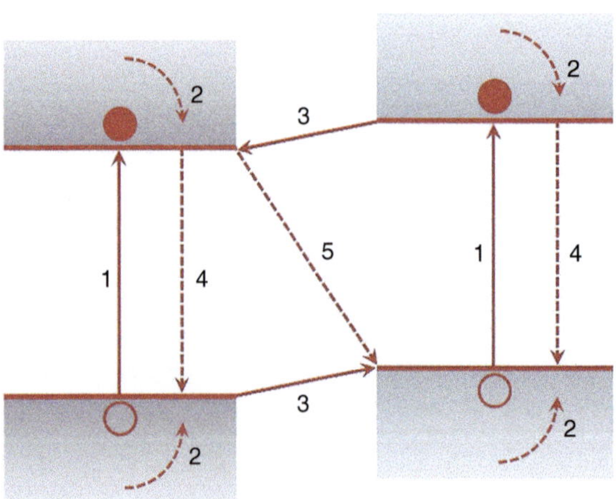

Figure 6.2 Absorption of a photon (1) by either electron donor or electron acceptor leads to charge separation (3) due to electron or hole transfer, respectively. Competing with the separation, the electron and hole can undergo recombination (4) or relaxation (2) inside either material. Following the separation, the charges can recombine at the interface (5). Source: Wei Wei.

a valence band hole. In accordance to the Fourier transform of the energy gap fluctuations, the 400 cm^{-1} phonon corresponding to the out-of-plane A_{1g} phonon mode of MoS_2 (with the frequency being 404.1 cm^{-1}) promotes this band edge electron–hole recombination. In spite of the not significantly populated shallow hole trap states induced by S adatom, the electron–hole recombination is eight orders of magnitude faster (49 ps), even faster than the case of S vacancy (225 ps). In such an anomalous case, the acceleration by S adatom can be explained by the strong nonadiabatic coupling. In comparison to the root-mean-square value of the nonadiabatic coupling of perfect MoS_2 (2.78 meV), the presence of S adatom leads to the larger value between VBM and CBM (8.44 meV). It can be rationalized by the fact that S adatom breaks the symmetry of two-dimensional MoS_2 plane and involves multiple phonon modes that can be coupled to the electronic subsystem. In this regard, structure and symmetry perturbation duo to the defects come to play and should be paid careful attention, since the induced vibration modes will couple strongly to the electronic system and affect the relaxation dynamics. In other words, defects in two-dimensional TMDCs including vacancies, antisite atoms, and adatoms definitely enhance the electron–hole recombination. In respect to the performance of TMDCs-based devices, which is closely related to the efficient charge separation, defects are strongly suggested to be avoided during the preparation of samples.

It is a consensus that vdW heterostructures of two-dimensional TMDCs present a new class of material systems, providing an exciting platform for new physics and novel applications. In light of the type-II band alignment, which facilitates efficient separation of photoexcited electrons and holes, vdW heterostructures of two-dimensional TMDCs show great potential in light–energy conversion applications such as photovoltaics and photocatalytic devices. It has been concluded that the interlayer charge transfer is at the center of the photoresponse of these heterostructures. In particular, the overlap of interlayer states is responsible for the charge carrier transfer, thus the photoexcitation charge carrier transfer dynamics certainly shows dependence on the interlayer stacking orders (twisting, translation, and spacing) and interactions. In recent results from TDDFT, it has been demonstrated that specific interlayer registry between MoS_2 and WS_2 monolayers can significantly modulate the interlayer charge transfer, in particular, it changes the timescale from 100 to 1000 fs. It was further unraveled that the transfer rate is governed by the coupling between specific interlayer states, instead of the overall interlayer coupling strength. In particular, the VBM states of MoS_2 and WS_2 (at the K point in the Brillouin zone) are referred to as the interlayer states and are not sensitive to the interlayer interaction. In can be expected that shorter interlayer distance, or stronger interlayer electronic coupling, will lead to faster charge transfer. It seems that, nevertheless, there is no obvious correlation between the charge carrier transfer dynamics and the interlayer coupling strength, which is against the normal expectation. In terms of formation energy and/or interlayer distance, the mechanical or electronic coupling strength is actually a total 1. It should be pointed out that the charge carrier transfer is just related to the coupling strength of the states where charge carrier transfer takes place. It can be confirmed by evaluating the coupling strength using dipole transition matrix element (M):

$$M = \langle \varphi_1 | \hat{Z} | \varphi_2 \rangle$$

here $\hat{Z}$ is the position operator along the vertical direction, and $\varphi_{1,2}$ is the interlayer state (the VBM of individual monolayer in MoS_2/WS_2 vdW heterostructures). It was further found that $1/\tau$ shows exponential dependence on M, or $1/\tau \propto e^M$, which is universal in controlling the interlayer charge transfer dynamics in MoS_2/WS_2 vdW heterostructures of different stacking orders and interlayer distance.

In MoS_2/WS_2 heterostructures, photoexcited ultrafast charge transfer can be demonstrated through photoluminescence mapping and femtosecond pump–probe spectroscopy. It has been found that within a timescale of 50 fs after optical excitation hole transfer from the MoS_2 layer to the WS_2 layer occurs, which is a remarkable rate for two-dimensional vdW bilayers. It is of interest that ultrafast charge transfer can enable two-dimensional vdW heterostructures-based devices more promise for light harvesting and optoelectronics.

In addition, in $MoSe_2/WSe_2$ heterostructures interlayer excitons were also observed by photoluminescence and photoluminescence excitation spectroscopy. It was found that the interlayer exciton lifetime reaches about 1.8 ns, one order of magnitude longer than intralayer excitons in a monolayer. It was also revealed that the luminescence intensity and energy of the interlayer excitons are highly adjustable by an applied vertical gate voltage. It is of interest that, according to these results, optically exciting the interlayer polarization opens up a new avenue for new physical phenomena such as interlayer exciton condensation and novel applications such as laser, photovoltaic devices, and light-emitting diodes based on two-dimensional materials.

In the case of type-II band alignment, long excited charge carrier lifetimes and suppressed electron–hole recombination across the interface can be expected. In $MoS_2/MoSe_2$ heterobilayer, for example, the VBM and CBM band offsets between MoS_2 and $MoSe_2$ are 0.37 and 0.63 eV, respectively, and, however, the band edge offsets reduce to 0.06 and 0.5 eV if the interlayer coupling is considered. In accordance to the results from many-body perturbation theory, exciton binding energies for two-dimensional TMDCs monolayers range from 0.5 to 1.1 eV, larger than the offsets being the driving force for carrier separation. It is an indication that the efficient dissociation of excitons into free carriers in constituents within the vdW heterobilayers should be prevented. In experiments, however, the efficient charge separation in heterostructures of two-dimensional TMDCs in terms of photoinduced electron/hole transfer can observed. In $MoS_2/MoSe_2$ heterostructures, the timescales for photoexcitation charge carrier transfer as well as electron–hole recombination are subpicosecond. In particular, the lifetimes of spatially indirect excitons (up to 240 ps) are longer than the intralayer excitons in constituent MoS_2 (100 ps) and $MoSe_2$ (125 ps), and the photoluminescence quenching in MoS_2/WSe_2 heterostructures also supports the photoinduced charge transfer. It is obvious that conflict exists and needs to be resolved. As for $MoS_2/MoSe_2$ vdW heterostructures, TDDFT and NAMD simulations can provide fundamental insights into excitation dynamics [4]. In this respect, quantum coherence at the interface is demonstrated to play an important role in correctly describing the photoexcitation dynamics in such vdW heterostructures. In case of heating the $MoS_2/MoSe_2$ vdW heterostructures to room temperature, the average interlayer distance will decrease, indicating enhanced donor–acceptor coupling because of the thermal fluctuations. As the finite-temperature atomic motions will distort the geometry of the two-dimensional sheet, thus additional interaction chances will be provided.

In accordance to the band offsets in $MoS_2/MoSe_2$ vdW heterostructures, one can acknowledge that MoS_2 will be the hole donor and electron acceptor, while $MoSe_2$ reverses. Although the delocalization of photoexcited states corresponds to the strong donor–acceptor interaction, it does not insure the ultrafast charge transfer. In the presence of coherent superposition, long lifetimes of these states will facilitate fast dynamics, which is a requirement of quantum dynamics. In this regard, pure-dephasing functions with timescales from Gaussian fitting are useful to characterize the photoinduced loss of electronic coherence. In particular, nonadiabatic coupling are predicted to be 0.440, 0.257, and 0.135 meV for $MoSe_2$, MoS_2, and $MoS_2/MoSe_2$ vdW heterostructures, respectively, and the corresponding recombination timescales are 63, 41, and 480 ps, respectively. It can be seen that strong nonadiabatic coupling and long coherence mean fast charge transfer. It is worth noting that electron–hole at $MoS_2/MoSe_2$ interface is much slower than the corresponding rates in individual components, which is associated with fast decoherence. In a way by influencing the relative energies and the localization of the donor/acceptor states and by creating nonadiabatic coupling, vibrational motions are strongly suggested to be taken into account to describe the charge carrier transfer and lead to energy loss to heat. It concludes that the high-frequency phonon modes resulting in large nonadiabatic coupling couple to charge carrier separation and will facilitate the charge transfer. In particular, delocalization of photogenerated states promotes the quantum coherence, and, therefore, helps to conquer the electron–hole attractive interaction. As a consequence, efficient charge separation can be recognized. In conclusion, only taking these factors into account the nonadiabatic dynamics for charge carrier separation, energy relaxation, and electron–hole recombination in light-energy conversion processes can be correctly described, thus providing insights into the design principles for effective photovoltaic and photocatalytic systems.

In solar cell and photocatalytic applications, significantly longer electron–hole recombination than charge reparation is demanded. In vdW heterostructures of two-dimensional TMDCs, the monolayer pristine properties are maintained and, importantly, many novel properties arise. In type-II MoS_2/WSe_2 vertical heterostructures, ultrafast electron transfer from WSe_2 into MoS_2 (within 470 fs) has been recently illustrated via time-resolved photoluminescence spectroscopy. In other experiments, the ultrafast charge carrier transfer and significantly slow recombination are also demonstrated, giving rise to the grounds for designing efficient light–energy conversion devices based on two-dimensional TMDCs materials. In respect to the photoexcitation dynamics in TMDCs vdW heterostructures, therefore, the importance of donor–acceptor coupling and fundamental band

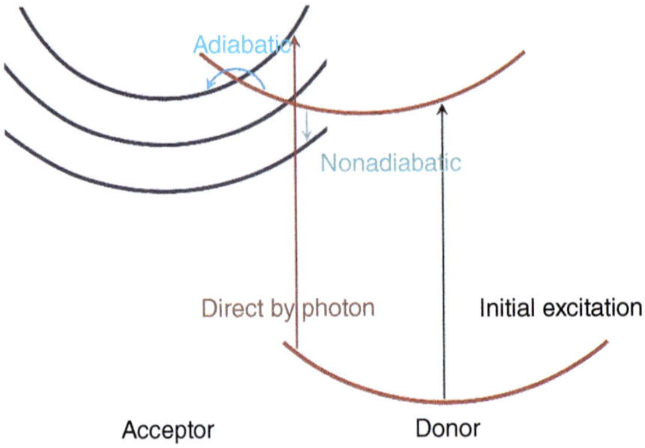

Figure 6.3 Scheme of the photoinduced electron transfer mechanism. Adiabatic electron transfer occurs by passing over a transition-state barrier. Nonadiabatic electron transfer occurs via a hop between donor and acceptor states. Photoexcitation can promote the electron directly from the donor material to a state that is localized on the acceptor, creating a charge-separated state. Source: Wei Wei.

should be emphasized. In order to mimic the time-resolved ultrafast pump–probe spectroscopy experiment, quantum–classical NAMD simulations in combination with TDDFT are usually the choice [5].

In general, the strong, finite temperature-induced interactions facilitate a number of pathways for the electron to be transferred, including direct inter-component excitation, adiabatic, and nonadiabatic mechanisms, Figure 6.3. In respect to the perovskite solar cells, the ultrafast interfacial injection guarantees efficient photoinduced charge separation that creates favorable conditions for operation.

In the case of MoS_2/WSe_2 vdW heterostructures, electron donor state is expanded into two components, and electron acceptor state is confined within MoS_2. As a result of wave functions overlap, such a situation enhances nonadiabatic coupling and facilitates charge carrier separation. In contrast, hole donor and acceptor states are, respectively, localized in each monolayer, thus decoupling each other and reducing nonadiabatic coupling. It should be kept in mind that strong donor–acceptor coupling leads not necessarily to ultrafast charge separation, but quantum coherence is also an important factor that affects the photoexcitation charge carrier dynamics. In principle, faster and more phonon modes participating in charge carrier transfer result in larger nonadiabatic coupling and accelerate charge carrier separation, since nonadiabatic coupling is proportional to the wave function overlap $\langle \tilde{\varphi}_m \mid \nabla_R \mid \tilde{\varphi}_k \rangle$ and to the velocities of nuclei $d\mathbf{R}/dt$. In calculations of TDDFT and NAMD, this can be evaluated by the spectral density calculated from Fourier transform of the fluctuations of the relative energy difference. In the case of MoS_2/WSe_2 vdW heterostructures, electron transfer (304 fs) is faster than hole transfer (443 fs) due to the reasons discussed earlier. In general, for MoS_2/WSe_2 vdW heterostructures, the weak coupling between electron and hole wave functions reduces the initial and final states interaction and inclines to suppress the recombination. In spite of the smaller band gap than individual monolayers, in vdW heterostructures the small nonadiabatic coupling due to the delocalized initial and final states as well as the fast decoherence due to phonon modes of high frequency will delay the electron–hole recombination. In particular, the electron–hole recombination timescale for MoS_2/WSe_2 vdW heterostructures is 496 ps, with a decoherence timescale being 6.8 fs.

It is of importance that, inspired by the new channel for hot electron relaxation in MoS_2/WSe_2 vertical bilayer, a new concept for design of light–energy conversion devices can be imagined. In vdW multilayers of distinct two-dimensional materials, photoexcited electrons diffuse vertically rather than decay thermally in one layer, with the hot holes transferring in an opposite direction to the hot electrons. In this regard, charge loss could be significantly inhibited, and high light–energy conversion efficiency could be expected. In Figure 6.4, such a new view for device design is schematically shown.

In 2014, in-plane heterostructures of two-dimensional TMDCs were first identified by the one-step or two-step chemical vapor deposition (CVD) method, which permits modification of the atomic composition of a single monolayer to manifest in-plane heterostructures [29–40]. In-plane heterostructures of two-dimensional TMDCs, such as MoS_2/WS_2, $MoSe_2/WSe_2$, $MoS_2/MoSe_2$, and MoS_2/WSe_2 mark the ultimate thickness limit for junctions between semiconducting materials. In a CVD growth for these in-plane heterostructures, the second TMDCs material is epitaxially grown from the edge of the first one with the formation of a seamless in-plane heterojunction, see Figure 6.5. In as-grown in-plane heterostructures, detailed atomic structure indicates a single hexagonal monolayer lattice and the

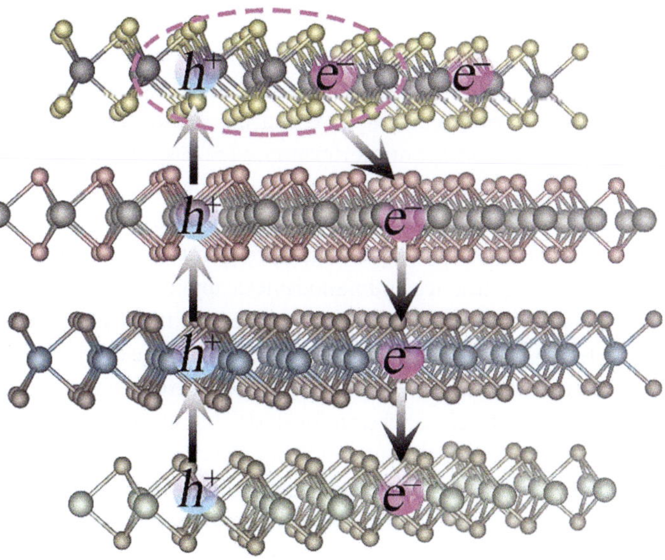

Figure 6.4 The vdW-coupled multilayers for photocatalytic and photovoltaic applications. The small ellipse suggests the intralayer exciton. In such an architecture, electrons and holes can be continuously transferred from one material to the most separated one, strongly reducing the undesirable recombination. Source: Wei Wei.

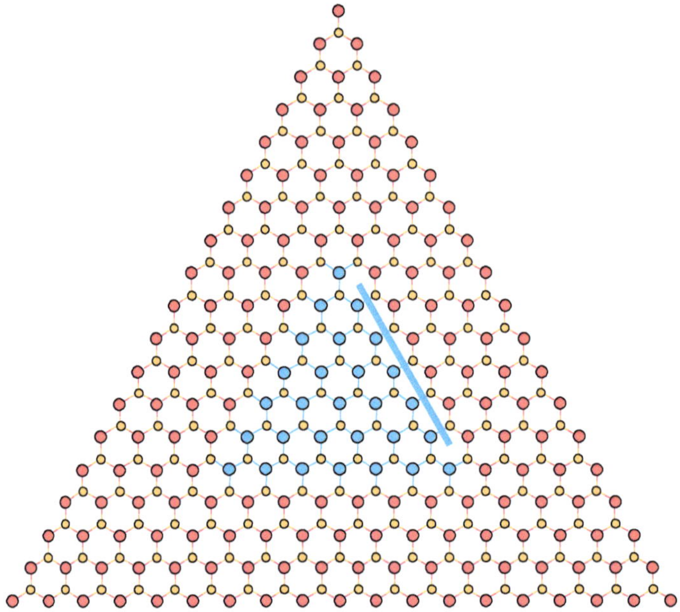

Figure 6.5 In experiments, in-plane heterostructures of two-dimensional TMDCs can be realized by CVD method; the second TMDCs material is epitaxially grown from the edge of the first one. The one-dimensional interface is denoted by a solid line. Source: Wei Wei.

same crystal orientation, confirming the atomic-sharp heterointerfaces. As the heterostructures are formed between two-dimensional materials, interface reduces to actually one-dimensional, referred to as interline. In the in-plane heterostructures composed of two-dimensional TMDCs, two TMDCs are linked by covalent bonds rather than the weak vdW forces in vertical vdW bilayers, ensuring the epitaxial quality and boosts the optical and electronic performance [41–50]. In particular, the intrinsic *p–n* junction behavior has been demonstrated in in-plane heterostructures of two-dimensional TMDCs. It can be expected that, as a consequence, they would comprise a significant platform for electronic engineering in two dimensions and open up new realms in materials science and nanodevices applications.

In in-plane MoS_2/WSe_2 heterostructures, electron states are mainly localized at the interface, while hole states are dominantly localized on WSe_2 side close to the interface. In this case, an exciton-like state forms due to the electron–hole Coulomb interactions, which is in favor of light emission for designing novel light-emitting devices. As a result of the strong electron–hole Coulomb attraction, strong nonadiabatic electron–phonon coupling appears because of the formation of new W–S chemical bonds along the interface, together with the increased quantum coherence. In light of this, the electron–hole recombination timescale is a factor of 2.5 faster (191 ps) than the vdW case (496 ps) [4].

In combination the real-time TDDFT with FSSH scheme, Zhao et al. developed the time-dependent ab initio NAMD code, that is, the Hefei-NAMD [51–55]. In particular, the photoexcited charge carrier dynamics in condensed matter systems can also be simulated. In accordance to this method, the charge carrier transfer dynamics across the interface, the excited spin-polarized hole dynamics, and the electron–hole recombination dynamics in different structures have been studied. In the framework of this method, photoexcitation charge carrier time-dependent dynamics could be investigated in energy, real, and reciprocal spaces. At the atomic level, studies based on the NAMD give rise to new insights into the ultrafast photoexcitation charge carrier dynamics in different condensed matter structures.

References

1 Wei, W., Huang, B., and Dai, Y. (2020). Photoexcited charge carrier behaviors in solar energy conversion systems from theoretical simulations. *WIREs Comput. Mol. Sci.* 10: e1441.

2 Li, L., Long, R., Bertolini, T., and Prezhdo, O.V. (2017). Sulfur adatom and vacancy accelerate charge recombination in MoS_2 but by different mechanisms: time-domain ab initio analysis. *Nano Lett.* 17: 7962–7967.

3 Li, L., Long, R., and Prezhdo, O.V. (2018). Why chemical vapor deposition grown MoS_2 samples outperform physical vapor deposition samples: time-domain ab initio analysis. *Nano Lett.* 18: 4008–4014.

4 Long, R. and Prezhdo, O.V. (2016). Quantum coherence facilitates efficient charge separation at a $MoS_2/MoSe_2$ van der Waals junction. *Nano Lett.* 16: 1996–2003.

5 Yang, Y., Fang, W.H., and Long, R. (2017). Disparity in photoexcitation dynamics between vertical and lateral MoS_2/WSe_2 heterojunctions: time-domain simulation emphasizes the importance of donor–acceptor interaction and band alignment. *J. Phys. Chem. Lett.* 8: 5771–5778.

6 Li, L., Long, R., and Prezhdo, O.V. (2017). Charge separation and recombination in two-dimensional MoS_2/WS_2: Time-domain ab initio modeling. *Chem. Mater.* 29: 2466–2473.

7 Long, R., Guo, M., Liu, L., and Fang, W. (2016). Nonradiative relaxation of photoexcited black phosphorus is reduced by stacking with MoS_2: a time domain ab initio study. *J. Phys. Chem. Lett.* 7: 1830–1835.

8 Wang, L., Long, R., and Prezhdo, O.V. (2015). Time-domain ab initio modeling of photoinduced dynamics at nanoscale interfaces. *Rev. Phys. Chem.* 66: 549–579.

9 Long, R., Prezhdo, O.V., and Fang, W. (2017). Nonadiabatic charge dynamics in novel solar cell materials. *WIREs Comput. Mol. Sci.* 7: e1305.

10 Alexey, V.A. and Prezhdo, O.V. (2015). Large-scale computations in chemistry: a bird's eye view of a vibrant field. *Chem. Rev.* 115: 5797–5890.

11 Jankowska, J., Long, R., and Prezhdo, O.V. (2017). Quantum dynamics of photogenerated charge carriers in hybrid perovskites: dopants, grain boundaries, electric order, and other realistic aspects. *ACS Energy Lett.* 2: 1588–1597.

12 Alexey, V.A. and Prezhdo, O.V. (2013). The PYXAID program for non-adiabatic molecular dynamics in condensed matter systems. *J. Chem. Theory Comput.* 9: 4959–4972.

13 Akimov, A.V. and Prezhdo, O.V. (2014). Advanced capabilities of the PYXAID program: integration schemes, decoherence effects, multiexcitonic states, and field-matter interaction. *J. Chem. Theory Comput.* 10: 789–804.

14 Tully, J.C. (1990). Molecular dynamics with electronic transitions. *J. Chem. Phys.* 93: 1061–1071.

15 Lian, C., Guan, M., Hu, S. et al. (2018). Photoexcitation in solids: first-principles quantum simulations by real-time TDDFT. *Adv.Theory Simul.* 1: 1800055.

16 Lian, C., Zhang, S.B., and Meng, S. (2016). Ab initio evidence for nonthermal characteristics in ultrafast laser melting. *Phys. Rev. B* 95: 184310.

17 Meng, S. and Kaxiras, E. (2008). Real-time, local basis-set implementation of time-dependent density functional theory for excited state dynamics simulations. *J. Chem. Phys.* 129: 054110.

18 Ma, W., Zhang, J., Yan, L. et al. (2016). Recent progresses in real-time local-basis implementation of time dependent density functional theory for electron–nucleus dynamics. *Comput. Mater. Sci.* 112: 478–486.

19 Ren, J., Kaxiras, E., and Meng, S. (2010). Optical properties of clusters and molecules from real-time time-dependent density functional theory using a self-consistent field. *Mol. Phys.* 108: 1829–1844.

20 Kidon, L., Wang, H., Thoss, M., and Rabani, E. (2018). On the memory kernel and the reduced system propagator. *J. Chem. Phys.* 149: 104105.

21 Wang, H. and Thoss, M. (2017). A multilayer multiconfiguration time-dependent Hartree simulation of the reaction-coordinate spin-boson model employing an interaction picture. *J. Chem. Phys.* 146: 114112.

22 Wang, H. and Thoss, M. (2016). Employing an interaction picture to remove artificial correlations in multilayer multiconfiguration time-dependent Hartree simulations. *J. Chem. Phys.* 145: 164105.

23 Wang, H. and Thoss, M. (2018). A multilayer multiconfiguration time-dependent Hartree study of the nonequilibrium Anderson impurity model at zero temperature. *Chem. Phys.* 509: 13–19.

24 Hong, J., Hu, Z., Probert, M. et al. (2015). Exploring atomic defects in molybdenum disulphide monolayers. *Nat. Commun.* 6: 6293.

25 Zhou, W., Zou, X., Najmaei, S. et al. (2013). Intrinsic structural defects in monolayer molybdenum disulfide. *Nano Lett.* 13: 2615–2622.

26 González, C., Biel, B., and Dappe, Y.J. (2016). Theoretical characterisation of point defects on a MoS_2 monolayer by scanning tunnelling microscopy. *Nanotechnology* 27: 105702.

27 Santosh, K.C., Roberto, C.L., Rafik, A. et al. (2014). Impact of intrinsic atomic defects on the electronic structure of MoS_2 monolayers. *Nanotechnology* 25: 375703.

28 Liu, D., Guo, Y., Fang, L., and Robertson, J. (2013). Sulfur vacancies in monolayer MoS_2 and its electrical contacts. *Appl. Phys. Lett.* 103: 183113.

29 Duan, X., Wang, C., Shaw, J.C. et al. (2014). Lateral epitaxial growth of two-dimensional layered semiconductor heterojunctions. *Nat. Nanotechnol.* 9: 1024–1030.

30 Bogaert, K., Liu, S., Chesin, J. et al. (2016). Diffusion-mediated synthesis of MoS_2/WS_2 lateral heterostructures. *Nano Lett.* 16: 5129–5134.

31 Yoo, Y., Degregorio, Z.P., and Johns, J.E. (2015). Seed crystal homogeneity controls lateral and vertical heteroepitaxy of monolayer MoS_2 and WS_2. *J. Am. Chem. Soc.* 137: 14281–14287.

32 Chen, K., Wan, X., Wen, J. et al. (2015). Electronic properties of MoS_2–WS_2 heterostructures synthesized with two-step lateral epitaxial strategy. *ACS Nano* 9: 9868–9876.

33 Zhang, X.Q., Lin, C.H., Tseng, Y.W. et al. (2015). Synthesis of lateral heterostructures of semiconducting atomic layers. *Nano Lett.* 15: 410–415.

34 Chen, K., Wan, X., Xie, W. et al. (2015). Lateral built-in potential of monolayer MoS_2–WS_2 in-plane heterostructures by a shortcut growth strategy. *Adv. Mater.* 27: 6431–6437.

35 Gong, Y., Lin, J., Wang, X. et al. (2014). Vertical and in-plane heterostructures from WS_2/MoS_2 monolayers. *Nat. Mater.* 13: 1135–1142.

36 Huang, C., Wu, S., Sanchez, A.M. et al. (2014). Lateral heterojunctions within monolayer $MoSe_2$–WSe_2 semiconductors. *Nat. Mater.* 13: 1096–1101.

37 Gong, Y., Lei, S., Ye, G. et al. (2015). Two-step growth of two-dimensional $WSe_2/MoSe_2$ heterostructures. *Nano Lett.* 15: 6135–6141.

38 Mahjouri-Samani, M., Lin, M.W., Wang, K. et al. (2015). Patterned arrays of lateral heterojunctions within monolayer two-dimensional semiconductors. *Nat. Commun.* 6: 7749.

39 Li, M.Y., Shi, Y., Cheng, C.C. et al. (2015). Epitaxial growth of a monolayer WSe_2–MoS_2 lateral p–n junction with an atomically sharp interface. *Science* 349: 524–528.

40 Zhao, J., Cheng, K., Han, N., and Zhang, J. (2018). Growth control, interface behavior, band alignment, and potential device applications of 2D lateral heterostructures. *WIREs Comput. Mol. Sci.* 8: e1353.

41 Sun, Q., Dai, Y., Ma, Y. et al. (2016). Ab Initio prediction and characterization of Mo_2C monolayer as anodes for lithium-ion and sodium-ion batteries. *J. Phys. Chem. Lett.* 7: 937–943.

42 Zhao, W., Li, Y., Duan, W., and Ding, F. (2015). Ultra-stable small diameter hybrid transition metal dichalcogenide nanotubes X-M-Y (X, Y = S, Se, Te; M = Mo, W, Nb, Ta): a computational study. *Nanoscale* 7: 13586–13590.

43 Er, D., Ye, H., Frey, N.C. et al. (2018). Prediction of enhanced catalytic activity for hydrogen evolution reaction in Janus transition metal dichalcogenides. *Nano Lett.* 18: 3943–3949.

44 Guan, Z., Ni, S., and Hu, S. (2018). Tunable electronic and optical properties of monolayer and multilayer Janus MoSSe as a photocatalyst for solar water splitting: a first-principles study. *J. Phys. Chem. C* 122: 6209–6216.

45 Ji, Y.J., Yang, M.Y., Lin, H.P. et al. (2018). Janus structures of transition metal dichalcogenides as the heterojunction photocatalysts for water splitting. *J. Phys. Chem. C* 122: 3123–3129.

46 Wang, J., Shu, H., Zhao, T. et al. (2018). Intriguing electronic and optical properties of two-dimensional Janus transition metal dichalcogenides. *Phys. Chem. Chem. Phys.* 20: 1857–18578.

47 Wei, W., Dai, Y., Sun, Q. et al. (2015). Electronic structures of in-plane two-dimensional transition-metal dichalcogenide heterostructures. *Phys. Chem. Chem. Phys.* 17: 29380–29386.

48 Wei, W., Dai, Y., and Huang, B. (2016). In-plane interfacing effects of two-dimensional transition-metal dichalcogenide heterostructures. *Phys. Chem. Chem. Phys.* 18: 15632–15638.

49 Wei, W., Dai, Y., and Huang, B. (2017). Straintronics in two-dimensional in-plane heterostructures of transition-metal dichalcogenides. *Phys. Chem. Chem. Phys.* 19: 663–672.

50 Wei, W., Dai, Y., Niu, C., and Huang, B. (2015). Controlling the electronic structures and properties of in-plane transition-metal dichalcogenides quantum wells. *Sci. Rep.* 5: 17578.

51 Zheng, Q., Chu, W., Zhao, C. et al. (2019). Ab initio nonadiabatic molecular dynamics investigations on the excited carriers in condensed matter systems. *WIREs Comput. Mol. Sci.* 9: e1411.

52 Zheng, Q., Saidi, W.A., Xie, Y. et al. (2017). Phonon-assisted ultrafast charge transfer at van der Waals heterostructure interface. *Nano Lett.* 17: 6435–6442.

53 Zheng, Q., Xie, Y., Lan, Z. et al. (2018). Phonon-coupled ultrafast interlayer charge oscillation at van der Waals heterostructure interfaces. *Phys. Rev. B* 97: 205417.

54 Zheng, Z., Huang, B., Qin, X. et al. (2011). Facile in situ synthesis of visible-light plasmonic photocatalysts M@TiO_2 (M= Au, Pt, Ag) and evaluation of their photocatalytic oxidation of benzene to phenol. *J. Mater. Chem.* 21: 9079–9087.

55 Zhao, C., Zheng, Q., Wu, J., and Zhao, J. (2017). Ab initio nonadiabatic molecular dynamics investigation on the dynamics of photogenerated spin hole current in Cu-doped MoS2. *Phys. Rev. B* 96: 134308.

7

Simulations for Photocatalytic Materials

7.1 Photocatalysis and Photocatalytic Reactions

It is estimated that 80% of the worldwide primary energy consumption is obtained from fossil fuels such as petroleum, oil, and natural gas. It is well known that solar energy is the most abundant and cleanest renewable energy source, and the sunlight reaching the Earth's surface on an hourly basis actually exceeds the annual global energy consumption [1–3], and thus the efficient harvesting and conversion of solar energy has been a worldwide priority target in the past years. As a result of an increasing focus on dwindling energy resources and environmental deterioration, solar power has emerged as a promising alternative to fossil fuels. However, due to variations in the degree of sunlight reaching the planet depending on geographical location, seasons, and time of day, the efficient utilization of solar energy requires that this energy be converted and stored in a cost-effective and environmentally benign fashion [1–4]. In this context, the use of sunlight to drive light–energy conversion via photocatalytic (photoelectrochemical) and photovoltaic applications is of paramount importance toward a sustainable future [5–19].

In photocatalytic H_2O splitting and CO_2 reduction, for example, the energy of solar photons can be stored in chemical bonds by using semiconductor photocatalysts like TiO_2. In particular, the direct semiconductor/electrolyte junction provides an effective driving force for photocatalytic (photoelectrochemical) reactions under irradiation, offering a more straightforward, cost-effective, and convenient way to achieve light–energy conversion of high efficiency [20–24]. It is however indicative that most photocatalysts present low solar energy utilization, as solar light is composed of ultraviolet, visible, and infrared components (accounting for 5, 43, and 52%, respectively). In addition, the long-term stability, high charge separation efficiency, low electron–hole recombination, and strong redox ability are still in need in current photocatalytic materials. As an example, overall water splitting using particulate photocatalysts has been considered as a low-cost technology with the potential to enable large-scale solar hydrogen production, because of the ready synthesis of the associated photocatalysts as well as the simple reactor and facility designs [25–27]. A technical/economic analysis has determined that a solar-to-hydrogen energy conversion efficiency of 5–10%, or even lower under some operational conditions, could allow photocatalytic overall water splitting to

Calculations and Simulations of Low-Dimensional Materials: Tailoring Properties for Applications, First Edition. Ying Dai, Wei Wei, Yandong Ma, and Chengwang Niu.

be economically viable for solar hydrogen production [25, 26]. It concludes that, however, the current solar-to-hydrogen values for overall water splitting using particulate photocatalysts on the laboratory scale are only approximately 1% [28]. In current stage, successful photocatalytic systems for the decomposition of H_2O into H_2 and O_2 are based on one of two approaches, with one of them involving splitting water with a single particulate photocatalyst via one-step excitation [29, 30]. It is now a consensus that, therefore, developing highly efficient photocatalytic overall water splitting systems is vital to achieving the solar-to-hydrogen required for scalability. In other fields of photocatalytic reactions, efficient photocatalysts are also needed to be explored.

It is therefore still very challenging to design novel photocatalysts that are abundant, stable, facile to fabrication, and present high light–energy conversion efficiency. It is a consensus that single-component photocatalysts cannot simultaneously take on wide light absorption range and strong redox ability partially due to, for instance, the fast recombination between conduction band electrons and valence band holes, which can be overcome by designing proper heterogeneous photocatalytic systems [31–33]. In case of heterogeneous photocatalytic materials, photoexcited electrons and holes are transferred to opposite components due to the type-II band alignment (Figure 7.1), reducing the charge carrier recombination and thus improving the light–energy conversion efficiency. In order to realize high light–energy conversion efficiency, a variety of strategies have been developed to improve the photoelectrochemical performance of semiconductor materials via addition of electron donor (hole scavenger), controlled incorporation of oxygen vacancies, noble metal loading, metal ion doping, anion doping, dye sensitization, and formation of composite semiconductors [34–36].

In last decades, photocatalysis has made a great advancement, as more and more materials are demonstrated to be photocatalytically active, and the efficiency has increased year after year. In 2004, graphene was born, and since then more and more two-dimensional materials are synthesized, transferred, and predicted. In the great application potentials for two-dimensional materials, such as transition

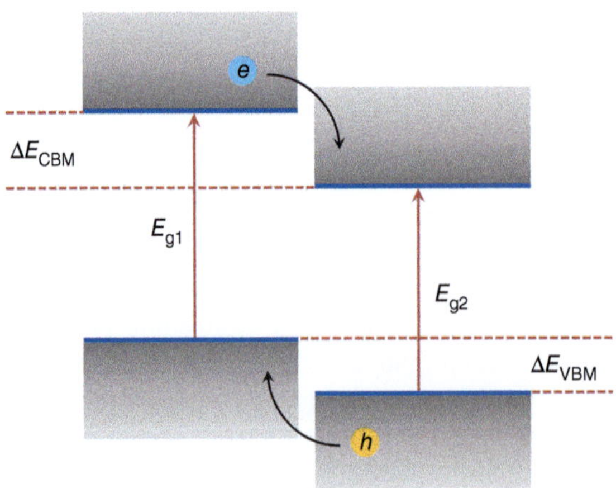

Figure 7.1 Type-II band alignment in heterogeneous catalysts; arrows indicate the transfer direction of electrons (*e*) and holes (*h*). Source: Wei Wei.

metal dichalcogenides (TMDCs), transition metal carbides and nitrides (MXenes), group-III and group-IV metal monochalcogenides, black phosphorus (phosphorene), and so on [37–51], photocatalysis is one of the most important aspects. In respect to potential photocatalytic applications, two-dimensional monolayer materials find their advantages like high specific surface area, strong light absorption, high charge carrier mobility, unique electronic structures, and chemical stability. In particular, the two-dimensional nature of materials minimizes the distance that photogenerated electrons and holes have to migrate before reaching the material surface, reducing the possibility of electron–hole recombination. As a result, photoelectrochemistry in two-dimensional materials has been extensively investigated, and novel light–energy conversion devices have been designed and/or proposed.

In respect to the two-dimensional materials, for example, graphitic carbon nitride (i.e. g-C_3N_4, see Figure 7.2) has been illustrated to be active for various photocatalytic reactions including water splitting, due to the strong light–matter interaction, appropriate band structure, high carrier mobility, large specific surface area, etc. [52–65] In application in photocatalysis, in particular, g-C_3N_4 has attracted great attention in the field of energy conversion and storage due to its unique layered structure, tunable band gap, metal-free characteristic, high physicochemical stability, and easy accessibility. It has been demonstrated that two-dimensional g-C_3N_4 nanosheets have the features of short charge/mass transfer path, abundant reactive sites, and easy functionalization, which are beneficial to optimizing their performance in different fields. In recent publications, the diversified applications of g-C_3N_4 in energy conversion and storage, including photocatalytic

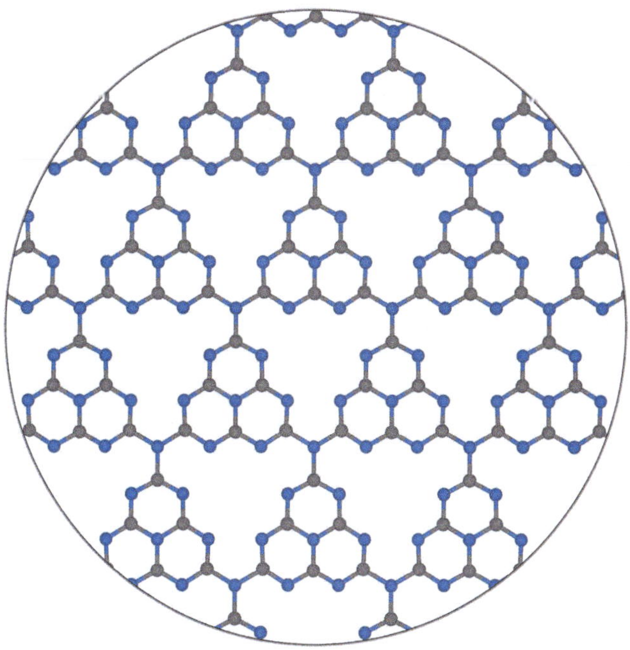

Figure 7.2 Two-dimensional *g*-C_3N_4 monolayer. Source: Wei Wei.

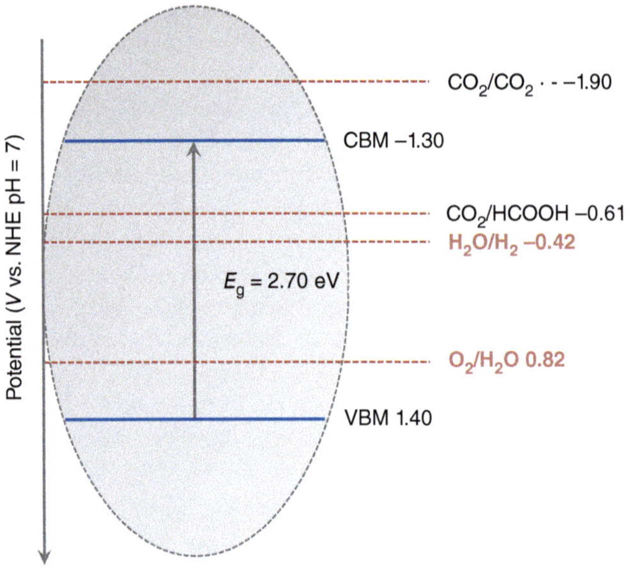

Figure 7.3 Band energy positions are shown for g-C_3N_4 and corresponding redox potentials for reactions.

H_2 evolution, CO_2 reduction, electrocatalytic H_2 evolution, O_2 evolution, O_2 reduction, alkali-metal ion batteries, lithium-metal batteries, lithium–sulfur batteries, metal-air batteries, and supercapacitors, have been discussed [66]. In Figure 7.3, band energy positions are shown for g-C_3N_4.

It is a consensus that solar energy bears great potential in the replacement of conventional fossil fuels to enable a more sustainable world. In order to harness the solar energy, on the other hand, solar cells made of semiconductor materials have been developed to convert sunlight into electricity based on the photovoltaic effects [67–76]. In this field, several new generation thin film solar cells, including organic solar cells, Schottky junction solar cells, dye-sensitized solar cells, quantum dot-sensitized solar cells, and perovskite solar cells, have been developed and considerable advances have already been made. In general, an efficient photovoltaic energy conversion process requires three basic attributes [10]: (i) solar light absorption with energy exceeding the band gap of semiconductor that results in the generation of charge carriers (electron–hole pairs or excitons); (ii) the charge carriers diffuse through the semiconductor and reach an energy barrier that permits one kind of charge carrier to pass but blocks the opposite charge carrier; and (iii) the separated charge carriers move through the semiconductor to an external circuit. In particular, among these the efficient charge separation plays a key role in solar energy capture and conversion. It is an indication that, therefore, efficient charge carrier separation requires the development of spatial variations in an electronic environment to provide the essential driving force. In conventional inorganic photovoltaic cells, the electric field at the interface plays the role of the driving force. As for the organic photovoltaic cells (excitonic solar cells) [77–80], sunlight absorption leads to the production of strongly bound excitons, and the band offset between donor and acceptor materials provides the pathway for dissociation of excitons at the heterostructures' interface. In this view, the generation, separation, and transport of charge carriers in a photovoltaic process give rise to a photocurrent in short-circuit operation and a photovoltage in open-circuit operation. However, difficult requirements of the currently employed epitaxial technique and the limitation of abundance of some elements have impeded the economic viability and the widespread application of the photovoltaic technology. In view of designing lightweight, flexible, and highly efficient photovoltaic devices, van der Waals (vdW) heterostructures of dissimilar two-dimensional materials arise [81–93]. In the light of strong light–matter interaction, spatial indirect geometry, and large driving force, vdW heterostructures based photovoltaic systems with high power conversion efficiency (PCE) have been attracting extensive attention. It should be emphasized again that understanding the photoexcitation dynamics of charge carriers is the foundation for designing more promising photovoltaic devices.

In Figure 7.4, photoelectric events such as photon absorption, electrons and holes separation and recombination are schematically shown. As a photocatalytic semiconductor is excited by photons with energy equal to or larger than the band gap, electrons in the valence band will be promoted to the conduction band, leaving holes in the valence band. In the following, photogenerated electrons and holes will separate and migrate to surface reaction sites, while electron–hole recombination occurs in the bulk or on the surface during the separation and migration. In particular, parameters to characterize this type of transport are the charge carrier diffusion length and the recombination rate. In general, especially for low-dimensional materials, Coulomb interaction can bind electron and hole together, that is, the formation of exciton. It can be deduced that, thus, small exciton binding energy is desirable for photocatalytic materials. In excitonic solar cells, the building block is a heterostructure formed by two different materials or the same materials but distinct phases, and

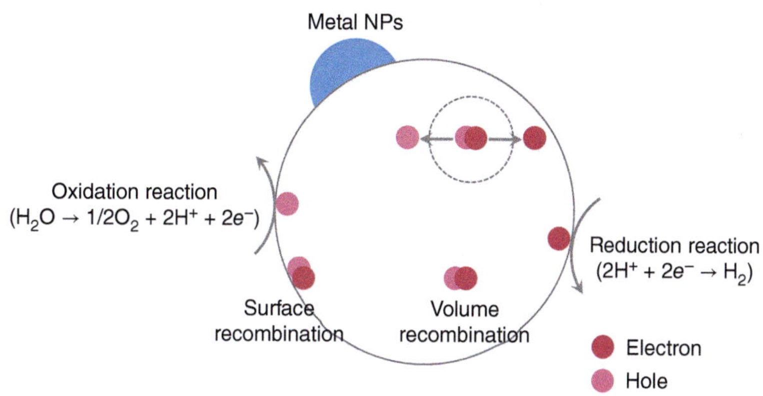

Figure 7.4 Photoelectric events in photocatalysts. Excited electrons and holes can recombine in bulk and on surface. Under light irradiation, metal nanoparticles, which can be used to enhance light absorption, show localized surface plasmon resonance. Source: Wei Wei.

photogenerated excitons will be dissociated in an ultrafast time scale at the heterointerface due to the discontinuities of the electron affinity and ionization potential across the interfaces, which generates free charge carriers and results in photoelectric conversion. It appears that the interfacial dissociation of excitons at a heterointerface into free electrons in one material and free hole on the other side of the interface is also the fundamentally important process in heterogeneous photocatalysts. In particular, electrons (holes) generated in donors will transfer to acceptors, creating a powerful photoinduced interfacial chemical potential energy gradient that drives the photoelectrochemical effects, even in the absence of a build-in electric potential. In general, such a charge carrier separation process is generally faster than the subsequently occurred electron–hole recombination. In practice, the resultant photocurrent can be collected from electrodes.

In principle, the photoinduced charge carrier behaviors constitute the foundation of photoelectronchemistry in photocatalytic and photovoltaic applications. It is of paramount importance to clearly understand these behaviors in improving and designing unprecedented photocatalytic and photovoltaic systems with commercially high light–energy conversion efficiency.

7.2 Photoresponsivity and Photocurrent from Simulations

It has been widely recognized that the first-principles calculations based on the density functional theory is successful in materials science, condensed matter physics, quantum chemistry, and theoretical biology. In these fields of simulation, the system is at equilibrium. It has been known that at nanoscale transport properties are sensitive to the chemical and atomic details of the materials and also to the external fields and quantum effects. In this sense, the electronic device operation is often under nonlinear and nonequilibrium conditions, and is beyond the scope of density functional theory from simulation point of view.

In order to include the microscopic physics for making quantitative predictions for charge/spin quantum transport, some theoretical methods have been developed. It becomes possible to predict a wide range of quantum transport properties of nanostructures from atomic first-principles without any phenomenological parameters. In combination of a real space self-consistent field (SCF) theory with the Keldysh nonequilibrium Green's function (NEGF) formalism, the Hamiltonian and electronic structures of the devices can be self-consistently calculated, and the nonequilibrium quantum statistical properties and density matrix can be calculated by NEGF, accounting for the open device transport boundary conditions and electrostatic boundary conditions by real space numerical techniques.

In practice, NEGF calculations can be performed based on an exact device model, which has two or more electrodes extending to electron reservoirs at infinity where bias voltage V_b is applied and current collected. In Figure 7.5, device model in NEGF calculations is schematically shown. It consists of three parts, the central scattering region and two electrodes. In a model of the two-probe open transport junction, the

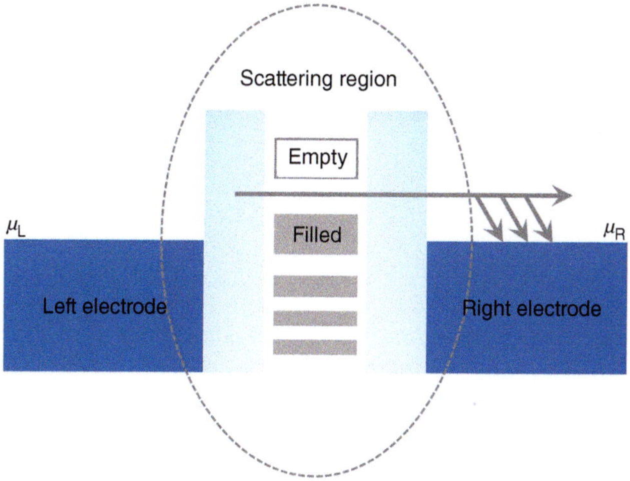

Figure 7.5 Schematic of a two-probe device model, with μ_L and μ_R being the electrochemical potentials of the left and right electrodes, respectively; $\mu_L - \mu_R = V_b$, where V_b is the bias voltage and e the electron charge. The dash-lined ellipse indicates the device scattering region that has some quantum levels, and the tunnel barriers indicate coupling of the scattering region to the semi-infinitely long electrodes. Source: Wei et al. [94]/Reproduced with permission of John Wiley and Sons.

scattering region must include several layers of the electrode atoms, while device electrodes or leads extend to $z = \pm\infty$ where bias voltages are applied and electric current collected. It is obvious that the electrode is half-infinite, providing the transport boundary conditions. In practice, the electrodes are usually metallic materials and hence maintain equal potential all the way near to the device scattering region, and V_b is dropped across the scattering region. On account of its own chemical potential μ of each electrode, the entire system is at nonequilibrium, and the potential or Hamiltonian of the scattering region has to be calculated by means of using the SCF method. Once the Hamiltonian is obtained, in equilibrium, the energy level can be calculated by Fermi–Dirac distribution function to construct the density matrix. In nonequilibrium, the density matrix can be constructed through NEGF. In the following, a self-energy term can be added to the Hamiltonian of the scattering region, and thus contributions from infinitely large number of electrons in electrodes can be integrated out. It can be known that calculations on infinitely large device model actually reduce to the one focusing on the scattering region, and inclusion of the dissipative self-energy terms makes the Hamiltonian to be non-Hermitian.

In combination of the Keldysh NEGF and an SCF theory, quantum transport calculations could be done. It should be kept in mind that NEGF–SCF theory is not a ground state theory because it is determined by a nonvibrational and nonequilibrium density matrix [95–110].

In this section, important equations and the main idea to perform NEGF–SCF calculations are presented. In general, the Green's function in NEGF–SCF theory takes

$$\boldsymbol{G}^{\mathrm{R,A}} = [ES - H - \Sigma^{\mathrm{R,A}}]^{-1}$$

where R and A stand for the retarded and advanced quantities, respectively. In the consideration of $\boldsymbol{G}^{\mathrm{A}} = (\boldsymbol{G}^{\mathrm{R}})^{\dagger}$, only $\boldsymbol{G}^{\mathrm{R}}$ needs to be calculated by inverting the matrix of the right side of the equation. On top of the obtained $\boldsymbol{G}^{\mathrm{R,\,A}}$, the Keldysh NEGF $\mathbf{G}^{<}$ can be obtained from the Keldysh equation:

$$\mathbf{G}^{<} = \mathbf{G}^{\mathrm{R}}\Sigma^{<}\mathbf{G}^{\mathrm{A}}$$

where

$$\Sigma^{<} = i\sum_{\alpha} f_{\alpha}\Gamma_{\alpha}$$

is justified for mean field theory in the central scattering region of the device, and

$$\Gamma_{\alpha} = i[\Sigma^{\mathrm{R}} - \Sigma^{\mathrm{A}}]$$

with f_{α} being the Fermi function of lead-α, and Γ_{α} is the linewidth function of the same lead.

In the linear combination of atomic orbital (LCAO) space, the nonequilibrium density matrix $\hat{\rho}$ reads:

$$\hat{\rho} = \frac{1}{2\pi}\left[\int_{-\infty}^{+\infty} dE\mathbf{G}^{<}(E)\right]$$

In case of using LCAO basis functions, the real space $\hat{\rho}$ can be expressed as:

$$\hat{\rho}(\mathbf{r}, \mathbf{r}') = \sum_{\mu\nu} \left\langle \zeta_{\mu}(\mathbf{r}) \mid \mathbf{G}^{<}_{\mu\nu} \mid \zeta_{\nu}(\mathbf{r}') \right\rangle$$

here $\mathbf{r}$ is associated with orbital ζ_{μ} and $\mathbf{r}'$ with ζ_{ν}. In the following, the charge density $\rho(\mathbf{r})$ is the diagonal elements of the $\hat{\rho}$, i.e.

$$\rho(\mathbf{r}) = \hat{\rho}(\mathbf{r}, \mathbf{r})$$

In practice, $\hat{\rho}$ reduces to

$$\hat{\rho} = \frac{1}{\pi}\mathrm{Im}\left[\int_{-\infty}^{\mu_{\mathrm{l}}} dE\mathbf{G}^{\mathrm{R}}(E)\right] + \frac{1}{2\pi}\left[\int_{\mu_{\mathrm{l}}}^{\mu_{\mathrm{r}}} dE\mathbf{G}^{<}(E)\right]$$

if the upper limit is truncated to the chemical potential of the right electrode μ_{r} (assuming that it is larger than the chemical potential of the left electrode μ_{l}) and for $E < \mu_{\mathrm{l}}$ all the energy levels of the scattering region are occupied at low temperature. In this context, the integrals can be calculated discretized and numerically.

In an open system with several electrodes, electrons at energy ε coming from all channels of an electrode α will be scattered into channels of another electrode β. In terms of a transmission coefficient $T_{\alpha\beta}(\varepsilon)$, the total scattering probability can be calculated through Green's function:

$$T_{\alpha\beta}(\varepsilon) = tr[G^{\mathrm{r}}(\varepsilon)\Gamma_{\alpha}(\varepsilon)G^{\alpha}(\varepsilon)\Gamma_{\beta}(\varepsilon)]$$

where $tr[\cdots]$ stands for the trace of $[\cdots]$, and the linewidth function of the electrode α is defined as:

$$\Gamma_{\alpha}(\varepsilon) \equiv i\left[\Sigma^{\mathrm{r}}_{\alpha}(\varepsilon) - \Sigma^{\mathrm{a}}_{\alpha}(\varepsilon)\right]$$

with $\Sigma^{\mathrm{r}}_{\alpha}(\varepsilon)$ and $\Sigma^{\mathrm{a}}_{\alpha}(\varepsilon)$ being the retarded and advanced self-energy of electrode α. In a common manner, the number of transmission channels is calculated by counting the number of the Bloch waves existing in the system, which propagate along a given direction.

In consideration of an open system with several electrodes, spin nonpolarized electric current in electrode β can by evaluated by Landauer formula:

$$I_\beta = \frac{2e}{h}\sum_{\alpha}\int d\varepsilon[f_\alpha(\varepsilon)T_{\alpha\beta}(\varepsilon) - f_\beta(\varepsilon)T_{\beta\alpha}(\varepsilon)] = \frac{2e}{h}\sum_{\alpha}\int d\varepsilon[f_\alpha(\varepsilon) - f_\beta(\varepsilon)]T_{\alpha\beta}(\varepsilon)$$

where $f_\alpha(\varepsilon)$ is the distribution function of the electrons in electrode α, and the factor 2 comes from the spin degeneracy. In case of defining the conductance as:

$$G_{\alpha\beta} = \frac{2e^2}{h}\int d\varepsilon \frac{f_\alpha(\varepsilon) - f_\beta(\varepsilon)}{eV_\beta - eV_\alpha}T_{\alpha\beta}(\varepsilon)$$

with V_β being the bias voltage applied on electrode β, the electric current in electrode β can be rewritten as:

$$I_\beta = \sum_{\alpha} G_{\alpha\beta}(V_\beta - V_\alpha)$$

In the case of spin-polarized transport, the spin current (spin-polarized charge current) for a two-probe system can be obtained as follows:

$$I_\sigma = \frac{e}{h}\int d\varepsilon T_\sigma(\varepsilon, V_{\mathrm{b}})[f_{\mathrm{L}}(\varepsilon) - f_{\mathrm{R}}(\varepsilon)]$$

In this equation, T_σ represents the spin resolved transmission coefficient, where $\sigma \equiv \uparrow, \downarrow$ is the spin index. It then gives the total charge current:

$$I = I_\uparrow + I_\downarrow$$

and the total spin current reads:

$$I_{\mathrm{S}} = I_\uparrow - I_\downarrow$$

In the case for the device being under irradiation of a beam of light, interactions between the electric field of the light and the electrons in the device can be treated as a small perturbation to the original Hamiltonian of the device without light. The whole Hamiltonian can therefore be expressed as:

$$H = H_0 + H_1 = H_0 + \frac{e}{m_0}\boldsymbol{A}\cdot\hat{\mathbf{p}}$$

here, H_0 is the unperturbed Hamiltonian, H_1 is the perturbation from the light, $\boldsymbol{A}$ is the electromagnetic vector potential of the light, and $\hat{\mathbf{p}}$ is the electronic momentum operator. In consideration of a single-mode monochromatic light, vector potential $\boldsymbol{A}$, which is second quantized in a volume V, can be written as:

$$\boldsymbol{A} = C_0\left(be^{-i\omega t}\mathbf{e}_{\mathrm{p}} + b^\dagger e^{i\omega t}\mathbf{e}^\dagger_{\mathrm{p}}\right)$$

where $\mathbf{e}_{\mathrm{p}}$ is a complex unit vector characterizing the light polarization, b and $b^\dagger$ are the bosonic annihilation and creation operators acting on photons, respectively, and

$C_0 = \sqrt{\hbar/2\omega\tilde{\varepsilon}V}$ with $\tilde{\varepsilon}$ the dielectric constant of the device material. In terms of atomic orbital basis $\zeta_\nu(\mathbf{r})$, electronic momentum operator $\hat{\mathbf{p}}$ is expressed as:

$$\mathbf{p}_{\mu\nu} = \int d\mathbf{r}\zeta_\mu^*(\mathbf{r})\hat{\mathbf{p}}\zeta_\nu(\mathbf{r})$$

As a consequence, Hamiltonian of the electron–photon interaction H_1 can be obtained:

$$(H_1)_{\mu\nu} = be^{-i\omega t}M_{\mu\nu} + b^\dagger e^{i\omega t}M_{\mu\nu}^\dagger$$

where

$$M_{\mu\nu} = (e/m_0)C_0\mathbf{e}_\mathrm{p} \cdot \mathbf{p}_{\mu\nu}$$

As the $(H_1)_{\mu\nu}$ is defined, self-energies due to the interactions with photon can be written within Born approximation as:

$$\Sigma^{<(\mathrm{ph})}(E) = MNG_0^<(E - \hbar\omega)M^\dagger + M^\dagger(N+1)G_0^<(E + \hbar\omega)M$$

$$\Sigma^{>(\mathrm{ph})}(E) = M^\dagger NG_0^>(E + \hbar\omega)M + M(N+1)G_0^>(E - \hbar\omega)M^\dagger$$

with N being the total number of photons within the volume V, and $G_0^{<,>,\mathrm{r},\mathrm{a}}$ is the unperturbed Green's function. In this situation, the linear component of Green's function due to the electron–photon interactions can be calculated as:

$$G^{<(\mathrm{ph})}(E) = G_0^\mathrm{r}(E)\Sigma^{<(\mathrm{ph})}(E)G_0^\mathrm{a}(E)$$

$$G^{>(\mathrm{ph})}(E) = G_0^\mathrm{r}(E)\Sigma^{>(\mathrm{ph})}(E)G_0^\mathrm{a}(E)$$

then the photocurrent moving into electrode L then can be obtained:

$$J_\mathrm{L}^{(\mathrm{ph})} = \frac{ie}{h}Tr\int \Gamma_\mathrm{L}\left[G^{<(\mathrm{ph})}(E) + f_\mathrm{L}(E)\left(G^{>(\mathrm{ph})} - G^{<(\mathrm{ph})}\right)\right]dE$$

which is the linear part of the response to light, namely, it is proportional to number of photon or the flux of the photon. In closing, the photocurrent response function thus can be defined as:

$$R_\mathrm{L}^{(\mathrm{ph})} = \frac{J_\mathrm{L}^{(\mathrm{ph})}}{eI_\omega}$$

and its decomposed form takes

$$R_\mathrm{L}^{(\mathrm{ph})} = R_\mathrm{L}^{(\mathrm{ph.ex.elec})} + R_\mathrm{L}^{(\mathrm{ph.de.elec})} + R_\mathrm{L}^{(\mathrm{ph.ex.hole})} + R_\mathrm{L}^{(\mathrm{ph.de.hole})}$$

In general, only the photoexcited current is of interest and, therefore, only the first terms in the expression of self-energies are taken into account in the practical calculations.

It is a consensus that solar energy is the most abundant and cleanest renewable energy source. In recent few years, vdW heterostructures composed of distinct two-dimensional materials are usually constructed to harness the solar energy, suggesting a new class of photocatalytic and photovoltaic devices. It should be pointed out here that charge transfer is the fundamental process that determines the

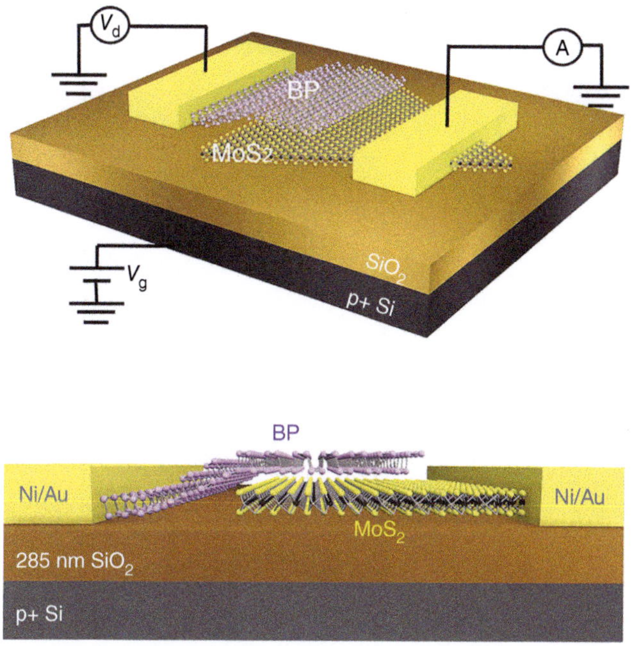

Figure 7.6 Schematics of the device structure for intuitive. Source: Deng et al. [111]/with permission of American Chemical Society.

performance of solar cells and photocatalytic devices. In Figure 7.6, experimentally constructed device is shown.

In recent publications, the NEGF–SCF method has been demonstrated to be successful in simulating the transport properties of two-dimensional materials-based devices. In particular, for InSe/InTe vdW heterostructures charge carrier can follow either the regular scheme (R-scheme) or the Z-scheme transfer path, depending on the coupling between the interlayer states at the band-edge positions [112]. It is of interest that the proposed R-scheme and Z-scheme transfer mechanism can be verified by quantum transport calculations based on the NEGF–SCF method, solidly evidencing the charge transfer mechanism.

In case of applying a tensile strain of 3%, coupling states appear within the InSe/InTe vdW heterostructures, providing an additional electron–hole recombination channel. In this case, the electron–hole recombination time is 1.2 ps, which is 3 orders of magnitude faster than the case of under a compressive strain of −3% (1.4 ns) and comparable to the case of strain free (5.8 ps). After generating electron–hole pairs by photoexcitation, charge carrier transfer can follow either the R-scheme or the Z-scheme. In the case of the former one, the work done by the build-in electric field force has to be overcome. In contrast, for the latter situation, charge carrier transfer is much easier since it is along the direction of electric field. In the presence of interlayer coupling states, band offsets between band edges are vanishing. As a result, photogenerated electrons from InSe recombine with holes

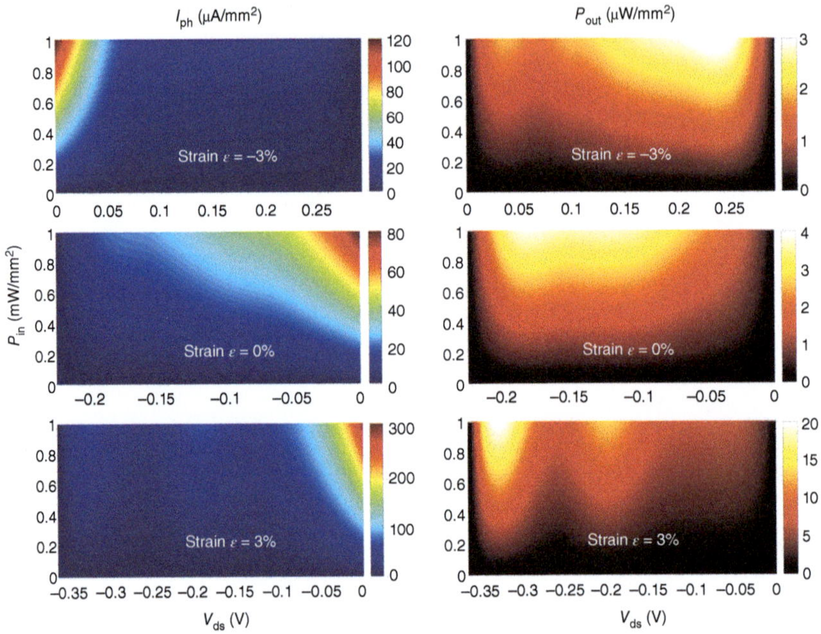

Figure 7.7 Photocurrent maps (I_{ph}), and output power density (P_{out}) with various bias voltages (V_{ds}) and incident light power densities (P_{in}) under different strains. Source: Jin et al. [112]/Reproduced with permission of American Chemical Society.

from InTe following the Z-scheme, whereas electrons (holes) from InTe (InSe) are swept to the right (left) electrodes guided by the build-in electric field. Thus, photocurrent from InTe to InSe could be collected. On the basis of NEGF–SCF simulations, the quantum transport process can be confirmed.

As shown in Figure 7.7, with linearly polarized light irradiating the scattering region, photocurrent (I_{ph}) direction for −3% compressively strained InSe/InTe vdW heterostructures is opposite to those of 3% tensely strained and strain-free structures. It therefore concludes that the R-scheme dominates the charge carrier transfer process for the model with −3% compressive strain, while Z-scheme plays a leading role for the models of strain free and 3% tensile strain.

In Figure 7.7, the photocurrent contour map for InSe/InTe vdW heterostructures with the incident light power density (P_{in}) varying from 0 to 1 mW mm^{-2} (i.e. AM1.5 illumination) is also shown. In accordance to

$$P_{out} = I_{ph} V_{ds}$$

the output power density P_{out} can be calculated, with V_{ds} being the bias voltage. It can be found that system of 3% tensile strain outputs higher power density, peaking at a maximum value $P_{out}^{max} = 21\ \mu W\ mm^{-2}$ when $P_{in} = 1\ mW\ mm^{-2}$, which is approximate 1 order of magnitude larger than the compressive system. In the mass, the photovoltaic performance can be evaluated by the photoresponse R_{ph}, the external quantum efficiency (EQE) and the energy conversion efficiency η

$$R_{ph} = \frac{I_{ph}}{eF_{ph}}$$

$$\text{EQE} = \frac{I_{sc}}{P_{in}} \frac{hc}{\lambda}$$

$$\eta = P_{out}/P_{in}$$

In these equations, e is the electron charge, F_{ph} is the photon flux defined as the number of photon per unit time per unit area, and h, c, λ are the Planck's constant, the speed of light, and the wavelength of light, respectively. In conclusion, results reveal that Z-scheme shows much better performance than R-scheme. In particular, InSe/InTe vdW heterostructures of 3% tensile strain exhibit $\eta = 2.08\%$, which is about 1 order of magnitude larger than MoS_2/WSe_2 vdW heterostructures ($\eta = 0.2\%$).

In vdW heterostructures of two-dimensional materials, build-in electric field will be established once the charge redistribution reaches equilibrium, causing band bending depending on the electron and hole transfer direction and affecting the photoexcited charge carrier separation. In case of R-scheme, a tunneling barrier is present at the interface and, thus, barrier height and tunneling probability come to play. It is an indication that if the photoinduced charge carrier transfer is impeded by the build-in electric field, the device performance will be poor.

In respect to photocatalytic water splitting into H_2 and O_2, metal-free graphitic carbon nitride (g-C_3N_4) has triggered intensive scientific interest because of its suitable band gap (2.7 eV), nontoxicity, high stability, and easy preparation [113–128]. It is proven however to suffer from fast electron–hole recombination, showing undesirable photocatalytic efficiency. In order to achieve high light-harvesting and light–energy conversion efficiency, g-C_3N_4 is usually combined with other materials to inhibit the recombination of photogenerated electron–hole pairs, such as CdS, $BiVO_4$, SnS_2, C_{60}, MoS_2, and InSe. In these heterostructures, build-in electric field due to charge redistribution and/or potential drop in the direction perpendicular to the interface shows influence on the photoexcited charge carrier separation. In the case of g-C_3N_4/InSe vdW heterobilayers, generated build-in electric field points from g-C_3N_4 to InSe. In contrast to the weak photocurrent I_{ph} in isolated g-C_3N_4, remarkable I_{ph} can be detected in g-C_3N_4/InSe vdW heterobilayers under visible light, maximizing 2.9 $\mu A\,mm^{-2}$, which is 2 orders of magnitude higher than g-C_3N_4/TiO_2 nanocomposites ($3.42 \times 10^{-2}\ \mu A\,mm^{-2}$) and 3 orders of magnitude higher than g-C_3N_4/carbon systems ($1.7 \times 10^{-3}\ \mu A\,mm^{-2}$) [129].

It is indicative that heterostructures with type-II band alignment and build-in electric field enhance the light–electricity conversion efficiency. In this view, thus, by selecting appropriate cocatalysts to improve the quantum yield is an appealing strategy for design light–energy devices. In general, the generated photocurrent can be a reference for target cocatalysts.

In addition, rectangular TiN monolayers are theoretically predicted to show enhanced auxeticity and ferroelasticity, and their metallic features can be modulated to semi-conductive via halogenation. In particular, halogenated TiNX (X = F, Cl, Br, I) monolayers present moderate direct band gap (1.2–1.6 eV), strong light absorption, small effective mass, and small exciton binding energy, hiding the potential in photovoltaic applications. In a recent publication, the photoresponsivities for TiNF, TiNCl, and TiNBr are calculated to be 0.31, 0.12, and 0.22 AW^{-1}, respectively [130], about 1 order of magnitude larger than phosphorene (0.068 AW^{-1}) and MoS_2 (0.016 AW^{-1}), manifesting themselves advanced in solar cells applications. In aspects of photocatalytic and photovoltaic applications, fast charge carrier separation plays a central role in the device performance. As a result, charge carrier mobility m^* and exciton binding energy E_b are usually calculated to evaluate the application potential of materials. In accordance to the following equation, the effective masses of electrons at CBM (m_e*) and holes at VBM (m_h*) can be calculated:

$$m^* = \hbar^2/(\partial^2 E_k/\partial k^2)$$

with k and E_k being the wave vector and the energy corresponding to the wave vector k, respectively. Then, the charge carrier mobility for two-dimensional materials μ_{2D} could be evaluated by [131–133]:

$$\mu_{2D} = \frac{2e\hbar^3 C}{3k_B T|m^*|^2 E_d^2}$$

where e is the electron charge, and k_B, T, and m^* are the Boltzmann constant, temperature, and carrier effective mass, respectively. In this equation,

$$C = [\partial^2 E/\partial \varepsilon^2]/S_0$$

is the elastic constant of the two-dimensional systems with E and S_0 representing the total energy and surface area of the equilibrium system, respectively, and E_d is the deformation potential constant describing the strain-induced band edge shift. In line with the hydrogenic model, exciton binding energy can be estimated:

$$E_b = \frac{\mu}{m_0 \varepsilon_r^2} R_H$$

here μ, m_0, and R_H are the exciton effective reduced mass, free electron mass, and Rydberg constant of a hydrogen atom (13.6 eV), respectively. In particular, the macroscopic static dielectric tensor ε_r equals the sum of electron contribution ε_∞ and ionic contribution ε_{vib}. It should be kept in mind that the hydrogenic model is a convenient approximation, and exact exciton binding energy in low-dimensional materials can be quantitatively obtained from *GW*+BSE approach. In regard to TiNX monolayers, charge carrier mobility is relatively high. In addition, the exciton binding energies for TiNF, TiNCl, and TiNBr are as small as 42, 19, and 17 meV, respectively, which are significantly smaller than MoS_2 (280 meV), g-C_3N_4 (728 meV), and phosphorene (78 meV) obtained at the same level of theory [130]. It is thus an indication that photogenerated excitons in TiNX monolayers are easy to dissociate into free charge carriers.

On the basis of quantum transport simulations, photocatalytic and photovoltaic device performance can be evaluated, providing insights into the design principle for light–energy and light–electricity conversion systems. As for excitonic solar cells, light absorption and charge carrier transfer being the main physical processes are dominated by the donor materials [134, 135]. It concludes that the promising donor should have suitable band gap (1.2–1.6 eV) and small exciton binding energy, or, in other words, strong light absorption and reduced charge carrier recombination. In this regard, group-IV monochalcogenides MX (M = Ge, Sn; X = S, Se) show potential as photovoltaic donor materials. It is similar to phosphorene; MX monolayers belong to *P*nma space group, with each X (M) atom connecting to two M (X) atoms. Under $0.1\,\mathrm{W\,cm^{-2}}$ power density illumination, MX monolayers show large photocurrent. As for GeS and SnS, the photocurrent along armchair direction maximizes 3.0 and $7.0\,\mathrm{mA\,mm^{-2}}$, respectively. In cases of GeSe and SnSe, the induced photocurrent can be 12.0 and $14.0\,\mathrm{mA\,mm^{-2}}$, respectively [131]. In respect to the photoresponsivities, 0.035, 0.075, 0.13, and $0.16\,\mathrm{A\,W^{-1}}$ are obtained for GeS, SnS, GeSe, and SnSe at the wavelength of 400–500 nm, respectively. In comparison to MoS_2 ($1.6\times10^{-2}\,\mathrm{A\,W^{-1}}$) and graphene ($5\times10^{-2}\,\mathrm{A\,W^{-1}}$), photoresponsivities in MX are strikingly large, and the EQE for MX correspondingly ranges from 10.27 to 30.32%.

In current stage, silicon-based solar cells show weak light absorption in the solar spectrum due to the indirect band gap of silicon (1.1 eV), and relatively low carrier mobility ($1400\,\mathrm{cm^2\,V^{-1}\,s^{-1}}$) also limits the efficient charge extraction. Thus, one of the most urgent tasks in photovoltaic market is searching for new materials to realize high photocurrent and PCE. In general, the quality of photovoltaic devices can be estimated by the maximum PCE [130, 136, 137]:

$$\eta = \frac{\beta_{\mathrm{FF}} V_{\mathrm{oc}} J_{\mathrm{sc}}}{P_{\mathrm{solar}}} = \frac{\beta_{\mathrm{FF}}(E_{\mathrm{g}} - \Delta E_{\mathrm{c}} - 0.3)\int_{E_{\mathrm{g}}}^{\infty} \frac{P(\hbar\omega)}{\hbar\omega} d(\hbar\omega)}{\int_0^{\infty} P(\hbar\omega) d(\hbar\omega)}$$

here $\beta_{\mathrm{FF}} = 0.65$ is the band-fill factor, E_{g} is the donor band gap, ΔE_{c} is the conduction band offset between donor and acceptor, and $(E_{\mathrm{g}} - \Delta E_{\mathrm{c}} - 0.3)$ is an estimation of the maximum open circuit voltage V_{oc} in eV. In addition, $P(\hbar\omega)$ stands for the AM1.5 solar energy flux (in $\mathrm{W\,m^{-2}\,eV^{-1}}$) at the photon energy $\hbar\omega$. In the dominator, the integral in numerator represents the short circuit current J_{sc} performed applying the limit EQE of 100%, and the integral in denominator denotes the incident solar irradiation P_{solar} ($10^3\,\mathrm{W\,m^{-2}}$).

In case of combining different TiNX monolayers, type-II donor–acceptor interfaces for excitonic solar cells with ultrahigh photovoltaic power energy conversion (PEC) form. In particular, the PEC is 18% for TiNF/TiNBr, 19% for TiNCl/TiNBr, and 22% for TiNF/TiNCl, which are far superior to the conventional polymer and fabricated two-dimensional solar cells. In fact, excitonic solar cells based on two-dimensional materials such as MoS_2, WS_2, graphene, h-BN, SiC_2, bilayer phosphorene, and organic–inorganic hybrid perovskites are potentially seen as the next generation of thin film solar cells. In respect to phosphorene that possesses a band gap in the visible region, potential application in the thin film excitonic solar cells can be expected. In a recent work, vdW heterostructures composed of phosphorene and several TMDCs were modeled and the viability as excitonic solar cells was evaluated. In particular, phosphorene could be the donor when paired with octahedral TMDCs (TiS_2 and ZrS_2) and the acceptor when paired with trigonal prismatic TMDCs (MX_2, where M = Mo, W; X = S, Se, Te). In these vdW heterostructures, ZrS_2/phosphorene and $MoTe_2$/phosphorene have the highest PCE of 12%, which is comparable to the theoretically proposed bilayer-phosphorene/MoS_2, PCBM/CBN, and g-SiC_2/GaN and higher than that achieved by existing excitonic solar cells (less than 9%) [137]. In addition, the PEC can be improved to 20% by 2% compressive strain applied on phosphorene, because strain tunes the CBM position of phosphorene.

In principle, efficient donor material for excitonic solar cells of high PEC should have both suitable direct band gap and high carrier mobility. In comparison to experiments, theoretical simulations can predict promising candidates with higher efficiency. As an example, among the five phases of arsenic–phosphorous (α, β, γ, δ, ε), α-AsP shows a direct band gap of 1.54 eV and high carrier mobility of 14380 $cm^2\ V^{-1}\ s^{-1}$, making itself a promising choice as a donor for excitonic solar cells. In case of employing GaN monolayer as the electron acceptor, PCE of the designed α-AsP/GaN type-II heterobilayer excitonic solar cells can be as high as 22.1%, which is competitive with the reported PCE to data for two-dimensional thin film solar cells [136].

In other fields, simulations based on NEGF–SCF theory make successes, proving itself a powerful tool in studies of electronics, spintronics, and optoelectronics. In triangulene-based spin-photovoltaic devices, for example, quantum transport calculations reveal that pure spin current can be generated without charge current. It is similar to a fragment of graphene; triangulene is made up of six hexagons of carbon jointed along their edges to form a triangle. As the triangulene molecule is irradiated with photon energy ranging from 1.8 to 2.5 eV, pure spin current can be observed. In case of applying a gate voltage of −1 V, the spin current can up to 59.3 $nA\ mm^{-2}$ [138]. In addition, the vanishing charge current suggests Joule heat free and low energy consuming for optoelectronic devices. In application of field effect transistors (FETs), NEGF–SCF method also succeeds [139, 140]. In particular, MoS_2 has been widely investigated as a channel material in FETs, and, however, high Schottky barrier and Fermi level pinning effects hamper the performance of MoS_2-based FETs. It is of interest that MXenes and their modified derivatives, Ta_2C, Ta_2CF_2, and $Ta_2C(OH)_2$, are found to be suitable electrodes for the MoS_2 channel with F and OH atomic layers playing a role as buffer layer to tune the interlayer coupling. In respect to the heterostructures composed of MoS_2 and these MXenes, Ohmic contacts form with vanishing Schottky barrier height. In a modeled FET device, however, Schottky barrier of 0.2 eV appears exclusively due to the lateral interface effects [133]. It thus sheds some light into the development of appropriate electrode materials and correct evaluation of the performance of FET devices.

7.3 Simulation for Localized Surface Plasmon Resonance

In recent years, people have seen the great progress in direct conversion of solar light to chemical energy using photocatalysts. It is an important approach using the light-trapping properties of plasmonic metal nanoparticles (NPs) (e.g. Ag and

Au) to improve the light–energy conversion efficiency in optoelectronics, photovoltaics, and photocatalyst. In the light of the coupling with the incident light at the frequency of local surface plasmon resonance (LSPR), metal NPs trigger the redistribution, localization, and enhancement of the electromagnetic field, thus the light absorption of semiconductor can be improved via both near-field electromagnetic field enhancement and the increase of optical path length by light scattering of the metal NPs. In regard to the plasmonic photocatalysts, both experiment and theoretical works are available, from where one can find more details [141–160].

In Figure 7.8, plasmon resonance absorption of metal NPs loaded on photocatalyst surface is simply shown. In case of light radiation, the electric field will displace the conduction electrons relative to the nuclei, inducing a large electric dipole; at the same time, a restoring force arises because of the Coulomb attraction between electrons and nuclei, which leads to a resonant oscillation of the conduction electrons at a certain frequency. In general, LSPR occurs when the plasmonic NPs are considerably smaller than the wavelength of the incident light, and the resultant electric dipole can create an intense electric field near the plasmonic NPs, with the magnitude ten to thousand times greater than that of incident light. It has been widely demonstrated that both the LSPR wavelength and absorption efficiency can be tuned by adjusting the sizes, shapes, compositions, and dielectric environments of plasmonic NPs and the corresponding intensity and distribution range of the near-field change. It should be pointed out that LSPR leads to energy transfer from NPs to semiconductors, providing versatile means for tuning the light–energy conversion efficiency. In principle, two competitive ways are responsible for the dephasing of the resonant oscillation, i.e. radiative emission of photons and nonradiative relaxation through electron–electron, electron–phonon,

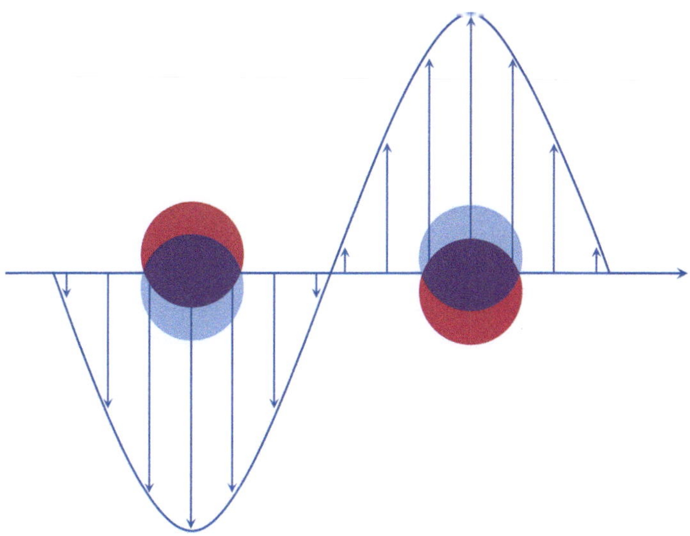

Figure 7.8 Schematic of plasmon resonance absorption of metal nanoparticles. Source: Wei Wei.

electron–surface, and electron–adsorbate scattering. In case of having larger energy than the Schottky barrier at the interface between the metal and semiconductor, energetic hot electrons migrate to the adjacent semiconductor, and hot electron injection mechanism is very similar to that of dye-sensitized solar cells [161–163].

In order to model the absorption and scattering, solving the Maxwell's equations for the target system is an effective method. It has already been known that, nevertheless, exact solutions to Maxwell's equations are only for special geometries such as spheres, spheroids, or infinite cylinder, so approximation methods are in general needed. In recent publications, the discrete dipole approximation (DDA) has been illustrated to be one of the powerful methods developed for calculating the absorption and electromagnetic scattering of isolated particles with complex geometry [164–170]. In particular, the DDA can treat inhomogeneous targets and anisotropic materials and has been extended to treat targets near substrates. On the basis of DDA, then, near-field calculations of light field intensity inside and in the vicinity of a scattering particle can be performed. In DDA, conceptually, the target of interest is approximated by an array of polarizable points. Once the polarizability tensors α_j are specified, Maxwell's equations can be solved accurately for the dipole array.

In case of illumination by monochromatic incident plane wave,

$$\boldsymbol{E}_{\mathrm{inc}}(\mathbf{r}, t) = \boldsymbol{E}_0 \exp(i\boldsymbol{k}_0 \cdot \boldsymbol{r} - i\omega t)$$

the polarization $\boldsymbol{P}_j$ of the dipoles in the target will oscillate coherently. In this situation, each dipole i will be affected by the incident wave together with the electric field at location i generated by all of the other point dipoles:

$$\boldsymbol{E}_i = \boldsymbol{E}_{\mathrm{inc},i} - \sum_{j \neq i} \tilde{\mathbf{A}}_{ij} \boldsymbol{P}_j$$

here, the vector of polarizations $\boldsymbol{P}_j$ must satisfy the system of equations:

$$\boldsymbol{P}_i = \alpha_i \left[\boldsymbol{E}_{\mathrm{inc}}(\boldsymbol{r}_i) - \sum_{j \neq i} \tilde{\mathbf{A}}_{ij} \boldsymbol{P}_j \right]$$

In case of N dipoles, this equation corresponds to a system of $3N$ linear equations. In consideration of $\mathbf{A}_{ij} \equiv \tilde{\mathbf{A}}_{ij} + \alpha_i^{-1} \delta_{ij}$, the incident plane wave induced electric field can be rewritten as:

$$\boldsymbol{E}_{\mathrm{inc},i} = \sum_{j \in \mathrm{target}} \mathbf{A}_{ij} \boldsymbol{P}_j$$

As $\boldsymbol{P}_j$ are obtained, the electric field located at i can be calculated as follows:

$$\boldsymbol{E}_i = \boldsymbol{E}_{\mathrm{inc},i} - \sum_{j \in \mathrm{target}} \tilde{\mathbf{A}}_{ij} \boldsymbol{P}_j$$

In this equation, the interaction matrix $\tilde{\mathbf{A}}_{ij}$ is spatially invariant and depends only on the displacement between i and j. In many applications, one is often interested in the near field outside the target, which can be directly evaluated from the last equation [165–170].

In practice, the LSPR of noble metal NPs has long been the subject of importance in surface-enhanced Raman spectroscopy, photovoltaics, and biosensing [171–174].

In particular, the incident light induced collective oscillation of conduction electrons in metal NPs arises largely from the dipole plasmon resonance (DPR) and a small contribution is from a higher-energy mode such as the quadrupole plasmon resonance (QDR). It has been demonstrated that the metal NPs LSPR frequencies can be easily tuned to cover the whole range of the visible light spectrum by controlling the size/shape and local dielectric environment [175–180], and the LSPR is characterized in high light absorption and strong electric field enhancement [181–189]. In general, energy transfer from metal NPs to semiconductors generating electrons and holes takes place in two possible ways: (i) direct electron transfer and (ii) via the electric field enhancement near the semiconductor surface.

In this section, Ag/AgCl system should be highlighted, which is the mostly studied system of the plasmonic metal/semiconductor photocatalysts from both experiments and simulations. In this system, the wide range of visible light absorption is ascribed to the plasmon absorption of Ag NPs, while the energy transfer is mediated by the mid-gap defect states in AgCl. It has been revealed that for spherical Ag NPs the LSPR wavelength (between 350 and 450 nm) is nearly independent of the sphere radius [152]. In another case, embedding of Ag NPs into AgCl lattice can however extend the visible light adsorption range. In case of Ag nanocubes of different sizes, high LSPR modes, i.e. QDR, are excited. In essence, LSPR absorption covers the range between 400 and 650 nm, and, notably, a several of absorption peaks of low wavelength appear. As the size of the nanocubes increases (up to such as 60 nm), absorption efficiency drops dramatically due to the resonant scattering. It therefore concludes from simulation results that Ag nanocubes of large size are not in favor of intense LSPR absorption. It has been argued that the LSPR absorption of Ag and Au nanorods can extend to visible (longer than 650 nm) and even infrared region by tuning the aspect ratio. As the aspect ratio increases from 1.5 to 4 for Ag nanorods on AgCl, the magnitude of absorption peak decreases gradually and the LSPR wavelength red shifts slightly for transversal LSPR mode. In general, the transversal LSPR is not very sensitive to the aspect ratio and the wavelength is nearly independent of the width of the nanorods. In respect to Au nanorods on TiO_2, simulation results are in agreement with experimental observations. In contrast, the longitudinal LSPR absorption depends strongly on the aspect ratio. As the aspect ratio increases, the LSPR wavelength of Ag nanorods substantially red shifts, ranging from 450 to 900 nm, and the peak magnitude increases nearly exponentially until the aspect ratio reaches 3.5, keeping constant when the aspect ratio increases further. It is of importance that the LSPR absorption efficiency for Ag nanorods is almost an order of magnitude greater than those of Ag nanospheres and nanocubes. It means that the dominating Ag NPs in experiments should be nanorods.

In the absence of LSPR, electric field effect on a semiconductor can be approximated by a periodic perturbation with the frequency as that of the incident electromagnetic wave. As the simplest example of the incident electromagnetic wave (with magnetic field neglected), the plane wave propagates in the z direction, the electric field (polarized in x direction) can be described as [153]:

$$\boldsymbol{E}(z,t) = E_0\hat{x}\cos(\omega z/c - \omega t)$$

where $\hat{x}$ is the unit vector along x direction, and E_0 and ω are the amplitude and angular frequency of the incident light electric field, respectively. In this case, an electron is perturbed by:

$$H' = e\boldsymbol{r} \cdot \boldsymbol{E}(z, t)$$

with $\boldsymbol{r}(x, y, z)$ being the coordinate of the electron, and the center of the atom bounding the electron is regarded as the origin of the coordinate system. As the atomic radius (10^{-10} m) is far smaller than the wavelength of the incident visible light (10^{-7} m), H' can be approximated as:

$$H' = eE_0 x \cos \omega t$$

that is, the well-known electric dipole approximation. In consequence, the transition probability per second from the occupied state to the empty state is given by:

$$w^0_{\mathrm{k}\to\mathrm{m}} = (2\pi e^2/\hbar)E_0{}^2 |x_{\mathrm{mk}}|^2 \delta(E_\mathrm{m} - E_\mathrm{k} - \hbar\omega)$$

where

$$x_{\mathrm{mk}} = \int \varphi^*_\mathrm{m}(\boldsymbol{r}) x \varphi_\mathrm{k}(\boldsymbol{r}) d\tau$$

and E_k and E_m are the eigenenergies of φ_k and φ_m states, respectively. In a same way, the effect of an LSPR induced electric field E_0^{LSPR} can also be described by using the electric dipole approximation. In regard to electrons of the semiconductor in the regions covered by the E_0^{LSPR}, the transition probability can be expressed as:

$$w^{\mathrm{LSPR}}_{\mathrm{k}\to\mathrm{m}} = (2\pi e^2/\hbar)\left(E_0^{\mathrm{LSPR}}(x, y, z)\right)^2 |x_{\mathrm{mk}}|^2 \delta(E_\mathrm{m} - E_\mathrm{k} - \hbar\omega)$$

In general,

$$\left(E_0^{\mathrm{LSPR}}(x, y, z)\right)^2 / E_0{}^2 = 10 - 1000$$

leading to $w^{\mathrm{LSPR}}_{\mathrm{k}\to\mathrm{m}}$ greater than $w^0_{\mathrm{k}\to\mathrm{m}}$ by the same factor. It also operates for the non-polarized natural visible light, just by replacing $|x_{\mathrm{mk}}|^2$ with

$$\left(|x_{\mathrm{mk}}|^2 + |y_{\mathrm{mk}}|^2 + |z_{\mathrm{mk}}|^2\right)/3$$

where

$$y_{\mathrm{mk}} = \int \varphi^*_\mathrm{m}(\boldsymbol{r}) y \varphi_\mathrm{k}(\boldsymbol{r}) d\tau$$

and

$$z_{\mathrm{mk}} = \int \varphi^*_\mathrm{m}(\boldsymbol{r}) z \varphi_\mathrm{k}(\boldsymbol{r}) d\tau$$

In the case of Ag/AgCl plasmonic photocatalytic models, simulation results confirm the significant LSPR induced electric field enhancement. In addition, the obvious dependence of the electric field enhancement on the size/shape of metal NPs can be expected. In particular, when increasing the size of Ag nanocubes, the electric field enhancement in AgCl increases in range and intensity by hundreds or even thousands of times.

In metal/semiconductor plasmonic photocatalysts, light absorption efficiency of the semiconductor can be strongly enhanced due to the LSPR of metal NPs. In addition, a new paradigm in light harvesting can be established, because the LSPR absorption of metal NPs can be manipulated. It has been widely accepted that hot electrons generated by the decay of the LSPR of metal NPs could be injected into the conduction band of the semiconductor, leaving holes in metal NPs. In general, the hot electron injection efficiency (HEIE) is low due to the charge transfer induced Schottky barrier, to large extent impairing the overall photocatalytic performance. It should be pointed out that conflict exists, for instance, in Au/TiO_2 composites [153]. In accordance to the Schottky model, the Schottky barrier between n-type TiO_2 and Au NPs is estimated to be 0.6 eV, significantly lower than the energy of hot electrons of 2 eV above the Fermi level. It is therefore reasonable that the relatively small Schottky barrier cannot account for the low HEIE. In a recent article, new insights were proposed to explain the low HEIE in Au/TiO_2 model plasmonic photocatalytic system. In such a model, Au atoms are assumed to occupy the surface oxygen vacancies. On one hand, electrons donated by bulk oxygen vacancies will populate TiO_2 surface to keep the Fermi level uniform, creating negatively charged surface layer. On the other hand, electrons from host conduction band will be trapped in unoccupied Au–Ti antibonding states, further increasing the negatively charged surface layer. In turn, the negatively charged surface layer can decrease HEIE because of the increased hot electron transfer barrier and extended space charge region. In combination with the upward band bending, large transfer barrier for hot electron and large space charge region appear, which can be used to explain the low HEIE. As a result, experimentally the surface oxygen vacancies should be strongly avoided.

In the determination of HEIE, the hot electron generation efficiency (HEGE), which is associated with the LSPR absorption efficiency and near-field enhancement, also plays an important role. It should be kept in mind that the LSPR absorption efficiency and near-field enhancement show dependence on the shape and size of metal NPs. In line with this, adjusting the morphology of metal NPs can improve the photocatalytic performance. In this view, simulations by solving the Maxwell's electrodynamics equations based on the DDA can provide guidelines for experiments. In respect to Au/TiO_2 system, incident light perpendicular to the metal/semiconductor interface cannot move the generated hot electrons toward the semiconductor due to the momentum conservation. If the size of metal NPs is considerably smaller than that of the semiconductor, the light absorption efficiency and near-field enhancement will be weak, not beneficial to the generation of a large number of hot electrons. It is therefore an indication that the sizes of metal NPs and semiconductor should be comparable, then more hot electrons can be generated. In addition, the intensity and distribution region of near field increase as the incident light varies from z to x direction, thus the near field enhanced energy transfer is increased. In Figure 7.9, Au/TiO_2 model and the near-field enhancement effect are shown.

In optoelectronics and photonics applications, the inherent monolayer thickness of two-dimensional materials poses a significant challenge for the interaction of

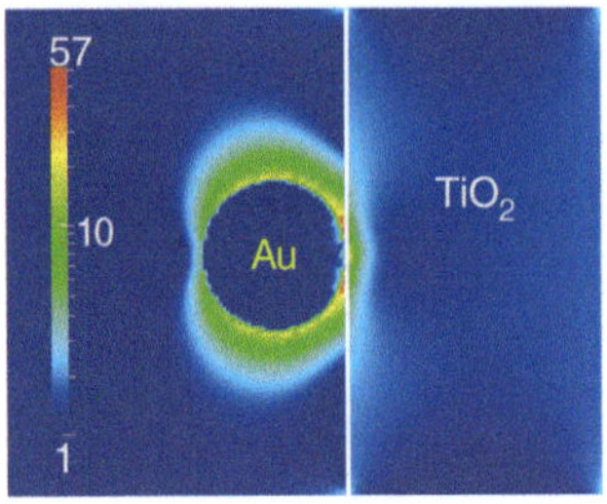

Figure 7.9 Near-field enhancement of Au@TiO_2 for incident light with the intersection angle of 0°. Source: Ma et al. [153]/Reproduced with permission of American Chemical Society.

light with the material, which therefore leads to poor light emission and absorption behavior, which is also applicable for two-dimensional TMDCs. It is fortunate to see that the light absorption can greatly benefit from the resonant electric field enhancement. In particular, plasmonic nanostructures are quite promising for boosting light absorption/emission as they enable very high electric field confinement and are easy to integrate with the merging two-dimensional TDMCs. In previous works, facilitated strong light–matter interactions by using plasmonic materials were confirmed in MoS_2. It has been illustrated that, for example, the surface plasmon enhanced photocurrent and photoluminescence can be realized in Au and core–shell NPs. In a recent work, the enhanced photoluminescence in large area (cm-sized) as-grown MoS_2 by chemical vapor deposition with designed plasmonic Ag NPs arrays was observed. In particular, as Ag nanodisc arrays with varying diameter sizes are fabricated onto monolayer MoS_2 film, spectral photoluminescence of plasmonic array/MoS_2 region is acquired. It shows a 12 times enhanced photoluminescence emission, which can be ascribed to both the excitation field enhancement at the pump wavelength and the efficient scattering at photoluminescence emission wavelengths [190]. It is therefore an indication that efficient light coupling to low-dimensional materials at nanoscale can break new ground in highly efficient optoelectronic devices.

It is confirmative that plasmonic materials provide a promising approach for light–energy conversion applications, due to the enhanced light–matter interactions, additional charge traps, and efficient charge transfer pathways. In a recent work, the importance of the intimate interface between plasmonic nanostructure semiconductor was demonstrated and emphasized based on the Au/MoS_2 model system [191]. As a prototype of photo-sensing devices, the Au/MoS_2 heterostructure constitutes Au NPs core that is encapsulated by a chemical vapor deposition-grown multilayer MoS_2 shell (Figure 7.10). In particular, the intimate and direct interfacing of Au and MoS_2 can be perfectly realized, which is important to the concept of plasmon-induced interfacial charge transfer transition, i.e. enabling the plasmon decay by directly exciting an electron into a strongly coupled semiconductor acceptor. It can be expected that such as Au/MoS_2 plasmonic system will show significantly enhanced visible light absorption due to the presence of LSPR, due to the enhanced local electric field induced by the LSPR of Au NPs. It can be found that the normalized electric field is mainly dispersed within the MoS_2 shell, with the field strength significantly enhanced compared to the bare shell. It is of paramount importance that the Au/MoS_2 model involves two types of plasmonic

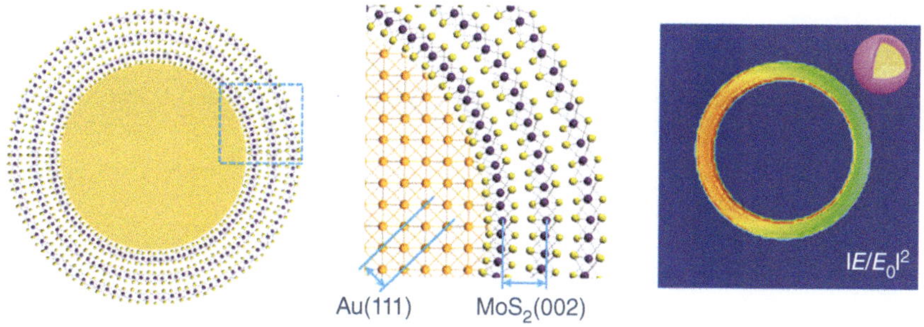

Figure 7.10 Schematic showing the Au core–multilayer MoS_2 shell structure, and atomic connection of the core–shell as highlighted. DDA-simulated electric field distribution on an individual Au@MoS_2 core–shell (as shown in the inset), and schematic demonstrating the possible photocarrier generation pathways in the Au@MoS_2 heterostructures. Source: Li et al. [191]/with permission of American Chemical Society.

photodetectors, i.e. an Au/MoS_2 field effect phototransistor and a Si-supported Au/MoS_2 *p–n* junction photodiode. As a phototransistor, Au/MoS_2 shows a photoresponsivity of 10 times higher than that of planar MoS_2 transistors. As a photodiode, Au/MoS_2 shows superior photoresponse and recovery ability with the photoresponsivity as high as 22.3 A W^{-1}, which is beyond the most distinguished values of previously reported similar gateless photodetectors.

It is known that excitons in two-dimensional TMDCs are highly confined in the in-plane direction, thus it is convenient to make the orientation of excitons align with the one of surface plasmons for effective dipole–dipole interactions. As a consequence, hybrid systems of two-dimensional TMDCs and plasmonic NPs could be ideal platforms for Fano resonances and plasmon–exciton coupling. In plasmonic systems incorporating dye molecules and quantum dots, both Fano resonances and Rabi splitting have been observed. As a result of the lack of efficient ways to tune the excitonic properties of dye molecules and quantum dots, two-dimensional TMDCs provide new choices and present superiorities. On the basis of the model system composing of single Au nanotriangles and monolayer WS_2, tunable Fano resonances and plasmon–exciton coupling in the monolayer was previously demonstrated [192]. In the light of tuning the exciton binding energy or the LSPR strength through the dielectric constant of surrounding solvents or the dimension of Au NPs, Fano resonances can be controlled. In addition, a transition from weak to strong plasmon–exciton coupling has been achieved. It appears that the strong field localization of the Au NPs and large transition dipole moment of the WS_2 exciton account for the large coupling strength of 50–170 meV occurring at room temperature.

It should be pointed out that open questions are still there, fundamental mechanism studies are required, and the effects of shape/size of these metal NPs on the LSPR absorption and the energy transfer process are still missing. It will provide insights into how to improve the device performance by unraveling these issues. In addition, photocarrier generation mechanisms could also be estimated.

References

1 Barber, J. (2009). Photosynthetic energy conversion: natural and artificial. *Chem. Soc. Rev.* 38: 185.

2 Lewis, N.S. (2007). Toward cost-effective solar energy use. *Science* 315: 798.

3 Gratzel, M. (2001). Photoelectrochemical cells. *Nature* 414: 338.

4 Wang, Z., Li, C., and Domen, K. (2019). Recent developments in heterogeneous photocatalysts for solar-driven overall water splitting. *Chem. Soc. Rev.* 48: 2109.

5 Maeda, K. and Domen, K. (2010). Photocatalytic water splitting: recent progress and future challenges. *J. Phys. Chem. Lett.* 1: 2655–2661.

6 Qu, Y. and Duan, X. (2013). Progress, challenge and perspective of heterogeneous photocatalysts. *Chem. Soc. Rev.* 42: 2568–2580.

7 Wang, Z., Liu, Y., Huang, B. et al. (2014). Progress on extending the light absorption spectra of photocatalysts. *Phys. Chem. Chem. Phys.* 16: 2758–2774.

8 Liu, G., Zhen, C., Kang, Y. et al. (2018). Unique physicochemical properties of two-dimensional light absorbers facilitating photocatalysis. *Chem. Soc. Rev.* 47: 6410–6444.

9 Wang, Q., Hisatomi, T., Jia, Q. et al. (2016). Scalable water splitting on particulate photocatalyst sheets with a solar-to-hydrogen energy conversion efficiency exceeding 1%. *Nat. Mater.* 15: 611–615.

10 Li, C., Cao, Q., Wang, F. et al. (2018). Engineering graphene and TMDs based van der Waals heterostructures for photovoltaic and photoelectrochemical solar energy conversion. *Chem. Soc. Rev.* 47: 4981–5037.

11 Lin, Q., Armin, A., Nagiri, R.C. et al. (2015). Electro-optics of perovskite solar cells. *Nat. Photonics* 9: 106–112.

12 Zhou, H., Chen, Q., Li, G. et al. (2014). Interface engineering of highly efficient perovskite solar cells. *Science* 345: 542–546.

13 Yang, W.S., Park, B.W., Jung, E.H. et al. (2017). Iodide management in formamidinium-lead-halide–based perovskite layers for efficient solar cells. *Science* 356: 1376–1379.

14 Dou, L., Andrew, B.W., Yu, Y. et al. (2015). Atomically thin two-dimensional organic-inorganic hybrid perovskites. *Science* 349: 1518–1521.

15 Wu, Y., Wang, P., Zhu, X. et al. (2018). Composite of $CH_3NH_3PbI_3$ with reduced graphene oxide as a highly efficient and stable visible-light photocatalyst for hydrogen evolution in aqueous HI solution. *Adv. Mater.* 30: 1704342.

16 Lv, X., Wei, W., Sun, Q. et al. (2017). Two-dimensional germanium monochalcogenides for photocatalytic water splitting with high carrier mobility. *Appl. Catal. B Environ.* 217: 275–284.

17 An, Y., Liu, Y., An, P. et al. (2017). Ni^{II} Coordination to Al-based metal-organic framework made from 2-aminoterephthalate for photocatalytic overall water splitting. *Angew. Chem. Int. Ed.* 56: 3036–3040.

18 Li, M., Dai, Y., Ma, X. et al. (2017). Insights into the effect of inner polarization and multiple Ag-O units on high-efficient Ag-based photocatalyst. *Appl. Catal. B Environ.* 205: 211–218.

19 Wang, G., Huang, B., Ma, X. et al. (2013). First near-infrared-light-driven photocatalyst $Cu_2(OH)PO_4$. *Angew. Chem. Int. Edit.* 52: 4810–4813.

20 Zhao, P., Ma, Y., Lv, X. et al. (2018). Two-dimensional III2-VI3 materials: promising photocatalysts for overall water splitting under infrared light spectrum. *Nano Energy* 51: 533–538.

21 Peng, R., Ma, Y., Huang, B., and Dai, Y. (2019). Two-dimensional Janus PtSSe for photocatalytic water splitting under the visible or infrared light. *J. Mater. Chem. A* 7: 603–610.

22 Li, X., Dai, Y., Ma, Y. et al. (2017). Landscape of DNA-like inorganic metal-free double helical semiconductors and potential applications in photocatalytic water splitting. *J. Mater. Chem. A* 5: 8484–8492.

23 Ma, X., Wu, X., Wang, H., and Wang, Y. (2018). A Janus MoSSe monolayer: a potential wide solarspectrum water-splitting photocatalyst with a low carrier recombination rate. *J. Mater. Chem. A* 6: 2295–2301.

24 Hisatomi, T., Kubota, J., and Domen, K. (2014). Recent advances in semiconductors for photocatalytic and photoelectrochemical water splitting. *Chem. Soc. Rev.* 43: 7520–7535.

25 Pinaud, B.A., Benck, J.D., Seitz, L.C. et al. (2013). Technical and economic feasibility of centralized facilities for solar hydrogen production via photocatalysis and photoelectrochemistry. *Energy Environ. Sci.* 6: 1983.

26 Fabian, D.M., Hu, S., Singh, N. et al. (2015). Particle suspension reactors and materials for solar-driven water splitting. *Energy Environ. Sci.* 8: 2825.

27 Goto, Y., Hisatomi, T., Wang, Q. et al. (2018). A particulate photocatalyst water-splitting panel for large-scale solar hydrogen generation. *Joule* 2: 509.

28 Wang, Q., Hisatomi, T., Jia, Q. et al. (2016). Scalable water splitting on particulate photocatalyst sheets with a solar-to-hydrogen energy conversion efficiency exceeding 1%. *Nat. Mater.* 15: 611.

29 Chen, S., Takata, T., and Domen, K. (2017). Particulate photocatalysts for overall water splitting. *Nat. Rev. Mater.* 2: 17050.

30 Kudo, A. and Miseki, Y. (2009). Heterogeneous photocatalyst materials for water splitting. *Chem. Soc. Rev.* 38: 253.

31 Zhou, P., Yu, J., and Jaroniec, M. (2014). All-solid-state Z-scheme photocatalytic systems. *Adv. Mater.* 26: 4920–4935.

32 Teixeira, I.F., Eduardo, C.M., Tsang, S.C., and Camargo, P.H. (2018). Carbon nitrides and metal nanoparticles: from controlled synthesis to design principles for improved photocatalysis. *Chem. Soc. Rev.* 47: 7783–7817.

33 Walter, M.G., Warren, E.L., McKone, J.R. et al. (2010). Solar water splitting cells. *Chem. Rev.* 110: 6446–6473.

34 Kumar, B., Llorente, M., Froehlich, J. et al. (2012). Photochemical and photoelectrochemical reduction of CO_2. *Annu. Rev. Phys. Chem.* 63: 541–569.

35 White, J.L., Baruch, M.F., Pander Iii, J.E. et al. (2015). Light-driven heterogeneous reduction of carbon dioxide: photocatalysts and photoelectrodes. *Chem. Rev.* 115: 12888–12935.

36 Sivula, K. (2015). Toward economically feasible direct solar-to-fuel energy conversion. *J. Phys. Chem. Lett.* 6: 975–976.

37 Rahman, M.Z., Kwong, C.W., Daveya, K., and Qiao, Z.S. (2016). 2D phosphorene as a water splitting photocatalyst: fundamentals to applications. *Energy Environ. Sci.* 9: 709–728.
38 Pang, J., Mendes, R.G., Alicja, B. et al. (2019). Applications of 2D MXenes in energy conversion and storage systems. *Chem. Soc. Rev.* 48: 72–133.
39 Zhang, X., Zhang, Z., Wu, D. et al. (2018). Computational screening of 2D materials and rational design of heterojunctions for water splitting photocatalysts. *Small Methods* 2: 1700359.
40 Singh, A.K., Mathew, K., Zhuang, H.L., and Hennig, R.G. (2015). Computational screening of 2D materials for photocatalysis. *J. Phys. Chem. Lett.* 6: 1087–1098.
41 Hu, W., Lin, L., Zhang, R. et al. (2017). Highly efficient photocatalytic water splitting over edge-modified phosphorene nanoribbons. *J. Am. Chem. Soc.* 139: 15429–15436.
42 Sun, J., Li, X., and Yang, J. (2018). The roles of buckled geometry and water environment in the excitonic properties of graphitic C_3N_4. *Nanoscale* 10: 3738–3743.
43 Fu, C.F., Luo, Q., Li, X., and Yang, J. (2016). Two-dimensional van der Waals nanocomposites as Z-scheme type photocatalysts for hydrogen production from overall water splitting. *J. Mater. Chem. A* 4: 18892–18898.
44 Zhou, L., Zhuo, Z., Kou, L. et al. (2017). Computational dissection of two-dimensional rectangular titanium mononitride TiN: Auxetics and promises for photocatalysis. *Nano Lett.* 17: 4466–4472.
45 Di, J., Xia, J., Li, H., and Liu, Z. (2017). Freestanding atomically-thin two-dimensional materials beyond graphene meeting photocatalysis: opportunities and challenges. *Nano Energy* 35: 79–91.
46 Li, X., Yu, J., Wageh, S. et al. (2016). Graphene in photocatalysis: a review. *Small* 12: 6640–6696.
47 Bonaccorso, F., Colombo, L., Yu, G. et al. (2015). Graphene, related two-dimensional crystals, and hybrid systems for energy conversion and storage. *Science* 347: 1246501.
48 Gao, G., O'Mullane, A.P., and Du, A. (2017). 2D MXenes: a new family of promising catalysts for the hydrogen evolution reaction. *ACS Catal.* 7: 494–500.
49 Wang, H., Jiang, S., Shao, W. et al. (2018). Optically switchable photocatalysis in ultrathin black phosphorus nanosheets. *J. Am. Chem. Soc.* 140: 3474–3480.
50 Guo, Y., Dai, B., Peng, J. et al. (2019). Electron transport in low dimensional solids: a surface chemistry perspective. *J. Am. Chem. Soc.* 141: 723–732.
51 Wang, H., Zhang, X., and Xie, Y. (2018). Photocatalysis in two-dimensional black phosphorus: the roles of many-body effects. *ACS Nano* 12: 9648–9653.
52 Shi, Y., Li, H., and Li, L.J. (2015). Recent advances in controlled synthesis of two-dimensional transition metal dichalcogenides via vapour deposition techniques. *Chem. Soc. Rev.* 44: 2744.
53 Wang, Y., Zhang, R., Zhang, Z. et al. (2019). Host–guest recognition on 2D graphitic carbon nitride for nanosensing. *Adv. Mater. Interfaces* 6: 1901429.

54 Wen, J., Xie, J., Chen, X., and Li, X. (2017). A review on g-C_3N_4-based photocatalysts. *Appl. Surf. Sci.* 391: 72.

55 Kumbhakar, P., Chowde Gowda, C., Mahapatra, P.L. et al. (2021). Emerging 2D metal oxides and their applications. *Mater. Today* 45: 142.

56 Wang, Y., Ren, B., Zhen Ou, J. et al. (2021). Engineering two-dimensional metal oxides and chalcogenides for enhanced electro- and photocatalysis. *Sci. Bull.* 66: 1228.

57 Huang, C., Li, C., and Shi, G. (2012). Graphene based catalysts. *Energy Environ. Sci.* 5: 8848.

58 Zhang, B.Y., Xu, K., Yao, Q. et al. (2021). Hexagonal metal oxide monolayers derived from the metal–gas interface. *Nat. Mater.* 20: 1073–1078.

59 Guan, G., Pan, J.H., and Li, Z. (2021). Innovative utilization of molecular imprinting technology for selective adsorption and (photo)catalytic eradication of organic pollutants. *Chemosphere* 265: 129077.

60 Heng, J.Z.X., Tang, K.Y., Regulacio, M.D. et al. (2021). Solar-powered photodegradation of pollutant dyes using silver-embedded porous TiO_2 nanofibers. *Nanomaterials* 11: 856.

61 Xu, J., Zhang, J., Zhang, W., and Lee, C.-S. (2017). Interlayer nanoarchitectonics of two-dimensional transition-metal dichalcogenides nanosheets for energy storage and conversion applications. *Adv. Energy Mater.* 7: 1700571.

62 Wang, X., Sun, G., Li, N., and Chen, P. (2016). Quantum dots derived from two-dimensional materials and their applications for catalysis and energy. *Chem. Soc. Rev.* 45: 2239.

63 Guo, Y., Wei, Y., Li, H., and Zhai, T. (2017). Layer structured materials for advanced energy storage and conversion. *Small* 13: 1701649.

64 Li, S., Sun, J., and Guan, J. (2021). Strategies to improve electrocatalytic and photocatalytic performance of two-dimensional materials for hydrogen evolution reaction. *Chin. J. Catal.* 42: 511.

65 Wang, Y., Liu, L., Ma, T. et al. (2021). 2D graphitic carbon nitride for energy conversion and storage. *Adv. Funct. Mater.* 31: 2102540.

66 Wang, Y., Yang, Y., Zhang, D. et al. (2020). Inter-overlapped MoS_2/C composites with large-interlayer-spacing for high-performance sodium-ion batteries. *Nanoscale Horiz.* 5: 1127.

67 Frost, J.M., Butler, K.T., Brivio, F. et al. (2014). Atomistic origins of high-performance in hybrid halide perovskite solar cells. *Nano Lett.* 14: 2584–2590.

68 Pattantyus-Abraham, A., Kramer, I.J., Barkhouse, A.R. et al. (2010). Depleted-heterojunction colloidal quantum dot solar cells. *ACS Nano* 4: 3374–3380.

69 Driscoll, K., Fang, J., Humphry-Baker, N. et al. (2010). Enhanced photoresponse in solid-state excitonic solar cells via resonant energy transfer and cascaded charge transfer from a secondary absorber. *Nano Lett.* 10: 4981–4988.

70 Shi, J., Li, Y., Li, Y. et al. (2018). From ultrafast to ultraslow: charge-carrier dynamics of perovskite solar cells. *Joule* 2: 879–901.

71 Yin, W.J., Yang, J.H., Kang, J. et al. (2015). Halide perovskite materials for solar cells: a theoretical review. *J. Mater. Chem. A* 3: 8926–8942.

72 Lewis, N.S. and Nocera, D.G. (2006). Powering the planet: chemical challenges in solar energy utilization. *Proc. Natl. Acad. Sci. U. S. A.* 103: 15729–15735.

73 Niu, G., Guo, X., and Wang, L. (2015). Review of recent progress in chemical stability of perovskite solar cells. *J. Mater. Chem. A* 3: 8970–8980.

74 Green, M.A., Ho-Baillie, A., and Snaith, H.J. (2014). The emergence of perovskite solar cells. *Nat. Photonics* 8: 506–514.

75 Gregg, B.A. (2005). The photoconversion mechanism of excitonic solar cells. *MRS Bull.* 30: 20–22.

76 Correa-Baena, J.P., Abate, A., Saliba, M. et al. (2017). The rapid evolution of highly efficient perovskite solar cells. *Energy Environ. Sci.* 10: 710–727.

77 Brian, A.G. (2003). Excitonic solar cells. *J. Phys. Chem. B* 107: 4688–4698.

78 Concina, I., Manzoni, C., Grancini, G. et al. (2015). Modulating exciton dynamics in composite nanocrystals for excitonic solar cells. *J. Phys. Chem. Lett.* 6: 2489–2495.

79 Bernardi, M., Palummo, M., and Grossman, J.C. (2012). Semiconducting monolayer materials as a tunable platform for excitonic solar cells. *ACS Nano* 6: 10082–10089.

80 Gan, L.Y., Zhang, Q., Cheng, Y., and Schwingenschlögl, U. (2014). Photovoltaic heterojunctions of fullerenes with MoS_2 and WS_2 monolayers. *J. Phys. Chem. Lett.* 5: 1445–1449.

81 Dai, J. and Zeng, X. (2014). Bilayer phosphorene: effect of stacking order on bandgap and its potential applications in thin-film solar cells. *J. Phys. Chem. Lett.* 5: 1289–1293.

82 Long, M., Liu, E., Wang, P. et al. (2016). Broadband photovoltaic detectors based on an atomically thin heterostructure. *Nano Lett.* 16: 2254–2259.

83 Scharber, M.C., Mühlbacher, D., Koppe, M. et al. (2006). Design rules for donors in bulk-heterojunction solar cells-towards 10% energy-conversion efficiency. *Adv. Mater.* 18: 789–794.

84 Hu, W., Lin, L., Yang, C. et al. (2016). Edge-modified phosphorene nanoflake heterojunctions as highly efficient solar cells. *Nano Lett.* 16: 1675–1682.

85 Bernard, M., Palummo, M., and Grossman, J.C. (2013). Extraordinary sunlight absorption and one nanometer thick photovoltaics using two-dimensional monolayer materials. *Nano Lett.* 13: 3664–3670.

86 Liu, Z., Lau, S.P., and Yan, F. (2015). Functionalized graphene and other two-dimensional materials for photovoltaic devices: device design and processing. *Chem. Soc. Rev.* 44: 5638–5679.

87 Chen, D., Zhang, H., Liu, Y., and Li, J. (2013). Graphene and its derivatives for the development of solar cells, photoelectrochemical, and photocatalytic applications. *Energy Environ. Sci.* 6: 1362–1387.

88 Tsai, H., Nie, W., Blancon, J.C. et al. (2016). High-efficiency two-dimensional ruddlesden–popper perovskite solar cells. *Nature* 536: 312–316.

89 Linghu, J., Yang, T., Luo, Y. et al. (2018). High-throughput computational screening of vertical 2D van der Waals heterostructures for high-efficiency excitonic solar cells. *ACS Appl. Mater. Interfaces* 10: 32142–32150.

90 Zhou, L., Zhang, J., Zhuo, Z. et al. (2016). Novel excitonic solar cells in phosphorene–TiO_2 heterostructures with extraordinary charge separation efficiency. *J. Phys. Chem. Lett.* 7: 1880–1887.

91 Mir, W.J., Swarnkar, A., Sharma, R. et al. (2015). Origin of unusual excitonic absorption and emission from colloidal Ag_2S nanocrystals: ultrafast photophysics and solar cell. *J. Phys. Chem. Lett.* 6: 3915–3922.

92 Polman, A., Knight, M., Garnett, E.C. et al. (2016). Photovoltaic materials: present efficiencies and future challenges. *Science* 352: aad4424.

93 Shi, E., Gao, Y., Finkenauer, B.P. et al. (2018). Two-dimensional halide perovskite nanomaterials and heterostructures. *Chem. Soc. Rev.* 47: 6046–6072.

94 Wei, W., Huang, B., and Dai, Y. (2020). Photoexcited charge carrier behaviors in solar energy conversion systems from theoretical simulations. *WIREs Comput. Mol. Sci.* 10: e1441.

95 Taylor, J., Guo, H., and Wang, J. (2011). Ab initio modeling of open systems: charge transfer, electron conduction, and molecular switching of a C_{60} device. *Phys. Rev. B* 3: 121104.

96 Brandbyge, M., Mozos, J.L., Ordejón, P. et al. (2002). Density-functional method for nonequilibrium electron transport. *Phys. Rev. B* 65: 165401.

97 Taylor, J., Guo, H., and Wang, J. (2001). Ab initio modeling of quantum transport properties of molecular electronic devices. *Phys. Rev. B* 63: 245407.

98 Perfetto, E., Uimonen, A.M., Leeuwen, R.V., and Stefanucci, G. (2015). First-principles nonequilibrium Green's-function approach to transient photoabsorption: application to atoms. *Phys. Rev. A* 92: 033419.

99 Stradi, D., Martinez, U., Blom, A. et al. (2016). General atomistic approach for modeling metal-semiconductor interfaces using density functional theory and nonequilibrium Green's function. *Phys. Rev. B* 93: 155302.

100 Esposito, E., Ochoa, M.A., and Galperin, M. (2015). Quantum thermodynamics: A nonequilibrium Green's function approach. *Phys. Rev. Lett.* 114: 080602.

101 Chen, X., Liu, Y., and Duan, W. (2018). Thermal engineering in low-dimensional quantum devices: a tutorial review of nonequilibrium Green's function methods. *Small Methods* 2: 1700343.

102 Yu, Z., Zhang, L., and Wang, J. (2017). First-principles investigation of transient spin transfer torque in magnetic multilayer systems. *Phys. Rev. B* 96: 075412.

103 Buddhiraju, S. and Fan, S. (2017). Theory of solar cell light trapping through a nonequilibrium Green's function formulation of Maxwell's equations. *Phys. Rev. B* 96: 035304.

104 Agarwalla, B.K., Kulkarni, M., Mukamel, S., and Segal, D. (2016). Tunable photonic cavity coupled to a voltage-biased double quantum dot system: diagrammatic nonequilibrium Green's function approach. *Phys. Rev. B* 94: 035434.

105 Wang, B., Wang, J., and Guo, H. (1999). Current partition: a nonequilibrium Green's function approach. *Phys. Rev. Lett.* 82: 398–401.

106 Chen, J., Hu, Y., and Guo, H. (2012). First-principles analysis of photocurrent in graphene PN junctions. *Phys. Rev. B* 85: 155441.

107 Zhang, L., Gong, K., Chen, J. et al. (2014). Generation and transport of valley-polarized current in transition-metal dichalcogenides. *Phys. Rev. B* 90: 195428.

108 Waldron, D., Haney, P., Larade, B. et al. (2006). Nonlinear spin current and magnetoresistance of molecular tunnel junctions. *Phys. Rev. B* 96: 166804.

109 Chen, X., Liu, D., Duan, W., and Guo, H. (2013). Photon-assisted thermoelectric properties of noncollinear spin valves. *Phys. Rev. B* 87: 085427.

110 Xie, Y., Chen, M., Wu, Z. et al. (2018). Two-dimensional photogalvanic spin-battery. *Phys. Rev. Appl.* 10: 034005.

111 Deng, Y., Luo, Z., Conrad, N.J. et al. (2014). Black phosphorus–monolayer MoS_2 van der Waals heterojunction p–n diode. *ACS Nano* 8: 8292–8299.

112 Jin, H., Li, J., Wei, Y. et al. (2018). Unraveling the mechanism of photoinduced charge-transfer process in bilayer heterojunction. *ACS Appl. Mater. Interfaces* 10: 25401–25408.

113 Wang, X., Maeda, K., Thomas, A. et al. (2009). Metal-free polymeric photocatalyst for hydrogen production from water under visible light. *Nat. Mater.* 8: 76–80.

114 Wang, X., Chen, X., Thomas, A. et al. (2009). Metal-containing carbon nitride compounds: a new functional organic-metal hybrid material. *Adv. Mater.* 21: 1609–1612.

115 Wang, X., Maeda, K., Chen, X. et al. (2009). Polymer semiconductors for artificial photosynthesis: hydrogen evolution by mesoporous graphitic carbon nitride with visible light. *J. Am. Chem. Soc.* 131: 1680–1681.

116 Liu, J., Liu, Y., Liu, N. et al. (2015). Metal-free efficient photocatalyst for stable visible water splitting via a two-electron pathway. *Science* 347: 970–974.

117 Zhang, G., Zang, S., and Wang, X. (2015). Layered $Co(OH)_2$ deposited polymeric carbon nitrides for photocatalytic water oxidation. *ACS Catal.* 5: 941–947.

118 Zhang, M., Yao, W., Lv, Y. et al. (2014). Enhancement of mineralization ability of C_3N_4 via a lower valence position by a tetracyanoquinodimethane organic semiconductor. *J. Mater. Chem. A* 2: 11432–11438.

119 Yang, C., Qin, J., Xue, Z. et al. (2017). Rational design of carbon-doped TiO_2 modified g-C_3N_4 via in-situ heat treatment for drastically improved photocatalytic hydrogen with excellent photostability. *Nano Energy* 41: 1–9.

120 Kong, H.J., Won, D.H., Kim, J., and Woo, S.I. (2016). Sulfur-doped g-C_3N_4/$BiVO_4$ composite photocatalyst for water oxidation under visible light. *Chem. Mater.* 28: 1318–1324.

121 Fu, J., Chang, B., Tian, Y. et al. (2013). Novel C_3N_4-CdS composite photocatalysts with organic-inorganic heterojunctions: in situ synthesis, exceptional activity, high stability and photocatalytic mechanism. *J. Mater. Chem. A* 1: 3083–3090.

122 Zhang, Z., Huang, J., Zhang, M. et al. (2015). Ultrathin hexagonal SnS_2 nanosheets coupled with g-C_3N_4 nanosheets as 2D/2D heterojunction photocatalysts toward high photocatalytic activity. *Appl. Catal. B Environ.* 163: 298–305.

123 Ma, X., Li, X., Li, M. et al. (2017). Effect of the structure distortion on the high photocatalytic performance of C_{60}/g-C_3N_4 composite. *Appl. Surf. Sci.* 414: 124–130.

124 Qian, X., Ding, J., Zhang, J. et al. (2018). Ultrathin molybdenum disulfide/carbon nitride nanosheets with abundant active sites for enhanced hydrogen evolution. *Nanoscale* 10: 1766–1773.

125 Hou, Y., Laursen, A.B., Zhang, J. et al. (2013). Layered nanojunctions for hydrogen-evolution catalysis. *Angew. Chem. Int. Ed.* 52: 3621–3625.

126 Li, J., Zhang, M., Li, Q., and Yang, J. (2017). Enhanced visible light activity on direct contact Z-scheme g-C_3N_4-TiO_2 photocatalyst. *Appl. Surf. Sci.* 391: 184–193.

127 Li, H., Li, F., Wang, Z. et al. (2018). Fabrication of carbon bridged g-C_3N_4 through supramolecular self-assembly for enhanced photocatalytic hydrogen evolution. *Appl. Catal. B Environ.* 220: 356–361.

128 Xu, Q., Cheng, B., Yu, J., and Liu, G. (2017). Making co-condensed amorphous carbon/g-C_3N_4 composites with improved visible-light photocatalytic H_2-production performance using Pt as cocatalyst. *Carbon* 118: 241–249.

129 Liang, Y., Long, C., Li, J. et al. (2018). InSe monolayer: promising cocatalyst of g-C_3N_4 for water splitting under visible light. *ACS Appl. Energy Mater.* 1: 5394–5401.

130 Liang, Y., Dai, Y., Ma, Y. et al. (2018). Novel titanium nitride halide TiNX (X = F, Cl, Br) monolayers: potential materials for highly efficient excitonic solar cells. *J. Mater. Chem. A* 6: 2073–2080.

131 Zhao, P., Yang, H., Li, J. et al. (2017). Design of new photovoltaic systems based on two-dimensional group-IV monochalcogenides for high performance solar cells. *J. Mater. Chem. A* 5: 24145–24152.

132 Zhao, P., Wei, W., Sun, Q. et al. (2017). In-plane heterostructures of Sb/Bi with high carrier mobility. *Nanotechnology* 28: 255201.

133 Zhao, P., Li, J., Wei, W. et al. (2017). Giant anisotropic photogalvanic effect in a flexible AsSb monolayer with ultrahigh carrier mobility. *Phys. Chem. Chem. Phys.* 19: 27233–27239.

134 Rodina, A.V., Dietrich, M., Göldner, A. et al. (2001). Free excitons in wurtzite GaN. *Phys. Rev. B* 64: 115204.

135 Zhou, L.J., Zhang, Y.F., and Wu, L.M. (2013). SiC_2 Siligraphene and nanotubes: novel donor materials in excitonic solar cells. *Nano Lett.* 13: 5431–5436.

136 Xie, M., Zhang, S., Cai, B. et al. (2016). A promising two-dimensional solar cell donor: black arsenic–phosphorus monolayer with 1.54 eV direct bandgap and mobility exceeding 14,000 $cm^2\ V^{-1}s^{-1}$. *Nano Energy* 28: 433–439.

137 Ganesan, V.D., Linghu, J., Zhang, C. et al. (2016). Heterostructures of phosphorene and transition metal dichalcogenides for excitonic solar cells: a first-principles study. *Appl. Phys. Lett.* 108: 122105.

138 Jin, H., Li, J., Wang, T., and Yu, Y. (2018). Photoinduced pure spin-current in triangulene-based nano-devices. *Carbon* 137: 1–5.

139 Jin, H., Li, J., Wan, L. et al. (2017). Ohmic contact in monolayer InSe–metal interface. *2D Mater.* 4: 025116.

140 Jin, H., Li, J., Dai, Y., and Wei, Y. (2017). Engineering the electronic and optoelectronic properties of InX (X = S, Se, Te) monolayers via strain. *Phys. Chem. Chem. Phys.* 19: 4855–4860.

141 Ma, X., Dai, Y., Yu, L., and Huang, B. (2016). Interface Schottky barrier engineering via strain in metal–semiconductor composites. *Nanoscale* 8: 1352–1359.

142 Ma, X., Dai, Y., and Huang, B. (2014). Origin of the increased photocatalytic performance of TiO_2 nanocrystal composed of pure core and heavily nitrogen-doped shell: a theoretical study. *ACS Appl. Mater. Interfaces* 6: 22815–22822.

143 Ma, X., Dai, Y., Guo, M., and Huang, B. (2013). Insights into the role of surface distortion in promoting the separation and transfer of photogenerated carriers in anatase TiO_2. *J. Phys. Chem. C* 117: 24496–24502.

144 Wang, P., Huang, B., Qin, X. et al. (2008). Ag@AgCl: a highly efficient and stable photocatalyst active under visible light. *Angew. Chem. Int. Ed.* 47: 7931–7933.

145 Wang, W., Huang, B., Ma, X. et al. (2013). Efficient separation of photogenerated electron–hole pairs by the combination of a heterolayered structure and internal polar field in pyroelectric $BiOIO_3$ nanoplates. *Chem. Eur. J.* 19: 14777–14780.

146 Zhang, R., Dai, Y., Lou, Z. et al. (2014). Layered photocatalyst $Bi_2O_2[BO_2(OH)]$ nanosheets with internal polar field enhanced photocatalytic activity. *CrystEngComm* 16: 4931–4934.

147 Cheng, H.F., Fuku, K., Kuwahara, Y. et al. (2015). Harnessing single-active plasmonic nanostructures for enhanced photocatalysis under visible light. *J. Mater. Chem. A* 3: 5244–5258.

148 Lou, Z., Wang, Z., Huang, B., and Dai, Y. (2014). Synthesis and activity of plasmonic photocatalysts. *ChemCatChem* 6: 2456–2476.

149 Wang, P., Huang, B., Dai, Y., and Whangbo, M.H. (2012). Plasmonic photocatalysts: harvesting visible light with noble metal nanoparticles. *Phys. Chem. Chem. Phys.* 14: 9813–9825.

150 Ma, X., Dai, Y., Yu, L., and Huang, B. (2014). Noble-metal-free plasmonic photocatalyst: hydrogen doped semiconductors. *Sci. Rep.* 4: 3986.

151 Zheng, Z., Huang, B., Qin, X. et al. (2011). Facile in situ synthesis of visiblelight plasmonic photocatalysts M@TiO_2 (M= Au, Pt, Ag) and evaluation of their photocatalytic oxidation of benzene to phenol. *J. Mater. Chem.* 21: 9079–9087.

152 Ma, X., Dai, Y., Yu, L. et al. (2014). Electron–hole pair generation of the visible-light plasmonic photocatalyst Ag@AgCl: enhanced optical transitions involving midgap defect states of AgCl. *J. Phys. Chem. C* 118: 12133–12140.

153 Ma, X., Dai, Y., Yu, L., and Huang, B. (2014). New basic insights into the low hot electron injection efficiency of gold-nanoparticle-photosensitized titanium dioxide. *ACS Appl. Mater. Interfaces* 6: 12388–12394.

154 Wang, P., Huang, B., Zhang, X. et al. (2009). Highly efficient visible-light plasmonic photocatalyst Ag@AgBr. *Chem. Eur. J.* 15: 1821–1824.

155 Wang, P., Huang, B., Lou, Z. et al. (2010). Synthesis of highly efficient Ag@AgCl plasmonic photocatalysts with various structures. *Chem. Eur. J.* 16: 538–544.

156 Ma, X., Dai, Y., Guo, M. et al. (2013). Insights into the adsorption and energy transfer of Ag clusters on the AgCl (100) surface. *Phys. Chem. Chem. Phys.* 15: 8722–8731.

157 Ma, X., Dai, Y., Guo, M., and Huang, B. (2012). The role of effective mass of carrier in the photocatalytic behavior of silver halide-based Ag@AgX (X=Cl, Br, I): a theoretical study. *ChemPhysChem* 13: 2304–2309.

158 Ma, X., Dai, Y., Lu, J. et al. (2012). Tuning of the surface-exposing and photocatalytic activity for AgX (X = Cl and Br): a theoretical study. *J. Phys. Chem. C* 116: 19372–19378.

159 Liang, X., Wang, P., Li, M. et al. (2018). Adsorption of gaseous ethylene via induced polarization on plasmonic photocatalyst Ag/AgCl/TiO_2 and subsequent photodegradation. *Appl. Catal. B Environ.* 220: 356–361.

160 Ma, X., Dai, Y., Yu, L., and Huang, B. (2016). Energy transfer in plasmonic photocatalytic composites. *Light Sci. Appl.* 5: e16017.

161 Grätzel, M. (2001). Photoelectrochemical cells. *Nature* 414: 338–344.

162 Hashimoto, K., Irie, H., and Fujishima, A. (2005). TiO_2 photocatalysis: a historical overview and future prospects. *Jpn. J. Appl. Phys.* 44: 8269.

163 Serpone, N. and Emeline, A.V. (2012). Semiconductor photocatalysis–past, present, and future outlook. *J. Phys. Chem. Lett.* 3: 673–677.

164 Baffoua, G. and Quidant, R. (2014). Nanoplasmonics for chemistry. *Chem. Soc. Rev.* 43: 3898–3907.

165 Draine, B.T. and Flatau, P.J. (1994). Discrete dipole approximation for scattering calculations. *J. Opt. Soc. Am. A* 11: 1491–1499.

166 Draine, B.T. and Flatau, P.J. (2008). Discrete-dipole approximation for periodic targets: theory and tests. *J. Opt. Soc. Am. A* 25: 2593–2703.

167 Flatau, P.J. and Draine, B.T. (2012). Fast near field calculations in the discrete dipole approximation for regular rectilinear grids. *Opt. Express* 20: 1247–1252.

168 Draine, B.T. (1988). The discrete dipole approximation and its application to interstellar graphite grains. *Astrophys. J.* 333: 848–872.

169 Flatau, P.J. and Draine, B.T. (1997). Improvements of the discrete dipole approximation method. *Opt. Lett.* 22: 1205–1207.

170 Flatau, P.J. (2004). Fast solvers for one dimensional light scattering in the discrete dipole approximation. *Opt. Express* 12: 3149–3155.

171 Stiles, P.L., Dieringer, J.A., Shah, N.C., and Van Duyne, R.P. (2008). Surface-enhanced Raman spectroscopy. *Annu. Rev. Anal. Chem.* 1: 601–626.

172 Michaels, A.M., Nirmal, M., and Brus, L.E. (1999). Surface enhanced Raman spectroscopy of individual rhodamine 6G molecules on large Ag nanocrystals. *J. Am. Chem. Soc.* 121: 9932–9939.

173 Moskovits, M. (1985). Surface-enhanced spectroscopy. *Rev. Mod. Phys.* 57: 783–826.

174 Lee, S., Anderson, L.J.E., Payne, C.M., and Hafner, J.H. (2011). Structural transition in the surfactant layer that surrounds gold nanorods as observed

by analytical surface-enhanced Raman spectroscopy. *Langmuir* 27: 14748–14756.
175 Atwater, H.A. and Polman, A. (2010). Plasmonics for improved photovoltaic devices. *Nat. Mater.* 9: 205–213.
176 Li, H., Yuan, K., Zhang, Y., and Wang, J. (2013). Synthesis of Au–SiO_2 asymmetric clusters and their application in ZnO nanosheet-based dye-sensitized solar cells. *ACS Appl. Mater. Interfaces* 5: 5601–5608.
177 Mahmoud, M.A. and El-Sayed, M.A. (2013). Different plasmon sensing behavior of silver and gold nanorods. *J. Phys. Chem. Lett.* 4: 1541–1545.
178 Anker, J.N., Hall, W.P., Lyandres, O. et al. (2008). Biosensing with plasmonic nanosensors. *Nat. Mater.* 7: 442–453.
179 Malinsky, M.D., Kelly, K.L., Schatz, G.C., and Van Duyne, R.P. (2001). Chain length dependence and sensing capabilities of the localized surface plasmon resonance of silver nanoparticles chemically modified with alkanethiol self-assembled monolayers. *J. Am. Chem. Soc.* 123: 1471–1482.
180 Kedem, O., Tesler, A.B., Vaskevich, A., and Rubinstein, I. (2011). Sensitivity and optimization of localized surface plasmon resonance transducers. *ACS Nano* 5: 748–760.
181 Mayer, K.M. and Hafner, J.H. (2011). Localized surface plasmon resonance sensors. *Chem. Rev.* 111: 3828–3857.
182 Khanal, B.P., Pandey, A., Li, L. et al. (2012). Generalized synthesis of hybrid metal–semiconductor nanostructures tunable from the visible to the infrared. *ACS Nano* 6: 3832–3840.
183 Awazu, K., Fujimaki, M., Rockstuhl, C. et al. (2008). Plasmonic photocatalyst consisting of silver nanoparticles embedded in titanium dioxide. *J. Am. Chem. Soc.* 130: 1676–1680.
184 Takai, A. and Kamat, P.V. (2011). Capture, store, and discharge. shuttling photogenerated electrons across TiO_2-silver interface. *ACS Nano* 5: 7369–7376.
185 Liu, L., Ouyang, S., and Ye, J. (2013). Gold-nanorod-photosensitized titanium dioxide with wide-range visible-light harvesting based on localized surface plasmon resonance. *Angew. Chem. Int. Ed.* 52: 6689–6693.
186 Mayer, K.M., Lee, S., Liao, H. et al. (2008). A label-free immunoassay based upon localized surface plasmon resonance of gold nanorods. *ACS Nano* 2: 687–692.
187 Lin, J., Li, H., Zhang, H., and Chen, W. (2013). Plasmonic enhancement of photocurrent in MoS_2 field-effect transistor. *Appl. Phys. Lett.* 102: 203109.
188 Sobhani, A., Lauchner, A., Najmaei, S. et al. (2014). Enhancing the photocurrent and photoluminescence of single crystal monolayer MoS_2 with resonant plasmonic nanoshells. *Appl. Phys. Lett.* 104: 031112.
189 Wang, M., Rajeeva, B.B., Scarabelli, L. et al. (2016). Molecular-fluorescence enhancement via blue-shifted plasmon-induced resonance energy transfer. *J. Phys. Chem. C* 120: 14820–14827.
190 Butun, S., Tongay, S., and Aydin, K. (2015). Enhanced light emission from large-area monolayer MoS_2 using plasmonic nanodisc arrays. *Nano Lett.* 15: 2700–2704.

191 Li, Y., JG, D.S., Murthy, A.A. et al. (2017). Superior plasmonic photodetectors based on Au@MoS_2 core–shell heterostructures. *ACS Nano* 11: 10321–10329.

192 Wang, M., Krasnok, A., Zhang, T. et al. (2018). Tunable Fano resonance and plasmon–exciton coupling in single Au nanotriangles on monolayer WS_2 at room temperature. *Adv. Mater.* 30: 1705779.

8

Simulations for Electrochemical Reactions

8.1 Single-atom Catalysts

In the field of electrochemical catalysis, single-atom catalysts (SACs) can be regarded a crucial concept for current catalyst study [1, 2], emerging as a new frontier of catalysis science. As for SACs, distributed metal atoms are anchored on support materials, indicative of ultrahigh atomic utilization ratio and high catalytic activity [3–18]. In applications of energy conversion and energy storage, developing efficient catalysts therefore becomes the key factor; the performance and the cost should be primarily considered. In spite of the experimental challenge in how to achieve the stabilization of single-metal atoms against migration and agglomeration, SACs receive great attention from the perspective of theoretical simulations, manifesting promising application potential in clean energy conversion reactions such as oxygen reduction reaction (ORR), hydrogen evolution reaction (HER), N_2 reduction reaction (NRR), and CO_2 reduction reaction (CO_2RR). It is beneficial for the decrease of catalyst cost due to the maximum atom utilization efficiency of SACs, which, at the same time, exhibits excellent performance.

In traditional heterogeneous catalysts that contain a mixture of metal particles of broad size distribution, only the metal particles with a suitable size distribution can serve as catalytic active sites, with the rests being either inert or triggering undesired side reactions. As a consequence, the metal utilization efficiency is low and the selectivity turns out to be poor, leading to high metal consumption along with cumbersome and expensive product purification and waste disposal. In contrast, catalysts with well-defined active sites and tunable coordination present excellent activity and exclusive selectivity for a specific reaction, which, however, show poor stability and bad recyclability. It is happy to see that SACs inherit the advantages of both heterogeneous and homogeneous catalysts, enabling the reasonable use of metal resources. In addition, for SACs the unique electronic structure and unsaturated active centers have been demonstrated to play a crucial role in catalytic activity for a variety of reactions. In a sense, SACs possess similarities to the homogeneous catalysts because of the homogeneity in active sites and the similar spatial and electronic interaction with reactive species. It concludes that electrocatalysts should meet two requirements: (1) a large number of active sites and (2) enhanced intrinsic

Calculations and Simulations of Low-Dimensional Materials: Tailoring Properties for Applications, First Edition. Ying Dai, Wei Wei, Yandong Ma, and Chengwang Niu.

activity of each active site. In this regard, SACs have the active atoms fully exposed thus effectively increasing the number of active sites. It has been well illustrated that the metal–support interaction determines the intrinsic activity of active sites, and there is an optimal metal–support match for a specific reaction.

In accordance with the preceding discussion, an appropriate support material is of importance for efficient SACs. In case of two-dimensional materials of large specific surface area, single metal or nonmetal atom will be stabilized by the lattice or the coordination environment. In both cases, strong covalent bonds form, and the single atom is doped into the lattice or anchored on the surface. In this context, the electronic properties of the single atom will be changed due to the interaction with the host, leading to the formation of new electronic states and thus possessing electrocatalytic activity. In Figure 8.1, change in electronic properties, for example, the density of states (DOS), is shown for single metal atom anchored on graphene. It can be seen that the DOS at the Fermi level continuously decreases as the number of layer reaches a monolayer (i.e. the graphene), while the DOS changes from a continuous one to be discrete when the bulk metal reduces to a single atom. In consideration of the metal–graphene interaction, new electronic states appear and dominate the catalytic activity. As summarized in the literature, two-dimensional materials confirmed SACs show several unique features: (1) exposed single atoms that adsorbed on or embedded in the two-dimensional supports serve directly as active sites, modulating the electronic properties of the coordination environment in intrinsically inactive two-dimensional materials; (2) single atoms confined in the two-dimensional structures are coordinatively unsaturated due to the vacuum on both sides of the two-dimensional structure, and thus achieve high catalytic activity; (3) the open structure on both sides of the two-dimensional plane enables expedited mass-transfer process, and theoretically ensures 100% exposure of single atoms to reactants and maximizes the catalytic reaction rates; (4) the local atomic and electronic structures can be accurately probed by advanced characterization techniques, which allow for further prediction of molecular reaction dynamics through theoretical calculation or even in situ observation in real space; (5) the well-defined motif

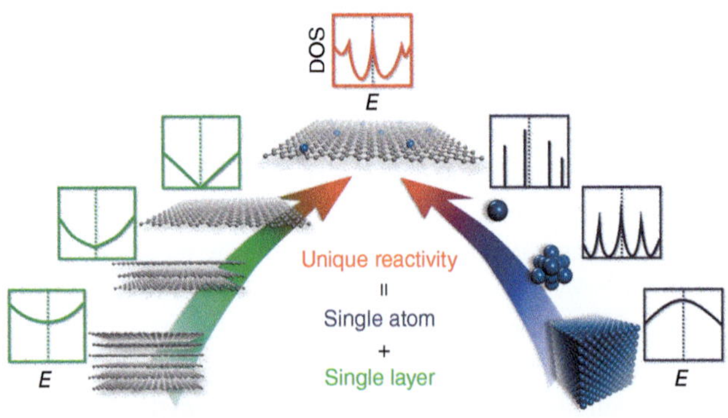

Figure 8.1 Electronic structure of single-atom catalysts supported on two-dimensional materials (here taking graphene as an example). Source: Wang et al. [2]/with permission of American Chemical Society.

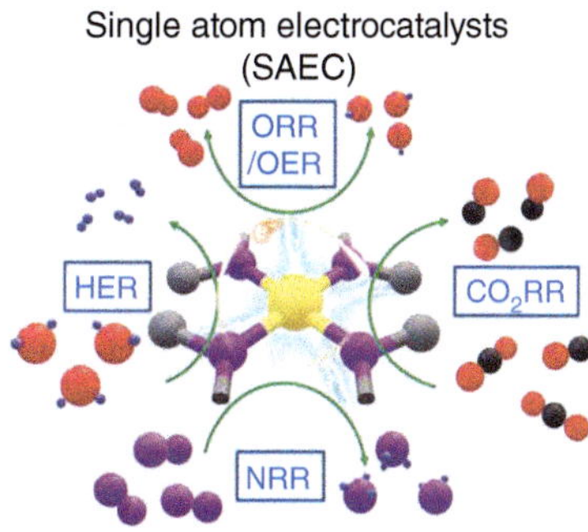

Figure 8.2 Transtion metal–N_4 active center in graphene. Source: Wang et al. [19]/with permission of American Chemical Society.

of the confined single atoms on two-dimensional materials provides a good platform suitable for model study on the interplay among geometric effect, electronic effect, and catalytic performance.

It is therefore conclusive that SACs based on two-dimensional materials have great significance in both theoretical and experimental studies of catalysis. In both experiments and simulations, transition metal–N_4–graphene being the active center (see Figure 8.2) has been attracting intensive attention, and its electrocatalytic activity has been demonstrated in series of electrochemical reactions [19]. In advanced calculations, the most stable configuration of the catalyst system can be accurately addressed, the thermodynamic and dynamic processes and corresponding energy profile can be obtained, and the physical and chemical origin of the electrocatalytic activity can be unraveled. In principle, theoretical simulations can definitely facilitate the comprehensive understanding on the structure–activity relationship, advancing the rational catalyst design of efficient catalysts for specific electrochemical reaction at atomic scale.

8.2 Stability of Catalyst

Prior to the electrochemical calculations, the model for catalyst should be carefully relaxed and the total energy can be obtained by the first-principles calculation on the basis of density functional theory. In particular, the stability and experimental accessibility should be demonstrated or discussed if the catalysts are theoretically predicted. In view of this, therefore, dynamic, thermodynamic, thermal, mechanical, and electrochemical stability should be first checked from calculations. It turns out to be dynamically stable if the phonon spectrum for a primitive cell of the catalyst shows no imaginary frequency over the whole Brillouin zone, which is a common strategy in many articles. In spite of the dynamic stability of a catalyst, it is hard to say that the material for catalyst can be realized experimentally. As for the thermodynamic stability, recent publications used ab initio molecular dynamics (AIMD) simulations to confirm it. It was concluded that the designed or predicted catalysts are thermodynamically stable if the free energy fluctuates within a small range (<1.0 eV) and the structure shows small distortion under certain temperature (such as 300 K or elevated 500 K).

In particular, for the so-called single-atom catalysts (SACs) with usually isolated transition metal (TM) atom adsorbed on a substrate (for example, graphene-like carbon nitride), binding energy verifies the stability of the catalysts

$$E_b = E_{m@sub} - E_{sub} - E_m$$

with $E_{m\,@\,sub}$, E_{sub}, and E_m being the total energies of the whole system, the substrate, and the single TM atom, respectively. In accordance with the definition, negative E_b suggests that catalysts are thermodynamically stable.

In the case of addressing the thermal stability, one can calculate the cohesive energy [20]:

$$E_{coh} = \frac{E_{bulk}}{n} - E_{atom}$$

where E_{bulk} and E_{atom} are the total energies of the unit cell in bulk phase and the total energy of an isolated atom, respectively, which can be obtained from the result of density functional theory, while n corresponds to atom number in a unit cell. In order to give an intuitive description, models for platinum in bulk phase (face-centered cubic, space group $m\bar{3}m$) and a single platinum atom for calculation are shown in Figure 8.3. In case of $\Delta E = E_b - E_{coh} < 0$, metal atoms tend to be fairly anchored on catalyst surface rather than aggregation, which is a severe problem for SACs.

In respect to the mechanical properties, elastic constants C_{11}, C_{12}, C_{22}, and C_{66} are always calculated to confirm whether the material (here in two dimensions) satisfies the Born criteria of mechanical stability, i.e. $C_{11}C_{12} > C_{12}^2$ and C_{11}, C_{22}, and $C_{66} > 0$ [21]. On account of the obtained elastic constants, additionally, Young's modulus

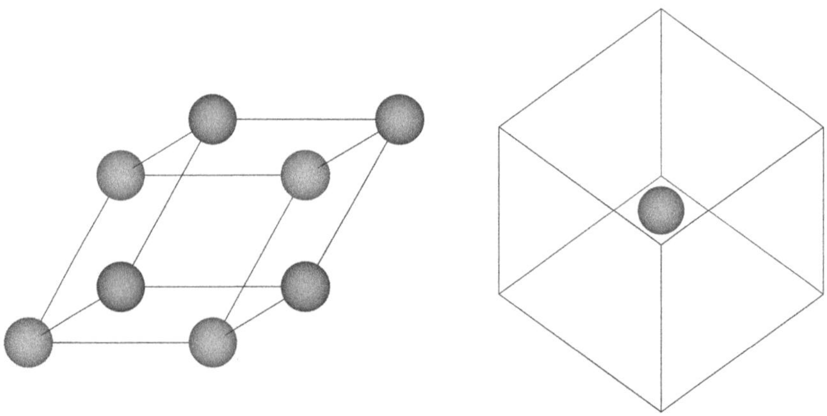

Figure 8.3 Models for obtaining the total energies of bulk metal (E_{bulk}) and single metal atom (E_{atom}) in practice calculations for the cohesive energy. It should be pointed out that the cell length for a single atom has to be sufficiently large to avoid the interactions due to the boundary condition. Source: Wei Wei.

$Y(\theta)$ and Poisson's ratio $\upsilon(\theta)$ can be calculated according to:

$$Y(\theta) = \frac{C_{11}C_{22} - C_{12}^2}{C_{22}\cos^4\theta + \left(\left(C_{11}C_{22} - C_{12}^2/C_{66}\right) - 2C_{12}\right)\cos^2\theta\sin^2\theta + C_{11}\sin^4\theta}$$

$$\upsilon(\theta) = \frac{C_{12}(\cos^4\theta + \sin^4\theta) - \left(C_{11} + C_{22} - \left(C_{11}C_{22} - C_{12}^2/C_{66}\right)\right)\cos^2\theta\sin^2\theta}{C_{22}\cos^4\theta + \left(\left(C_{11}C_{22} - C_{12}^2/C_{66}\right) - 2C_{12}\right)\cos^2\theta\sin^2\theta + C_{11}\sin^4\theta}$$

with θ being the angle with respect to the in-plane x axis. The Young's modulus of graphene of 340 N m^{-1} [22] could be a reference used to illustrate the flexibility of two-dimensional materials, and the Poisson's ratio of most of the two-dimensional materials is in the range from 0 to 0.5 [23].

In consideration that surface fluctuation of catalysts will largely affect the long-range order of the two-dimensional crystals, which results in significant change in electronic properties and the activity for a specific electrochemical reaction, in-plane stiffness can also be evaluated to see the mechanical stability:

$$C_{2D} = \frac{(\partial^2 E_{total}/\partial \varepsilon^2)}{S_0}$$

where E_{total} represents the total energy per unit cell, ε stands for the applied uniaxial strain, and S_0 is the area of the optimized unit cell. As examples, in-plane stiffness of MoS_2 monolayer is 125 N m^{-1} [24] and for phosphorene is 21–92 N m^{-1} [25, 26], and large result indicates more favorable mechanical stability.

In electrolyte environment, dissolution potential can be calculated to check the electrochemical stability of catalysts:

$$U_{diss} = U^0_{M^{z+}/M} - \frac{E_b - E_b}{ZF}$$

with $U^0_{M^{z+}/M}$ being the standard electrode potential of metal, E_b and E_b stand for the binding energy and cohesive energy, respectively; Z is the electron number involved in the dissolution, and F is the Faraday's constant.

The dissolution potential can be alternatively calculated as follows:

$$U_{diss} = U^0_{M^{z+}/M} - \frac{E_f}{ZF}$$

here E_f is the formation energy. In accordance with the equations, negative dissolution potential suggests that the catalyst is eletrochemically stable.

8.3 Electrochemical Reactions

8.3.1 Hydrogen Evolution Reaction (HER)

In recent years, global energy consumption and environmental problems are becoming more and more severe, which can be mainly ascribed to the burning of non-renewable fossil fuels. In order to overcome these problems, developing renewable energy has attracted increasingly extensive interest, and it has become one of the most urgent target in worldwide research community. In light of the advantages of high energy density, abundant source, and free of harmful by-products, hydrogen is regarded as a promising energy source to replace fossil fuels [27–29]. In modern society, electrochemical water splitting turns out to be the most effective and environmentally friendly method for hydrogen generation [30, 31]. In current stage, precious metal platinum-based materials are considered to be the best candidate catalysts for HER [32, 33]. It is however obvious that large-scale application is limited due to the natural scarcity and high cost, and it is therefore significant and urgent to explore alternatives to platinum-based electrocatalysts of high performance and low cost to meet the needs of the practical applications. As a result, finding the most effective catalysts is at the heart for electrochemical HER and the realization of industrial hydrogen generation, storage, and transfer.

In principle, HER is the elemental reaction for electrochemical reactions, generating H_2 gas and, importantly, is the basis for the proton–electron transfer process. In the process of experimentally electrochemical HER, which occurs at the cathode, usually the oxygen evolution reaction (OER) accompanies at the anode, see Figure 8.4.

In acidic condition, reactions at cathode and anode follow, respectively:

$$2H^+ + 2e^- \rightarrow H_2$$

$$H_2O \rightarrow \frac{1}{2}O_2 + 2H^+ + 2e^-$$

with the former reaction being a typical HER process and the latter an OER half-reaction, while in alkaline condition, HER and OER reactions occur as:

$$2H_2O + 2e^- \rightarrow 2OH^- + H_2$$

$$2OH^- \rightarrow \frac{1}{2}O_2 + H_2O + 2e^-$$

at cathode and anode, respectively. It is obvious that there is an electrochemical overall water splitting reaction:

$$H_2O \rightarrow H_2 + \frac{1}{2}O_2$$

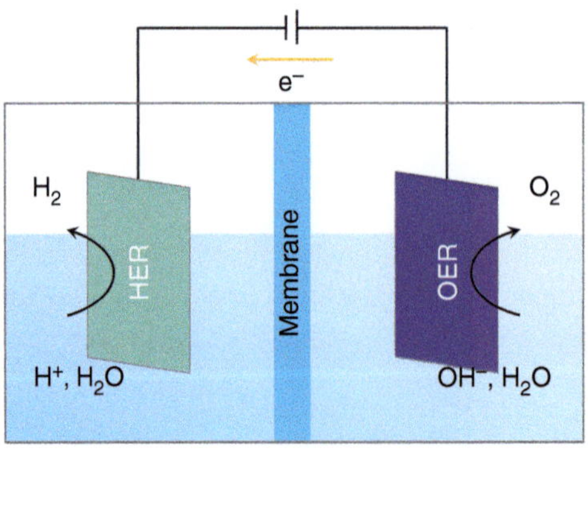

Figure 8.4 In a typical HER process, H_2 gas generation pathway differs depending on the electrolyte condition (acidic or alkaline), and at anode OER accompanies, corresponding to an overall water splitting reaction. In both cases, electron transfer involves. It is obvious that design of efficient catalysts is one of the most important targets for large-scale H_2 gas production. Source: Wei Wei.

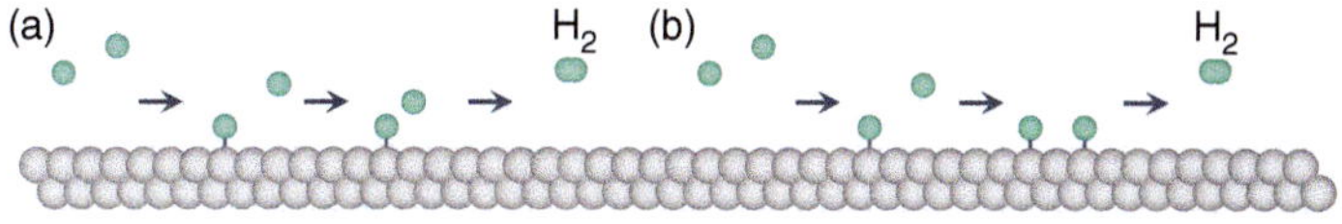

Figure 8.5 Schematic for a complete HER process. (a) Heyrovsky mechanism; (b) Tafel mechanism. In (a), an absorbed hydrogen atom (*H) is attached by another hydrogen atom, and then a H_2 molecule releases from the catalyst surface, while in (b), two absorbed hydrogen atoms react and form H_2 molecule. In both pathways, electron transfer involves, and the whole process is thermodynamic and dynamic. Source: Wei Wei.

In a typical electrochemical HER, three steps are involved: (1) atomic H adsorption on the catalyst surface (Volmer reaction); (2) formation of absorbed $^*H_2$; (3) release of H_2 molecule, with step (a) being extremely important since the energy of the intermediate H^* plays a decisive role in the reaction barrier during the following reaction:

$$H^+ + e^- + {}^* \rightarrow H^*$$

here * stands for the active site on catalyst surface and H^* is the intermediate. In respect to step (2), two pathways are possible, i.e. Tafel and Heyrovsky pathways, see Figure 8.5.

In most literatures, for catalysts the HER activity is commonly evaluated by calculating the Gibbs free energy change [34]:

$$\Delta G_{H^*} = \Delta E_{H^*} + \Delta E_{ZPE} - T\Delta S_{H^*} - neU + \Delta G_{pH}$$

where ΔE_{H^*} stands for the hydrogen adsorption energy that can be obtained from density functional theory calculations; ΔE_{ZPE} and ΔS_{H^*} are the corresponding changes in zero-point energy (ZPE) and entropy between the adsorbed hydrogen (H^*) and free-standing hydrogen in gas phase (H_2), respectively, which can be obtained from frequency calculation at $T = 298.15$ K. In this equation, n is the number of transferred electrons and U is the electrode potential. In our calculations, ΔG_{pH} is related to the pH by

$$\Delta G_{pH} = \ln 10 \times K_B T \times pH$$

with being the Boltzmann constant. In particular, ZPE and vibrational entropy of gas molecules could be looked up from the NIST database (https://janaf.nist.gov), while those of the adsorption intermediates can be obtained by [35]

$$E_{ZPE} = \frac{1}{2}\sum_i h\nu_i$$

$$S(T) = \sum_{i=1}^{3N}\left[-R\ln\left(1 - e^{-h\nu_i/k_B T}\right) + \frac{N_A h\nu_i}{T}\frac{e^{-h\nu_i/k_B T}}{1 - e^{-h\nu_i/k_B T}}\right]$$

here R is the ideal gas constant, h is Planck's constant, k_B is Boltzmann constant, T is the absolute temperature, ν_i is the vibrational frequency, N is the amount of adsorbed atoms, and N_A is Avogadro's number. In Table 8.1, entropies for common molecules are summarized.

Table 8.1 Entropies for common molecules under standard condition $p^0 = 0.1$ MPa and $T = 298.15$ K, and the unit is J K^{-1} mol^{-1}.

	H_2	H_2O	H_2O_2	CO	CO_2
S^0	130.680	188.834	232.991	197.653	213.795
	CH_4	N_2	NH_3	NH_2	NO
S^0	186.251	191.609	192.774	194.707	210.758

In case of calculating the ΔG_{H^*} for HER, the zero-point energy can be obtained by

$$\Delta E_{ZPE} = E_{ZPE}^{H} - \frac{1}{2}E_{ZPE}^{H_2}$$

where E_{ZPE}^{H} is the zero-point energy of the atomic hydrogen on the catalyst surface, and $E_{ZPE}^{H_2}$ is the zero-point energy of H_2 in gas phase.

In this context, ΔG_{H^*} can be expressed as:

$$\Delta G_{H^*} = \Delta E_{H^*} + 0.24\,\text{eV}$$

ΔE_{H^*} is the differential hydrogen adsorption energy, which is defined by

$$\Delta E_{H^*} = E_{H^*} - E_{\text{catalyst}} - \frac{1}{2}E_{H_2}$$

where E_{H^*}, E_{catalyst}, and E_{H_2} represent the total energies of the hydrogen adsorbed system, catalyst, and H_2 in gas phase, respectively.

In the volcano curve, the exchange current is based on the Nørskov's assumption [36], that is, if $\Delta G_{H^*} \leq 0$, the following expression for the i_0 at pH = 0 is used:

$$i_0 = -ek_0\frac{1}{1+\exp(-\Delta G_{H*}/k_B T)}$$

If $\Delta G_{H^*} \geq 0$, the exchange current is calculated by [37]

$$i_0 = -ek_0\frac{1}{1+\exp(\Delta G_{H*}/k_B T)}$$

where k_0 is the rate constant; however, there are no experimental data available, thus k_0 is usually set to 1.

In calculations, the HER polarization curve can be simulated using the turnover frequency (TOF) and the number of active sites, where [38]

$$\text{TOF} = \frac{N_{H_2}}{N_s}$$

In this equation, N_{H_2} is the total number of H_2 molecules per second, and N_s is the total number of active sites per unit area. As suggested in previous report [36, 39], the TOF of catalyst can be benchmarked with an Arrhenius-type equation using the hydrogen adsorption energy and an active site density of N_{as}, then the current density j can be obtained by

$$j = 2qN_{as} \times \text{TOF}$$

where $q = 1.6 \times 10^{-19}$ C is the elementary charge and factor 2 originates from the number of atoms per H_2.

In accordance to the computational hydrogen electrode (CHE) model [40], the chemical potential of one proton and electron pair ($H^+ + e^-$) is equivalent to one-half of the chemical potential of H_2 molecule under standard conditions (pH = 0 and $U = 0$) [41]. It should be kept in mind that the calculation gives only the hydrogen generation reaction energy thermodynamically, and ΔG_{H^*} is always used to evaluate the catalyst activity. It has been concluded that, importantly, for an ideal HER catalyst the ΔG_{H^*} should be close to zero, known as the Sabatier principle [42]. In accordance to the Sabatier principle, strong hydrogen adsorption is, however, not conducive to the hydrogen desorption.

In regard to the diffusion barrier and the corresponding minimum energy pathway, the climbing image nudged elastic band (CINEB) method can be used to determine the dynamic process for reactions on catalyst surface.

8.3.2 Oxygen Evolution Reaction (OER)

In general, the OER performance can be described by the reaction-free energy (ΔG_1, ΔG_2, ΔG_3, ΔG_4) for four elementary steps:

$$H_2O(l) + {*} \rightarrow {}^*OH + H + e^-$$
$${}^*OH \rightarrow {}^*O + H + e^-$$
$$H_2O(l) + {}^*O \rightarrow {}^*OOH + H + e^-$$
$${}^*OOH \rightarrow {*} + O_2(g) + H + e^-$$

In practice calculations, the CHE is applied to calculate the Gibbs free energy change (ΔG) for each elemental step in the OER process, which can be obtained from the following equation:

$$\Delta G = \Delta E + \Delta E_{ZPE} - T\Delta S + Ue + \Delta G_{pH}$$

where ΔE is the electronic energy difference between the product and reactant of each elemental step in OER process. ΔE_{ZPE} and ΔS are the differences in zero-point energy and entropy, respectively; e is the number of transferred electrons and U stands for the applied electrode potential. ΔG_{pH} is the correction to pH, which can be expressed by

$$\Delta G_{pH} = \text{In}10 \times K_B T \times \text{pH}$$

with K_B being the Boltzmann constant and the value of pH was set to zero in this work. In consideration that the external potential should be equal to the equilibrium potential (i.e. $U_{eq} = 1.23$ V), therefore, the theoretical overpotential of OER can be obtained from the following equation [43]:

$$\eta_{OER} = \max\{\Delta G_1, \Delta G_2, \Delta G_3, \Delta G_4\}/e - 1.23$$

8.3.3 Oxygen Reduction Reaction (ORR)

In respect to ORR, the elementary reaction steps are described as follows:

$$^* + O_2(g) + H + e^- \rightarrow {}^*OOH$$

$$^*OOH + H + e^- \rightarrow H_2O(l) + {}^*O$$

$$^*O + H + e^- \rightarrow {}^*OH$$

$$^*OH + H + e^- \rightarrow H_2O(l) + {}^*$$

with the corresponding reaction-free energy of each elementary step expressed as ΔG_I, ΔG_{II}, ΔG_{III}, ΔG_{IV}, where $\Delta G_I = -\Delta G_4$, $\Delta G_{II} = -\Delta G_3$, $\Delta G_{III} = -\Delta G_2$, and $\Delta G_{IV} = -\Delta G_1$. In analogy to OER, the theoretical overpotential of ORR can be written as:

$$\eta_{ORR} = \max\{\Delta G_I, \Delta G_{II}, \Delta G_{III}, \Delta G_{IV}\}/e + 1.23$$

8.3.4 Nitrogen Reduction Reaction (NRR)

As one of the most highly produced inorganic chemicals, ammonia (NH_3) has been widely used to produce fertilizer, plastics, fibers, and intermediates for pharmaceuticals. Nowadays, large-scale industrial synthesis of NH_3 mainly depends on the Haber–Bosch (H–B) process [44], for which, however, energy-extensive consumption, emissions of massive greenhouse gas, and extreme reaction conditions have always been the huge challenges in this process [45–49]. It is therefore highly demanded to develop new alternative technology or search for environmentally friendly and cost-effective catalysts to replace the H–B process. In the past few years, inspired by biological nitrogen (N_2) fixation, electrochemical N_2 conversion to NH_3 at ambient conditions has currently become a research hotspot in the field of electrochemistry, with the alluring merits of energy efficiency, greenness, and sustainability [50–52]. In the process of electrochemical NRR for NH_3, catalysts absolutely play the pivotal role. In current stage, NRR catalysts usually have the TM atoms being the active centers, such as Ru(0001) [53, 54]; metal oxides MoO_3 [55], MnO [56], and VO_2 [57]; as well as the SACs [58–61]. As for NRR electrocatalysts, the high overpotential and low Faraday efficiency (FE) are inevitably huge challenges [62–64], and, therefore, exploring and constructing highly active, efficient, and durable NRR catalysts are becoming increasingly important. As for SACs, it has been illustrated that the optimal TM–ligand match and local electronic property are closely related to the N_2 adsorption and activation [65–67]. In fact, however, achieving 100% individual atom dispersion is less realistic and thus the limited active sites, and the poor stability is not conducive to the recycling of the SACs. In this sense, TM-based catalysts with intrinsic TM-terminated surfaces of high stability and high activity are more promising.

In a photocatalytic process, photons are absorbed as the driving force to propel the N_2 activation and reduction; however, the efficiency is still far from satisfactory because of the weak binding strength of N_2 and inefficient electron transfer from

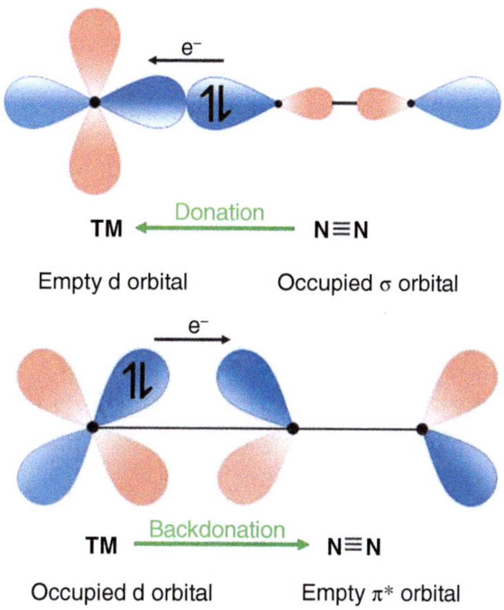

Figure 8.6 The donation–backdonation mechanism for N_2 electrochemical activation. Source: Zhang et al. [73]. Reproduced with permission of The Royal Society of Chemistry.

photocatalyst to N_2 ($N_2 + e^- \rightarrow N_2^{\bullet -}$), therefore the N≡N bond is difficult to be activated [68]. In order to improve the NRR activity, photocatalysts should facilitate the chemisorption of N_2 to guarantee the sufficient activation of the inert N≡N triple bond. On account of the coexisted empty and occupied d orbitals, TM atoms can not only accept the lone-pair electrons of N_2 to strength the TM–nitrogen bond but also donate electrons into the antibonding orbitals of N_2 to weaken the N≡N bond [69–72], see Figure 8.6. In Figure 8.7, orbitals of N_2 molecule are shown. In accordance with such a concept, nonmetal elements that can also function as TM atoms to activate N_2 are theoretically filtered.

In general, the elementary reactions of ammonia synthesis are:

$$N_2(g) + 2^* \leftrightarrow {}^{**}N_2$$

$$H_2(g) + 2^* \leftrightarrow 2H^*$$

$${}^{**}N_2 + H^* \leftrightarrow {}^{**}N_2H + {}^*$$

$${}^{**}N_2H + H^* \leftrightarrow {}^{**}N_2H_2 + {}^*$$

$${}^{**}N_2H_2 + H^* \leftrightarrow {}^{**}N_2H_3 + {}^*$$

$${}^{**}N_2H_3 + H^* \leftrightarrow 2{}^*NH_2 + {}^*$$

$${}^*NH_2 + H^* \leftrightarrow {}^*NH_3 + {}^*$$

$${}^*NH_3 \leftrightarrow NH_3(g) + {}^*$$

where * represents a metal active site.

It has been accepted that N_2 can be electrochemically reduced to NH_3 by six successive proton-coupled electron transfer reactions, following three typical reaction mechanisms, i.e. distal, alternating, and enzymatic pathways [74–77], see

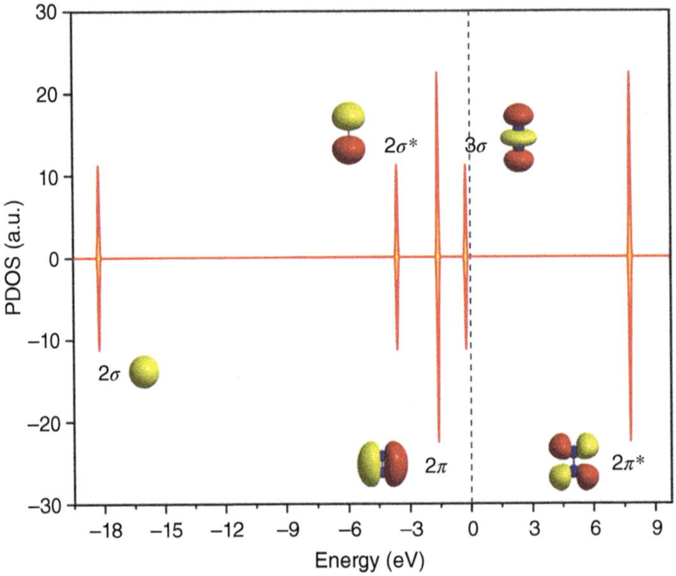

Figure 8.7 Orbitals of N_2 molecule by calculating projected density of states (PDOS). Source: Zhang et al. [73]. Reproduced with permission of The Royal Society of Chemistry.

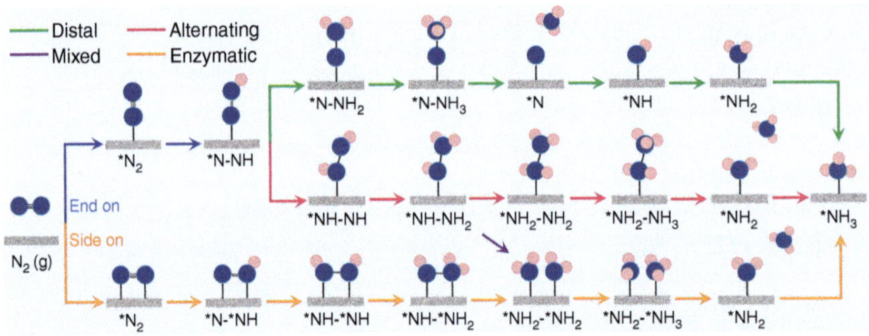

Figure 8.8 Schematic depiction of distal, alternating, enzymatic, and mixed mechanisms for N_2 reduction to NH_3. Source: Zhang et al. [73]. Reproduced with permission of The Royal Society of Chemistry.

Figure 8.8. In particular, if N_2 is initially adsorbed in the end-on mode, it will be reduced to NH_3 by two pathways (distal and alternating). As for the side-on mode, N_2 molecule is mainly reduced to ammonia by enzymatic mechanism. In the distal pathway, the proton–electron pair ($H^+ + e^-$) will first attack the terminal N atom of N_2 molecule with end-on configuration. After three protonation cycles, the first NH_3 molecule is released, and then by completing the next three-step protonation cycle N_2 will be reduced to NH_3 and released. In case of the alternating mechanism, six proton-electron pair ($H^+ + e^-$) alternately occur between two N atoms. In the enzymatic pathway, similar to the alternation mechanism, six proton-electron pair ($H^+ + e^-$) attack the two N atoms alternatively but occur in the side-on N_2 adsorption model. In order to screen out the optimal reduction pathway, all these possible routes for the selected candidates with high catalytic performance should be systematically evaluated.

In order to evaluate the reaction rate, each elementary step can be written as:

$$r_1 = k_1 p_{N_2} \theta_*^2 - k_{-1} \theta_{**N_2}$$

$$r_2 = k_2 p_{H_2} \theta_*^2 - k_{-2} \theta_{*H}^2$$

$$r_3 = k_3 \theta_{**N_2} \theta_{*H} - k_{-3} \theta_{**N_2H} \theta_*$$

$$r_4 = k_4 \theta_{**N_2H} \theta_{*H} - k_{-4} \theta_{**N_2H_2} \theta_*$$

$$r_5 = k_5 \theta_{**N_2H_2} \theta_{*H} - k_{-5} \theta_{**N_2H_3} \theta_*$$

$$r_6 = k_6 \theta_{**N_2H_3} \theta_{*H} - k_{-6} \theta_{*NH_2}^2 \theta_*$$

$$r_7 = k_7 \theta_{*NH_2} \theta_{*H} - k_{-7} \theta_{*NH_3} \theta_*$$

$$r_8 = k_8 \theta_{*NH_3} - k_{-8} p_{NH_3} \theta_*$$

where k_i is the rate constant for step i, p denotes the partial pressure of gas N_2 and NH_3, and θ denotes the coverage of the adsorbed species (N_xH_y). With the assumption that r_4 is the rate-determining step (RDS), under the quasi-equilibrium approximation (QEA) [78–80], the rates of all the other steps equal to zero ($r_1 = r_2 = r_3 = r_5 = r_6 = r_7 = 0$), and therefore the coverage of reaction species can be obtained:

$$\theta_{**N_2} = K_1 p_{N_2} \theta_*^2$$

$$\theta_{*H} = \sqrt{K_2 p_{H_2} \theta_*^2}$$

$$\theta_{**N_2H} = K_3 K_1 \sqrt{K_2 p_{H_2}} p_{N_2} \theta_*^2$$

$$\theta_{**N_2H_2} = \frac{p_{NH_3}^2 \theta_*^2}{K_8^2 K_7^2 K_6 K_5 K_2^2 P_{H_2}^2}$$

$$\theta_{**N_2H_3} = \frac{p_{NH_3}^2 \theta_*^2}{K_8^2 K_7^2 K_6 \sqrt{K_2 p_{H_2}}^{\,3}}$$

$$\theta_{*NH_2} = \frac{p_{NH_3} \theta_*}{K_8 K_7 \sqrt{K_2 p_{H_2}}}$$

$$\theta_{*NH_3} = \frac{p_{NH_3} \theta_*}{K_8}$$

where $K_i = k_i/k_{-i}$ is the equilibrium constant for step i, which can be expressed by

$$K_i = e^{-\Delta G_i/k_B T}$$

$$k_i = \frac{k_B T}{h} e^{-\Delta G_{TS}/k_B T}$$

where ΔG_i and ΔG_{TS} are free energies of reaction and activation, respectively; k_B is Boltzmann constant, and h is the Planck's constant. In this situation, the sum of coverage of all the reaction species equals to one:

$$\sum_i \theta_i = 1$$

In combination of these equations, a quadratic equation on θ_* can be solved analytically with the conversion ratio of NH_3 fixed at 10% [77], and then the rate of slow step r_3 can be obtained.

In general, the limiting potential defined as the lowest negative potential at which the reaction became exergonic is used to evaluate the intrinsic activity of NRR catalysts [81–88]. As the NRR is a complicated process, efficient screening descriptors are therefore desirable. As Figure 8.9 shows, for this purpose, a "Five-Step" strategy is proposed [89]: (1) the catalysts should possess high thermodynamic stability ($\Delta E_b < 0$ eV) and feasibility ($E_f < 0$ eV); (2) N_2 should be sufficiently activated with the $\Delta G_{*N_2} < -0.3$ eV; (3) and (4) to guarantee low energy cost, the ΔG of the first and last hydrogenation step (the most likely limiting steps) should be as low as possible with ΔG_{*N_2H} and $\Delta G_{*NH_3} < 0.55$ eV (the best catalyst Ru); (5) to guarantee the high selectivity of NRR, the maximum ΔG for NRR should be much lower than that for the competing HER. In these screening criteria, ΔE_b is the difference between binding energy of TM atoms on substrate and cohesive energy of TM atoms, E_f is

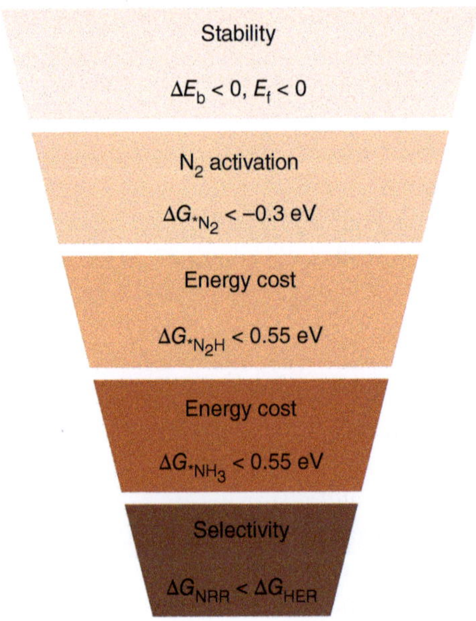

Figure 8.9 The proposed "Five-Step" strategy for screening NRR candidate catalysts. Source: Lv et al. [65]/with permission of American Chemical Society.

the formation energy, and ΔG_{*N_2}, ΔG_{*N_2H}, and ΔG_{*NH_3} are the Gibbs free energy changes of intermediates of N_2, N_2H, and NH_3, respectively.

8.3.5 Electrocatalytic Activity Evaluated from the First-principles Calculations

In a recent report, a new allotrope of graphyne, referred to as H4,4,4-graphyne, was proposed, which is composed of rectangular carbon rings and acetylenic rings in a hexagonal lattice [90]. In light of its porous structure, metal atoms can be anchored within the hole, serving as SACs. In addition to the dynamic, thermodynamic, and thermal stability, H4,4,4-graphyne shows unique electronic properties, i.e. the double Dirac points, indicative of superhigh carrier mobility and favorable electric conductivity for electrochemical charge transfer [90]. In this section, we will show how to evaluate the electrochemical HER, OER and ORR activities of SACs supported on H4,4,4-graphyne from simulations. In practice, a series of TM atoms (Sc, Ti, V, Cr, Mn, Fe, Co, Ni, Mo, Pd, Ir, and Pt) supported on the two-dimensional H4,4,4-graphyne monolayer were screened to identify the bifunctional catalytic activity for HER/OER and OER/ORR. In this section, background of electrochemical HER, OER, and ORR is briefly introduced, and the parameters for the first-principles calculations based on density functional theory are provided.

In recent years, as clean and sustainable energy hydrogen economy has been attracting increasing attention, which alleviates the global environmental problem and energy crisis. In many critical processes such as petroleum, spaceflight, metallurgy, and light industry, hydrogen is widely used [91–94], and, as discussed earlier, water electrolysis is an environmentally friendly manner to generate hydrogen. In respect to electrochemical HER, high-performance catalysts are playing a vital role. It has been well recognized that noble metals and noble-metal oxides based catalysts (e.g. Pt, RuO_2, and IrO_2) as the state-of-the-art electrocatalysts are up till now dominating in the water splitting reactions [95–98], which are however too expensive and scarce to be used on a large scale for hydrogen production. In order to seek out low-cost, high performance, and durable catalysts, therefore, a variety of emerging materials are extensively studied for water splitting, for example, Mxenes [99], TM oxides [100, 101], phosphides [102, 103], and some sulfides [104, 105].

It has already been known that the electrochemical water splitting constitutes two half-cell reactions (i.e. HER and OER). It is difficult for the unifunctional catalysts to present catalytic performance of HER and OER simultaneously under the same conditions [106, 107]. In comparison with unifunctional catalysts, bifunctional catalysts with simplified preparation process for water splitting are highly appealing [108].

In 2019 graphyne was successfully synthesized experimentally [109], in which the porosity of graphyne with sp- and sp^2-hybridized carbon atoms provides space to locate the single metal atoms, behaving as the SACs [110, 111]. It has been demonstrated that graphyne can support single metal atoms strongly at the center of the acetylenic ring, being an efficient catalyst, for example, for OER [112]. In addition, single metal atoms supported on nitrogen doped graphyne show catalytic activity for NRR [113], and single metal atoms supported graphdiyne suggest desirable catalytic performance for CO_2RR [114]. It is therefore conclusive that porous carbon-based monolayer can be a platform for studying the relationship of structure–electronic property–activity of electrocatalysts, thus advancing the design principle for catalysts. It is of interest that more and more carbon-based materials in two dimensions beyond graphene have been realized experimentally, providing new possibilities for model study on electrocatalysts.

In experiments, cyclobutadiene (C_4H_4) and acetylene (C_2H_2) as reactants can produce H4,4,4-graphyne by dehydrogenation. In particular, the dehydrogenation process of C_4H_4 and C_2H_2

$$3C_4H_4 + 6C_2H_2 \rightarrow H4{,}4{,}4 - \text{graphyne} + 12H_2$$

is an exothermic reaction, which indicates that the reaction is thermodynamically allowed [90]. In addition, a number of experiments have shown that two-dimensional carbon materials of porosity such as graphdiyne [115], γ-graphyne [116], and graphtetrayne [117] could be synthesized, enriching the graphyne family. It should be highlighted that, additionally, the successful synthesis of 4–6 carbophene [118] (4-carbon and 6-carbon rings in ratio of 1 : 1) further consolidates the feasibility of synthesizing H4,4,4-graphyne with rectangular carbon rings and triple bonds of carbon. It should be emphasized here that the holes of large size in graphdiyne provide stable anchoring sites also for metal dimer, metal trimer, and even metal clusters, which show probably better catalytic activity and selectivity than the single atom counterparts, deserving further investigations [114].

In principle, for SACs the TM–substrate coordination and local electronic properties are determinative in activating the adsorbed intermediates for various electrochemical reactions, such as HER, OER, ORR, and NRR [119, 120]. In the process of binding, charge transfer occurs between TM atoms and the substrate, accompanied with changes in local structure and electronic properties. It is thus conclusive that a key in the design principle for SACs with high stability and excellent catalytic activity is to find a suitable match between the metal and support for a specific reaction. In respect to the supports for SACs, two-dimensional materials such as graphene, graphene-like carbon nitrides (such as g-CN and g-C_3N_4), and borophene are increasingly drawing interest, which is ascribed to the large specific surface area and large number of sites for anchoring metal atoms [121, 122]. As examples, SACs such as Ni/g-CN [96], Co@GY/GY [123], Pt@MoS_2 [124], and Ni@β_{12}-BM [125] have been theoretically predicted to be HER and OER catalysts of high performance.

In case of H4,4,4-graphyne, transition atoms Sc, Ti, V, Cr, Mn, Fe, Co, Ni, Mo, Pd, Ir, and Pt are taken into account to evaluate their catalytic activity by the first-principles calculations. In the following, calculation parameters are provided. In practical calculations, geometry optimization should be first conducted, then the material stability and the electronic properties. In consideration of the magnetic atoms, spin-polarized density functional theory as implemented by Vienna ab initio simulation package (VASP) is employed [126, 127], with the Perdew–Burke–Ernzerhof (PBE) in framework of generalized gradient approximation (GGA) describing the exchange–correlation functional [128, 129]. In order to describe the electron–ion interactions, the projector-augmented wave

(PAW) method [130, 131] is used. In terms of van der Waals (vdW) interactions, the vdW–D2 corrections are carried out in all calculations [132]. In addition, cut-off energy for the plane-wave basis is set to be 550 eV and all structures will cease to be optimized when the convergence of total energy and force were less than 10^{-5} eV and 0.01 eV Å^{-1}, respectively. In cases of geometry optimization and electronic self-consistent calculations, Monkhorst–Pack k-point meshes of $3 \times 3 \times 1$ and $5 \times 5 \times 1$ are adopted, respectively. In consideration of the periodicity of the structure, a vacuum space of 20 Å was applied normally to the surface plane to avoid the interactions between periodic images. In addition, AIMD simulations [133] at 500 K for 5 ps with a time step of 1 fs were performed to evaluate the thermodynamic stability of the catalysts. In particular, a canonical ensemble is simulated using the algorithm of Nosé, controlling the frequency of the temperature oscillations during the simulations.

In regard to HER, the activity is usually assessed by means of the reaction-free energy (ΔG_{H^*}) in accordance with

$$\Delta G_{H^*} = \Delta E_{H^*} + \Delta E_{ZPE} - T\Delta S_{H^*}$$

where ΔE_{H^*} stands for the hydrogen adsorption energy obtained from calculations based on density functional theory, ΔE_{ZPE} and ΔS_{H^*} are the corresponding changes in zero-point energy and entropy between the adsorbed hydrogen (H^*) and hydrogen in gas phase (H_2), respectively, calculated from frequency calculation at $T = 298.15$ K.

In analogy to HER, for OER the Gibbs free energy change for each element step is:

$$\Delta G_{H^*} = \Delta E + \Delta E_{ZPE} - T\Delta S - neU + \Delta G_{pH}$$

with ΔE being the electronic energy difference between the product and reactant of each element step in OER process, and ΔE_{ZPE} and ΔS are the differences in zero-point energy and entropy, respectively; e is the electron number transferred and U stands for the applied electrode potential; ΔG_{pH} corresponds to the correction to pH value. In general, for calculations the pH value is usually set to zero.

As shown in Figure 8.10, the hexagonal framework of H4,4,4-graphyne (from top view) is composed of rectangular carbon rings and acetylenic rings, with the unit cell denoted by a rhombus. In a unit cell of H4,4,4-graphyne, 24 carbon atoms are connected via —C—C— single bonds and -C≡C- triple bonds. In accordance with the ground-state calculation, the optimized lattice constant of hexagonal H4,4,4-graphyne is 11.82 Å, and the corresponding bond lengths are $l_1 = 1.247$, $l_2 = 1.351$, $l_3 = 1.453$, and $l_4 = 1.489$ Å, which are in good agreement with the first-principles results by others [90].

In Figure 8.10, five possible anchoring sites for TM atoms (here Sc, Ti, V, Cr, Mn, Fe, Co, Ni, Mo, Pd, Ir, and Pt) on H4,4,4-graphyne are considered: center of the acetylenic ring (H_1), center of the rectangular ring (H_2), center of the 24-membered carbon ring (H_3), corner of the acetylenic ring (C_1), and corner of 24-membered carbon ring (C_2). It can be found that after full structure optimization the system undergoes severe distortion as TM atoms anchored at H_2, H_3, and C_2 sites. In addition, TM atoms located at C_1 site would be spontaneously converged to the H_1 site.

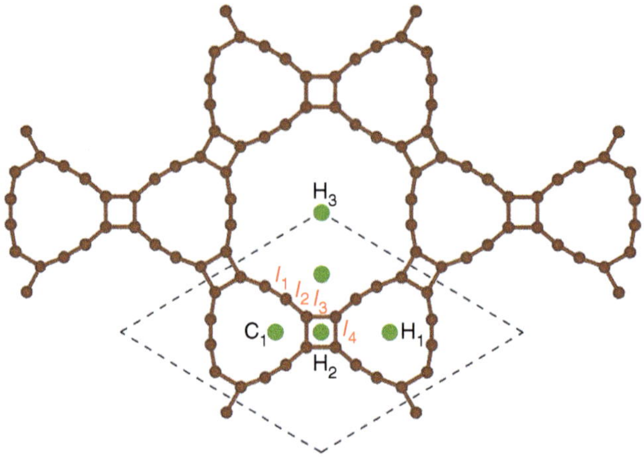

Figure 8.10 Structure of H4,4,4-graphyne from top view; bond lengths are denoted. The possible adsorption sites for transition metal atoms are marked. Source: Zhang et al. [30]. Reproduced with permission of The Royal Society of Chemistry.

In cases for V, Cr, and Mo, adsorption of these atoms on H4,4,4-graphyne at all possible sites will cause considerable deformation, and thus will not be discussed later.

In order to address the binding strength of TM atoms on H4,4,4-graphyne, binding energy is first calculated according to

$$E_b = E_{TM@H4,4,4\text{-}GY} - E_{H4,4,4\text{-}GY} - E_{TM}$$

where $E_{TM@H4,4,4\text{-}GY}$, $E_{H4,4,4\text{-}GY}$, and E_{TM} are the total energies of TM atom adsorbed H4,4,4-graphyne, H4,4,4-graphyne, and TM atom, respectively. In order to calculate the energy for a single transition atom, i.e. the E_{TM}, a cubic cell of $20 \times 20 \times 20$ Å is used.

In Table 8.2, binding energies for the rest of TM atoms are summarized, indicating that they prefer to be stably fixed at the H_1 site and the binding energies are smaller than −1.0 eV (except for Ti). In accordance with the definition of the binding energy, negative value illustrates strong interaction between the TM atom and the sp-hybridized carbon atoms of the acetylenic ring, and thus it concludes that the TM atoms are energetically favorable to be firmly anchored on the substrate.

Table 8.2 Binding energy (E_b) for transition metal atoms of consideration adsorbed on H4,4,4-graphyne.

TM atoms	Sc	Ti	Mn	Fe	Co
E_b	−3.70	0.34	−1.47	−2.30	−3.43
TM atoms	Ni	Pd	Ir	Pt	Co_2
E_b	−3.56	−1.88	−3.51	−3.05	−3.45

Source: Zhang et al. [30]. Reproduced with permission of The Royal Society of Chemistry.

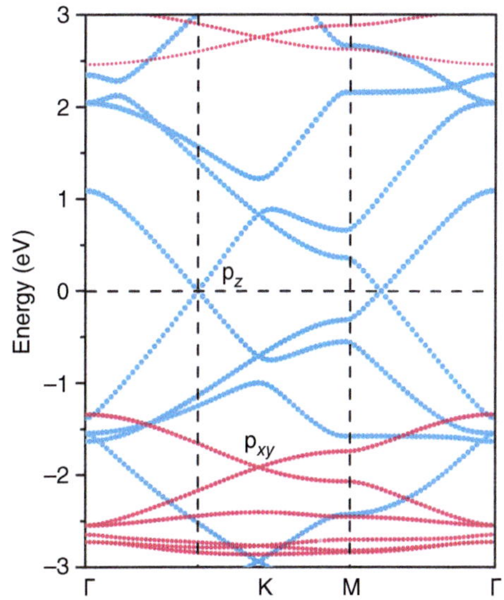

Figure 8.11 Orbital-projected band structure for H4,4,4-graphyne with two Dirac points along Γ–K and Γ–M directions, indicative of ultrahigh carrier mobility and desirable electric conductivity for charge transfer. The Fermi level is set to zero; contributions from carbon $p_{x,y}$ and p_z orbitals are denoted. Source: Zhang et al. [30]. Reproduced with permission of The Royal Society of Chemistry.

In addition, like the graphdiyne, the uniform large holes provide more high active sites to strongly anchor metal atoms and effectively inhibit the formation of clusters. In order to see whether the single metal atoms will aggregate into metal clusters, the migration energy can be calculated by the CINEB method. As an example, the diffusion barrier for one Co atom moving to another stable site is as high as 1.10 eV, which will prevent the Co atoms from agglomerating into clusters, indicating the kinetic stability of the supported Co atoms on the H4,4,4-graphyne surface.

In addition to firmly bind the TM atoms, desirable electronic properties are important for the activity of SACs. In Figure 8.11, the band structure of pristine H4,4,4-graphyne exhibits double Dirac points along Γ–K and Γ–M directions, indicative of ultrahigh carrier mobility and desirable electric conductivity for charge transfer. In order to have a more intuitive view on the interaction between TM atoms and the substrate, band structures are also calculated for the TM adsorbed H4,4,4-graphye. In most cases, the system preserves at least one Dirac point, and some of them change from semimetallic to be metallic. It is an indication that the desirable electric conductivity for electrocatalytic reactions could be maintained, and thus can enhance the activity. In the case of Co@H4,4,4-graphyne (see Figure 8.12), for example, inversion symmetry breaking results in gap opening at one Dirac point, and it rationalizes the intact double Dirac points Co_2@H4,4,4-graphyne of inversion symmetry. It is therefore indicative that the coordination between metal atoms and substrate could effectively adjust the electronic structure, which plays a crucial role in the catalytic performance of SACs.

In Figure 8.13a, the Gibbs free energy change of adsorbed H^* (ΔG_{H^*}) for different TM atoms is presented, with that of pristine H4,4,4-graphyne also shown for comparison. In the case of pristine H4,4,4-graphyne, hydrogen adsorption prefers the sp-hybridized carbon sites with an unsatisfied $\Delta G_{H^*} = 0.40$ eV. As for hydrogen adsorption on TM atoms, hydrogen atom tends to be located just above the metal

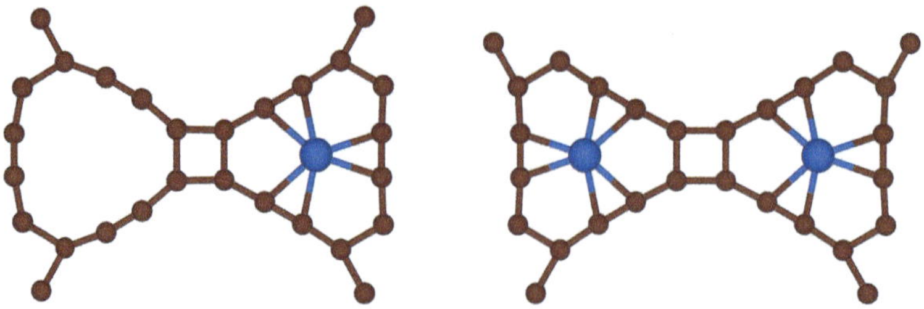

Figure 8.12 Transition metal (the bigger spheres) adsorption on H4,4,4-graphyne within a hollow site. It can be found that introduction of transition metal atoms shrinks the lattice. Source: Zhang et al. [30]. Reproduced with permission of The Royal Society of Chemistry.

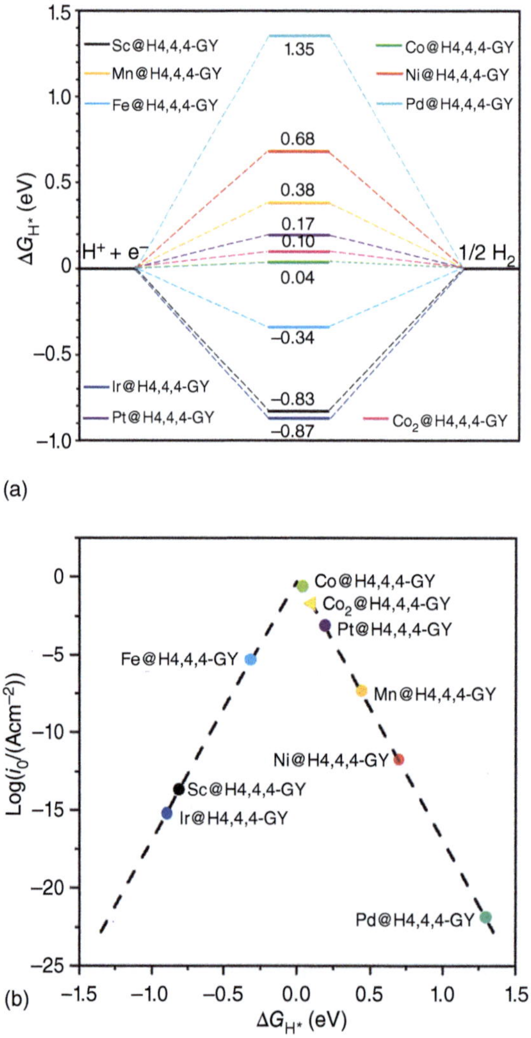

Figure 8.13 (a) Gibbs free energy change of hydrogen adsorption for transition metal atoms of interest supported on H4,4,4-graphyne (TM@H4,4,4-GY) under standard conditions. (b) Volcano curve of exchange current density (i_0) related to the Gibbs free energy change (ΔG_{H^*}) of hydrogen adsorption. Source: Zhang et al. [30]. Reproduced with permission of The Royal Society of Chemistry.

atom. In cases of Sc/Fe/Ir@H4,4,4-graphyne, the excessive negative ΔG_{H*} values (−0.83/−0.34/−0.87 eV, respectively) are not conducive to the desorption of H_2 molecule from the catalyst surface. On the contrary, the large positive ΔG_{H*} values for Mn/Ni/Pd@H4,4,4-graphyne indicate that a large amount of energy should be injected in the process of protonation. It has already been known that the ideal change of Gibbs free energy for HER should be close to zero (Sabatier principle) [134, 135]. In accordance with this, thus Co@H4,4,4-graphyne and Pt@H4,4,4-graphyne could be promising HER catalysts of high catalytic performance, with the ΔG_{H*} being as low as 0.04 and 0.17 eV, respectively. In comparison with currently commercialized Pt (0.09 V) [136], the corresponding overpotential of Co@H4,4,4-graphyne is even lower.

In the premise of the desirable catalytic HER activity of Co@H4,4,4-graphyne, its efficiency is further explored by increasing the active sites (within the same lattice, two Co atoms are considered, Co_2@H4,4,4-graphyne, see Figure 8.12). It can be found that Co_2@H4,4,4-graphyne still shows acceptable catalytic performance with a low overpotential of 0.1 V. In my opinion, the difference in the overpotentials for systems constituting one and two Co atoms originates partially from the change in electronic properties. In addition, AIMD simulation results at 500 K lasting for 5 ps confirm that Co@H4,4,4-graphyne, Co_2@H4,4,4-graphyne, and Pt@H4,4,4-graphyne are thermodynamically stable, since the structures do not show obvious distortion and the TM atoms are still firmly anchored on H4,4,4-graphyne. It can be concluded that, therefore, these SACs could be used as catalysts for HER with high activity and high stability.

In addition, the exchange current density (i_0) can be used as a descriptor to evaluate the performance of the catalysts. In accordance with the Nørskov's assumption [136], a linear relationship between i_0 and the Gibbs free energy change ΔG_{H*} can be obtained, as shown in Figure 8.13b. It can be seen from the volcano curve that Sc@H4,4,4-graphyne, Fe@H4,4,4-graphyne, and Ir@H4,4,4-graphyne on the left branch show too strong hydrogen adsorption and thus difficult for H_2 molecule desorption. On the other hand, Mn@H4,4,4-graphyne, Ni@H4,4,4-graphyne, and Pd@H4,4,4-graphyne on the right branch are not conducive to the hydrogen adsorption. It is obvious that Co@H4,4,4-graphyne, Co_2@H4,4,4-graphyne, and Pt@H4,4,4-graphyne are positioned near the peak of the volcano curve, indicative of promising HER performance.

It is well known that H_2O molecule can be converted to O_2 by four-step elementary reactions, that is, the well-accepted OER process. In the first step, as Figure 8.14 shows, H_2O molecule is dissociated into H^+ and $^*OH$ with the help of catalyst. In the second step, $^*OH$ is further decomposed into H^+ and $^*O$, and the third step is $^*O$ reaction with another H_2O molecule and generates $^*OOH$ species. In the last step, the final product O_2 molecule is formed and desorbs from the surface of catalyst. In Figure 8.15, the Gibbs free energy change (ΔG) for each elemental step for TM atoms deposited on H4,4,4-graphyne in OER process is presented, with the corresponding potential-determining step (PDS) and overpotential denoted.

In respect to Sc@H4,4,4-graphyne, Mn@H4,4,4-graphyne, Fe@H4,4,4-graphyne, and Ir@H4,4,4-graphyne, the negative ΔG_{*OH} illustrate too strong binding between

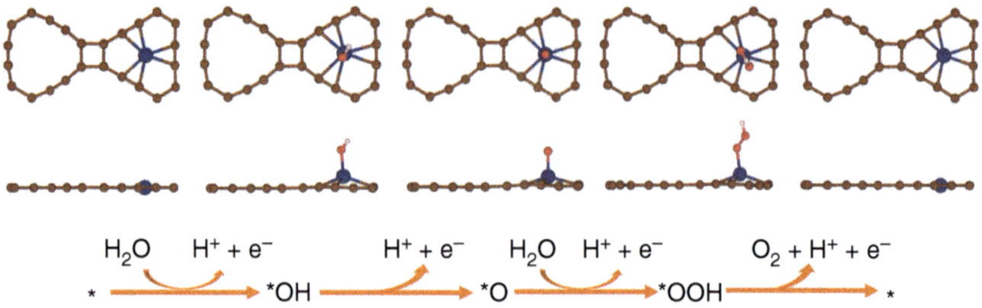

Figure 8.14 Schematic of the four-step elementary reactions for the OER process, where * represents the metal active site on transition metal atoms supported H4,4,4-graphyne. Source: Zhang et al. [30]. Reproduced with permission of The Royal Society of Chemistry.

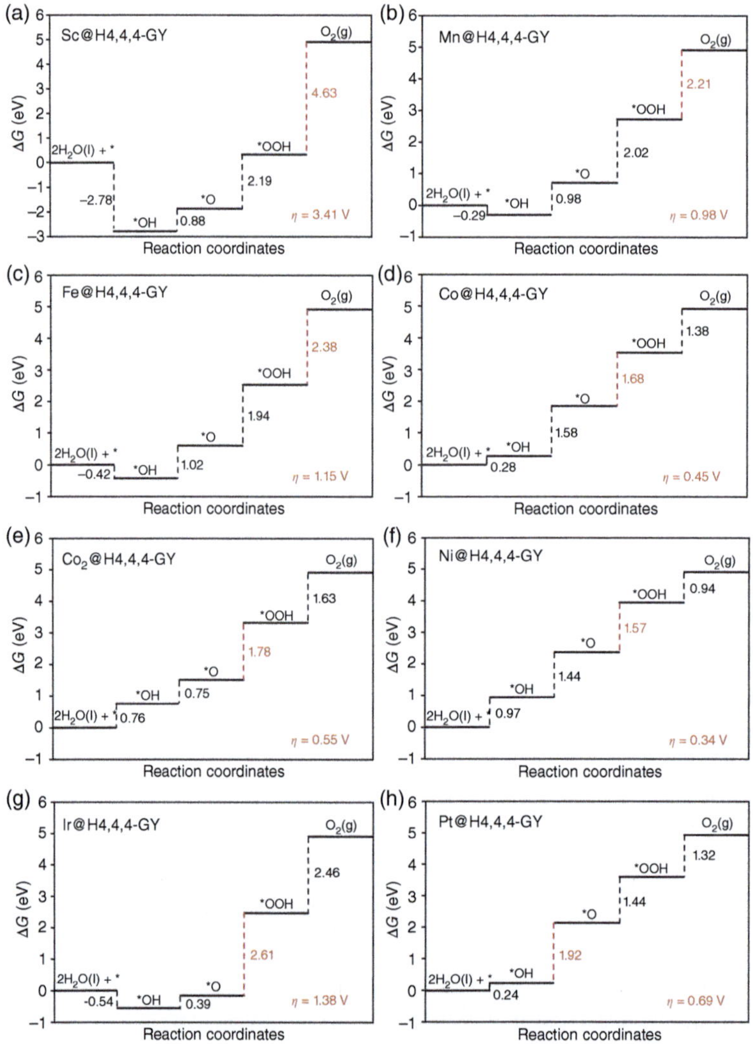

Figure 8.15 Gibbs free energy diagram of transition metal atoms on H4,4,4-graphyne (TM@H4,4,4-GY) for OER at $U = 0$, where the potential-determining step (PDS) of each elementary reaction is marked in burgundy (a) Sc, (b) Mn, (c) Fe, (d) Co, (e) Co_2, (f) Ni, (g) Ir, and (h) Pt. Source: Zhang et al. [30]. Reproduced with permission of The Royal Society of Chemistry.

the reaction intermediate (*OH) and the substrate, which is unfavorable to the decomposition of *OH in next reaction. In this case, the PDS usually occurs in the process of *OOH production (ΔG_3) or O_2 desorption (ΔG_4).

In Figure 8.15, the overpotentials are labeled. In particular, the values are 3.41, 0.98, 1.15, and 1.38 V for Sc, Mn, Fe, and Ir on H4,4,4-graphyne, which indicate that a large amount of energy is needed to drive the OER process. In regard to Co@H4,4,4-graphyne, Ni@H4,4,4-graphyne, and Pt@H4,4,4-graphyne, the Gibbs free energy change for the four-step chemical reaction is uphill with low overpotentials. It is noteworthy that the overpotentials of Co@H4,4,4-graphyne (0.45 V), Ni@H4,4,4-graphyne (0.34 V), and Pt@H4,4,4-graphyne (0.69 V) are comparable to or even lower than that of commercially available OER catalyst IrO_2 (0.55 V) [137], indicating that these systems could be considered as potential catalysts for OER.

In general, the catalytic OER performance can usually be described by the Gibbs free energy change for three intermediates, i.e. ΔG_{*OH}, ΔG_{*O}, and ΔG_{*OOH}. In order to efficiently screen out the highly active SACs for OER, a relationship for the Gibbs free energy change of the intermediates was established. As shown in Figure 8.16a, the scaling relationship between ΔG_{*OOH} and ΔG_{*OH} can be described according to

$$\Delta G_{*OOH} = 1.07\Delta G_{*OH} + 3.71$$

It can be found that a quite fine linear relationship between ΔG_{*OOH} and ΔG_{*OH} is obtained with a coefficient of determination (R^2) of 0.99.

In the four-step elementary reactions of OER, the difference in Gibbs free energy can be expressed simply by two descriptors based on the scaling relationship:

$$\Delta G_1 = \Delta G_{*OH}$$

$$\Delta G_2 = \Delta G_{*O} - \Delta G_{*OH}$$

$$\Delta G_3 = \Delta G_{*OOH} - \Delta G_1 - \Delta G_2$$

$$\Delta G_4 = 4.92 - \Delta G_{*OOH}$$

In accordance with the relationships mentioned earlier, remarkably, the overpotential can be presented visually by the relationship between three descriptors in the OER volcano plot. In Figure 8.16b, the volcano curve is shown, which is composed of four regions and each region corresponds to the Gibbs free energy difference between products and reactants. It can be found that for Sc/Mn/Fe/Ir@H4,4,4-graphyne the negative ΔG_{*OH} cause zone 4 or zone 3 to be the PDS in the process of accelerating OER. It is clear that Co@H4,4,4-graphyne, Ni@H4,4,4-graphyne, and Pt@H4,4,4-graphyne with relatively low overpotential (0.45, 0.34, and 0.69 V, respectively) are located near the volcano peak. In particular, Co@H4,4,4-graphyne and Ni@H4,4,4-graphyne might exhibit higher catalytic activity compared with other 2D electrocatalytic materials such as Co@g-CN [96], NiIT [138], and Ni/B_{36} [139].

In order to have an intensive view on the catalytic performance of SACs toward OER, the electronic properties of the catalysts should be examined to find the

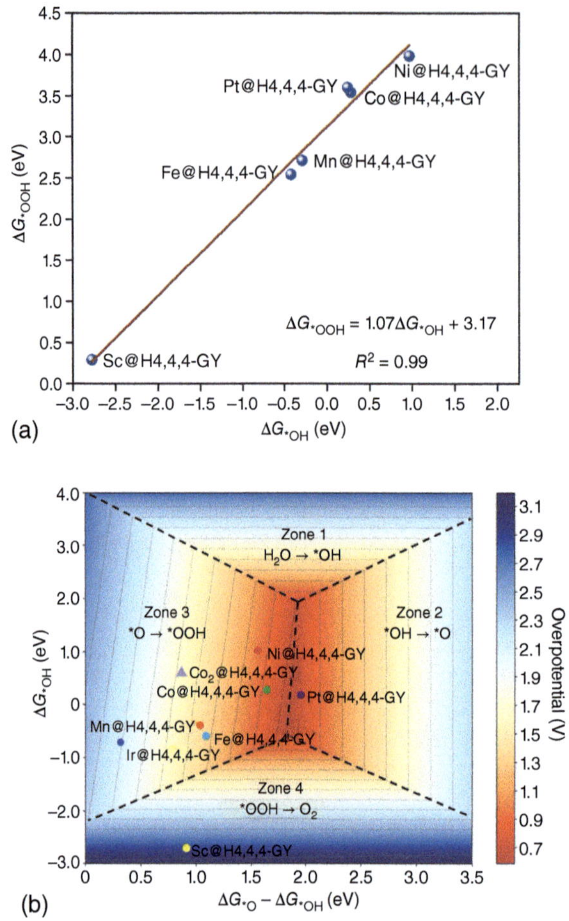

Figure 8.16 (a) Scaling relationship for the Gibbs free energy change of adsorbed *OOH (ΔG_{*OOH}) and *OH (ΔG_{*OH}). (b) OER activity volcano showing the ΔG_{max} as a function of Gibbs free energies change of the reaction intermediates; color bar represents the OER overpotential. Transition metal atom adsorbed H4,4,4-graphyne is denoted as TM@H4,4,4-GY. Source: Zhang et al. [30]. Reproduced with permission of The Royal Society of Chemistry.

physical origin. In general, *d* band center of metal atom is usually used as a descriptor to evaluate the catalytic activity. In particular, the *d* band centers of TM atoms could be calculated by analyzing the projected density of states (PDOS). In case of TM atom considered in this work, the *d* band centers of TM atoms move to low energy side with respect to the Fermi level with increased *d* electron occupation. In order to evaluate the correlation between the *d* band center and the catalytic performance for OER process, the relationship between the Gibbs free energy change of reaction intermediates and the *d* band center is constructed. As indicated in Figure 8.17a, the negative relationship suggests that the binding strength between the intermediates and the metal decreases with the increase of the *d* band center. In this case, Sc/Mn/Fe@H4,4,4-graphyne with the *d* band center closing to the Fermi level demonstrates relatively strong interaction with intermediates, resulting in high overpotentials for OER. On contrary, Co/Ni/Pt@H4,4,4-graphyne with more negative *d* band center imply that the binding strength between TM atoms and intermediates become weaker, thus exhibiting lower overpotentials. It should be emphasized that Co@H4,4,4-graphyne not only shows efficient catalytic

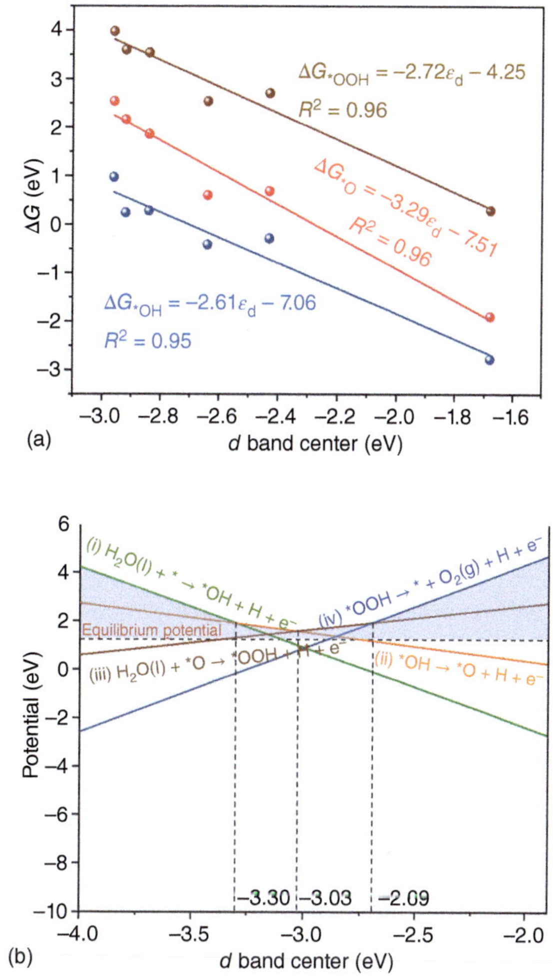

Figure 8.17 (a) Scaling relationships between Gibbs free energy change of each intermediate (*OH, *O, and *OOH) and *d* band center of transition metal atoms supported on H4,4,4-graphyne. (b) Potential of OER as a function of the *d* band center; the shaded area represents the overpotential. Source: Zhang et al. [30]. Reproduced with permission of The Royal Society of Chemistry.

performance for HER but also has desirable activity for OER. In other words, Co@H4,4,4-graphyne could be the appealing candidate as bifunctional catalyst for electrochemical water splitting.

In consideration of the perfect linear relationship between the Gibbs free energy change of reaction intermediates and the *d* band center of the TM atoms of interest, the relationship between *d* band center and Gibbs free energy change corresponding to each step of elementary reaction should be further studied to visualize the impact of *d* band center on the OER electrocatalytic activity. As shown in Figure 8.17b, it can be found that low and high *d* band centers signify weak and strong interactions between metal atoms and intermediates, respectively; too weak or too strong interaction usually leads to high overpotentials. In the former case, the first step is the PDS because few proton–electron pairs are transferred from TM atoms to the intermediates. In the latter case, while the interaction between the intermediate (*OOH) and the active sites is so strong that it is difficult for O_2 molecule desorption. As a result, the fourth step is generally the PDS. As an example, Sc@H4,4,4-graphyne with relatively high *d* band center shows large overpotential in the last elementary reaction step. It is of interest to find that when *d* band center is in a moderate range (from −3.30 to −2.69 eV), the binding strength between the intermediate (*OOH) and TM atoms will be neither too weak nor too strong. It can be seen that Co@H4,4,4-graphyne (−2.84 eV), Co_2@H4,4,4-graphyne (−3.18 eV), Ni@H4,4,4-graphyne (−2.96 eV), and Pt@H4,4,4-graphyne (−2.92 eV) with *d* band center located within this range demonstrate better catalytic performance for OER. It is worth noting that the *d* band center of TM atoms is adjustable experimentally by constructing SACs with different TM atom ratio. As a rational descriptor, thus *d* band center could be reasonably regulated to improve the catalytic performance of OER.

As the inverse process of OER, ORR plays a crucial role in fuel cells and metal–air batteries, which can also be described by the four-step elementary reactions. In this case, the corresponding relation of the four-step elementary reactions can be expressed by two descriptors ΔG_{*O} and $\Delta G_{*O} - \Delta G_{*OH}$, and the overpotential can be calculated by $\eta = 1.23 - \min\{\Delta G_4, \Delta G_3, \Delta G_2, \Delta G_1\}/e$. In the SACs considered in this work, Ni@H4,4,4-graphyne demonstrates desirable catalytic performance with a lower overpotential of 0.29 V in the process of ORR.

In conclusion, porous H4,4,4-graphyne shows potential to be a metal-free substrate for SACs. In particular, Co@H4,4,4-graphyne and Pt@H4,4,4-graphyne could be efficient bifunctional catalysts for water splitting with low overpotentials of 0.04/0.45 and 0.17/0.69 V for HER/OER, respectively. In addition, Ni@H4,4,4-graphyne as bifunctional catalyst also exhibits desirable catalytic activity for OER/ORR with low overpotentials of 0.34/0.29 V, even superior to commercial IrO_2 and RuO_2. It is of paramount importance that the TM atom–substrate coordination and local electronic properties play a crucial role in the electrocatalysis processes. In line with literatures, *d* band center as an effective descriptor could be adopted to optimize the catalytic performance of the catalysts, and OER catalytic performance can be significantly improved by adjusting the *d* band center to an appropriate value (−3.03 eV in this work).

8.3.6 Simulations for Nitrogen Reduction Reaction

In past few years, emerging two-dimensional TM borides (MBenes), as the boron analog of MXenes, has aroused extensive concern due to their structural diversity and fascinating properties [140–143]. As potential electrocatalysts of high performance, these TM borides with large specific surface area and high active center density exhibit intrinsic basal-plane activity and selectivity for electrochemical reactions [144–148]. In addition, TM borides also promise good stability since electron-donating metal atoms will be bonded with electron-deficient boron atom via covalent bonds. In order to achieve the optimal efficiency for diverse TM borides as electrochemical catalysts for NRR, it is necessary to explore their composition space and establish a rational design principle to satisfy the demand for high activity and high selectivity.

In this work, new structures of TM atom decorated honeycomb hexagonal borophene that are different from previous results are predicted. In particular, TM atoms are deposited over the B–B bonds of the hexagonal boron ring from both sides, forming the TM-shrouded borophene. It is of interest to find that the new structures are energetically more stable than those reported previously, giving large significance for the unrevealed properties and holding great promise in applications. On account of the charge transfer (from the metal to the boron), TM atoms are firmly anchored on the surface and the electronic properties are altered significantly by boron. In view of this, TM atom terminated borides with large specific surface area and high active center density may hide the great potential as effective catalysts toward electrochemical reactions.

In this work, 13 two-dimensional TM borides in formula of TMB_2 (TM = Ti, V, Cr, Mn, Fe, Co, Nb, Mo, Tc, Ru, W, Re, and Os) with high stability and great experimental accessibility are explored, and the potential of these TMB_2 as NRR catalysts is examined by the extensive first-principles calculations. In order to address the feasibility of these TMB_2 as NRR catalysts, three key factors are taken into account: (1) the activation of N_2, (2) the energy barrier of PDS, and (3) the selectivity toward HER and NRR [149–156]. In accordance with the preceding screening criteria, ReB_2 is identified as a highly appealing catalyst toward NRR with high activity and high selectivity, with a record-low limiting potential of −0.05 V and FE of 100%. In this work, we not only identify an efficient NRR electrocatalyst in particular, paving a way for sustainable NH_3 production, but also explain the chemical and physical origin of the activity, advancing the design principle for catalysts for various reactions in general.

In previous studies, layered OsB_2 and RuB_2 in bulk phase were synthesized experimentally, confirming the crystal structure and lattice constants by means of the powder X-ray diffraction [157, 158]. In this work, the predicted TMB_2 structrues are actually identical to the free-standing OsB_2 and RuB_2 monolayers. In analogy to MXenes and MBenes, TMB_2 could probably be exfoliated from the corresponding bulk counterparts. In particular, the exfoliation of non-van der Waals materials has been developed experimentally [159], providing new choice for obtaining new two-dimensional structures. As shown in Figure 8.18a of the new TMB_2 monolayer, the smallest repeating unit cell is a rectangle one, containing four B and two TM atoms. In particular, the inner borophene wrinkles like phosphorene and the transition atoms are coordinated by six boron atoms. In this work, 29 transition atoms (3d, 4d, and 5d) were then taken into account to see the possibility to stabilize the honeycomb hexagonal borophene, see Figure 8.19, and 13 among them were confirmed to be stable and form the new family of two-dimensional materials, the TMB_2 (TM = Ti, V, Cr, Mn, Fe, Co, Nb, Mo, Tc, Ru, W, Re, and Os). In respect to the material stability, it will be discussed later.

As Figure 8.18b shows, the band structure of representative ReB_2 indicates that TM borides are metallic, and the others are also metallic. It should be highlighted that the metallicity signifies the high electric conductivity, which is necessary for the charge transfer in electrochemical reactions. In addition, charge density difference indicates that electrons transfer from TM atom to electron-deficient B, achieving the stabilization of TMB_2 via covalent bonds.

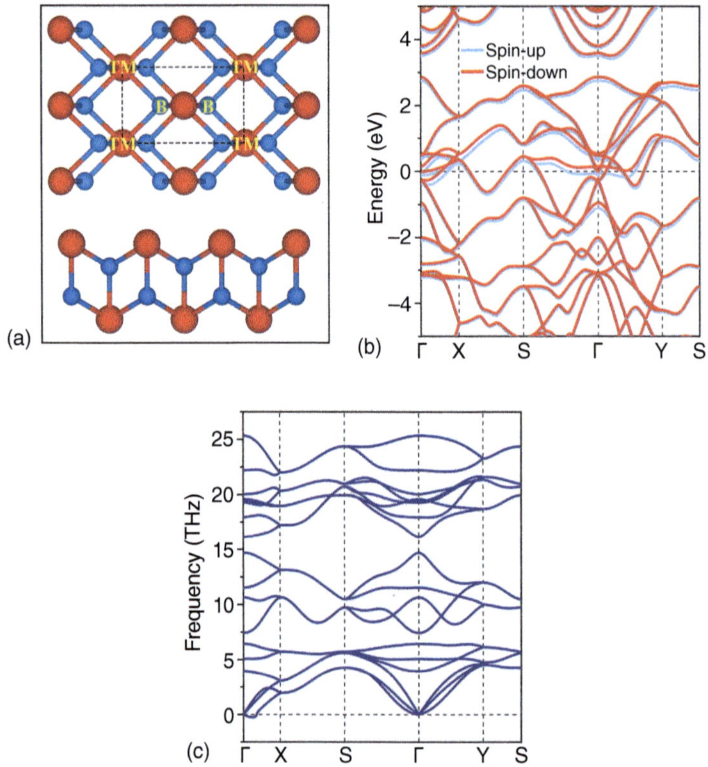

Figure 8.18 (a) Optimized structure of TMB_2, with the rectangle denoting the unit cell. (b) Band structure and (c) phonon spectrum of ReB_2. Source: Zhang et al. [73]. Reproduced with permission of The Royal Society of Chemistry.

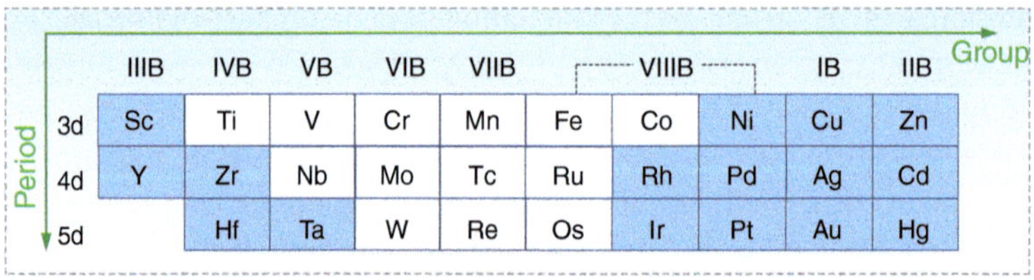

Figure 8.19 Twenty-nine transition metal atoms used to construct TMB_2 structure; green and blue backgrounds indicate stable and unstable TMB_2 candidates, respectively. Source: Zhang et al. [73]. Reproduced with permission of The Royal Society of Chemistry.

In respect to the stability, phonon dispersions of these TMB_2 show no imaginary frequency over the Brillouin zone, thus confirming the dynamic stability, see Figure 8.18c for the phonon spectrum for the representative ReB_2. In Figure 8.20a, cohesive energies are provided for 13 TMB_2 calculated according to

$$E_{\text{coh}} = \frac{(4nE_{\text{B}} + 2nE_{\text{TM}} - 2nE_{\text{TMB}_2})}{6n}$$

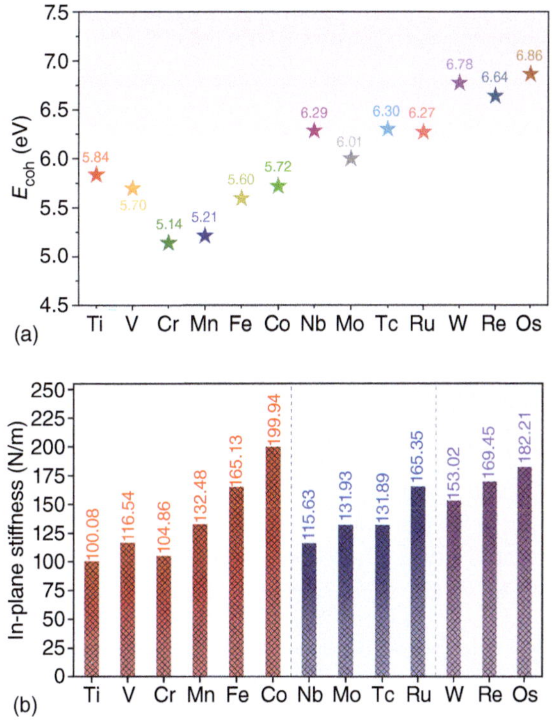

Figure 8.20 (a) Calculated cohesive energy of 13 TMB_2 monolayers. (b) In-plane stiffness of the 13 stable TMB_2 monolayers. Source: Zhang et al. [73]. Reproduced with permission of The Royal Society of Chemistry.

where E_B, E_{TM}, and E_{TMB_2} are the total energies of B atom, TM atom, and TMB_2, respectively. It can be seen that all the TMB_2 structures possess larger cohesive energy than MnB (4.80 eV atom^{-1}), Mn_2C (4.42 eV atom^{-1}), and MnN_2 (3.45 eV atom^{-1}) [45], suggesting good thermodynamic stability. In consideration that surface fluctuation can significantly affect the long-range order of two-dimensional crystals, the mechanical stability of the TMB_2 can be evaluated by calculating the in-plane stiffness:

$$C_{2D} = \frac{(\partial^2 E_{total}/\partial \varepsilon^2)}{S_0}$$

where E_{total} represents the total energy of TMB_2, ε stands for the applied uniaxial strain, and S_0 is the area of the optimized unit cell. In Figure 8.20b, results are summarized. It can be found that the values (100.1–199.9 N m^{-1}) are comparable or even higher than that of MoS_2 monolayer (125 N m^{-1}) [24] and phosphorene (21–92 N m^{-1}) [25, 26], indicating that our TMB_2 structures are mechanically more stable.

As a consequence of the interaction between TM atom-d and B-p orbitals, the electronic structures of exposed metal atoms will be adjusted and then will possibly serve as active centers for various electrochemical reactions such as HER, OER, and NRR. It is of interest that the characteristic geometry of the TMB_2 guarantees the large reaction area and high active center density. In comparison with the defect-based active sites, TMB_2 as electrocatalysts thus indicates irresistible superiority.

In principle, the activity of exposed metal atoms can be qualitatively evaluated by the Fukui functions:

$$F^{+}(\boldsymbol{r}) = \rho^{N+1}(\boldsymbol{r}) - \rho^{N}(\boldsymbol{r})$$

$$F^{-}(\boldsymbol{r}) = \rho^{N}(\boldsymbol{r}) - \rho^{N-1}(\boldsymbol{r})$$

In the equations, $\rho^{N+1}(\boldsymbol{r})$ and $\rho^{N-1}(\boldsymbol{r})$ are the total charge densities of the system with one extra electron and hole doped, respectively; $\rho^{N}(\boldsymbol{r})$ is that of the neutral system. As shown in Figure 8.21a and b, Fukui functions indicate that $F^{+}(\boldsymbol{r})$ and $F^{-}(\boldsymbol{r})$ are dominantly localized over the TM atoms, which means that the TM atoms are ready to obtain charge from electron donors and donate charge to acceptor species, and thus intrinsically active for NRR. It has already been identified that N_2 capture, $^*N_2H$ formation, and the protonation $^*NH_2 \rightarrow {}^*NH_3$ are the usuall critical steps for an NRR process [160–164]. As the starting point for NRR, especially, N_2 adsorption and activation initiate the subsequent reduction reactions. On the catalyst surface, N_2 adsorption takes two patterns, i.e. end-on and side-on adsorption. In comparison to the gas-phase N_2 molecule (1.11 Å), N–N bond length is elongated when N_2 is adsorbed on these TMB_2, indication of the effective activation. As Figure 8.21c and d shows, charge density difference for N_2 adsorption on ReB_2 clearly indicates

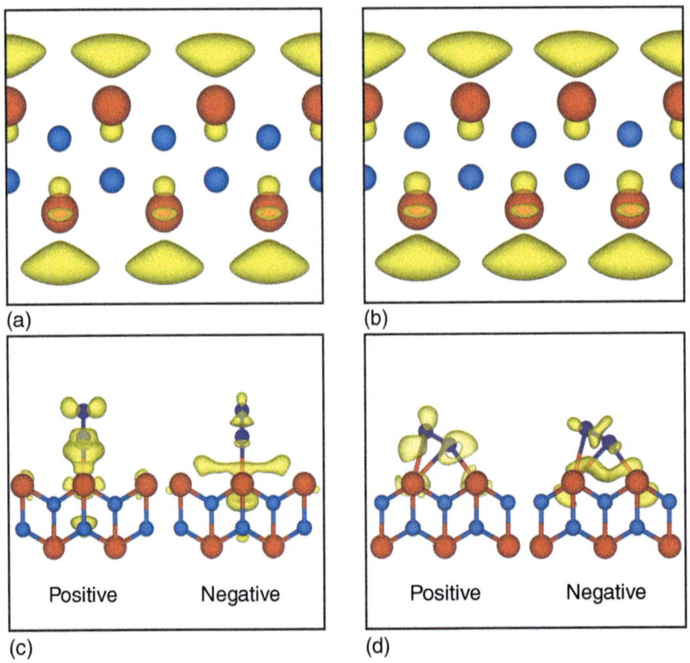

Figure 8.21 Fukui function of (a) $F^{+}(r)$ and (b) $F^{-}(r)$ of representative ReB_2; the isosurface value is 0.001 $e/Å^3$. Charge density difference of N_2 adsorption on ReB_2 in (c) end-on and (d) side-on pattern; the isosurface value is set to 0.004 $e/Å^3$. Electron accumulation and depletion are shown in yellow and cyan, respectively. B, N, and Re atoms are represented by light blue, blue, and red spheres, respectively. Source: Zhang et al. [73]. Reproduced with permission of The Royal Society of Chemistry.

the simultaneous charge accumulation and depletion on both N_2 and TM atoms. In particular, N_2 lone-pair electrons will be donated to TM atom-d orbitals, and at the same time TM atoms are going to transfer *d* electrons to N_2 antibonding orbital. In the case of ReB_2, for example, Bader charge calculation indicates that about 0.32 *e* and 0.62 *e* accumulate around N_2 for end-on and side-on adsorption configurations, respectively.

As a prerequisite for driving the NRR, N_2 activation plays a crucial role in the subsequent reactions ($N_2 \rightarrow NH_3$). As shown in Figure 8.22a, both side-on and end-on N_2 adsorptions on all of these 13 TMB_2 suggest negative Gibbs free energy change ($\Delta G_{*N_2} < 0$), demonstrating the spontaneous N_2 adsorption and the effective activation of N_2. It should be pointed out that moderate ΔG_{*N_2} is beneficial for the overall process of electrochemical NRR. In general, the PDS usually occurs in the first ($*N_2 + H^+ + e^- \rightarrow *N_2H$) and the last protonation ($*NH_2 + H^+ + e^- \rightarrow *NH_3$) of the six hydrogenation steps. In the former case, the strong N≡N triple bond will be broken by hydrogenation, which means that this thermodynamic process needs a mass of energy injection. In this view, the change of Gibbs free energy in the first protonation can be used as a descriptor to screen high-performance catalysts. In the latter case, the conversion of stably adsorbed $*NH_2$ to much less stable $*NH_3$ is thermodynamically unfavorable, along with positive Gibbs free energy change. In $*NH_2$, the sp^3 hybrid orbitals are partially occupied while fully filled in $*NH_3$. As a result, $*NH_2$ tends to combine TM atom-d orbitals strongly, whereas the interaction between $*NH_3$ and TM atom turns out to be relatively weak. On the basis of this, therefore, Gibbs free energy change in the last protonation can also be recognized as a descriptor for catalytic NRR properties.

In the present work, ΔG_{PDS} smaller than the benchmark Ru(0001) surface (0.98 eV) [165–167] is identified as a criterion to screen out the candidates of high-performance NRR catalysts. As shown in Figure 8.22b and c, in the case of end-on mode, the ΔG_{PDS} of CrB_2, ReB_2, and OsB_2 are smaller than that of stepped Ru(0001), while for VB_2, MnB_2, FeB_2, CoB_2, NbB_2, MoB_2, TcB_2, RuB_2, and WB_2, the ΔG_{PDS} exceed 0.98 eV. In the case of side-on pattern, for CrB_2, MnB_2, FeB_2, TcB_2, RuB_2, ReB_2, and OsB_2, the ΔG_{PDS} of the first and last protonation are smaller than 0.98 eV, while the rest show larger ΔG_{PDS}. It is therefore conclusive that (1) CrB_2, ReB_2, and OsB_2 with N_2 end-on adsorption and (2) CrB_2, MnB_2, FeB_2, TcB_2, RuB_2, ReB_2, and OsB_2 with N_2 side-on adsorption can be initially selected to further evaluate the NRR activity.

In general, the six successive proton-coupled electron transfer reactions during the N_2 conversion to NH_3 follow three typical reaction mechanisms, i.e. distal, alternating, and enzymatic [168–171]. In order to address the optimal reduction pathway, all these possible routes are evaluated for the selected catalysts. In addition to the three typical mechanisms, two mixed mechanisms for NRR on TMB_2 are also found. It should be emphasized that $*NH_3$ is easily converted to NH_4^+ under acidic conditions and desorbs from the surface of the catalyst, which is not an electrochemical process [172–174]. In this case, calculations for this process will not be involved.

In case of N_2 end-on adsorption, one will see that the largest ΔG_{max} occurs in the first protonation step ($*N_2 + H^+ + e^- \rightarrow *N_2H$), 1.17, 0.34, and 0.89 eV for CrB_2,

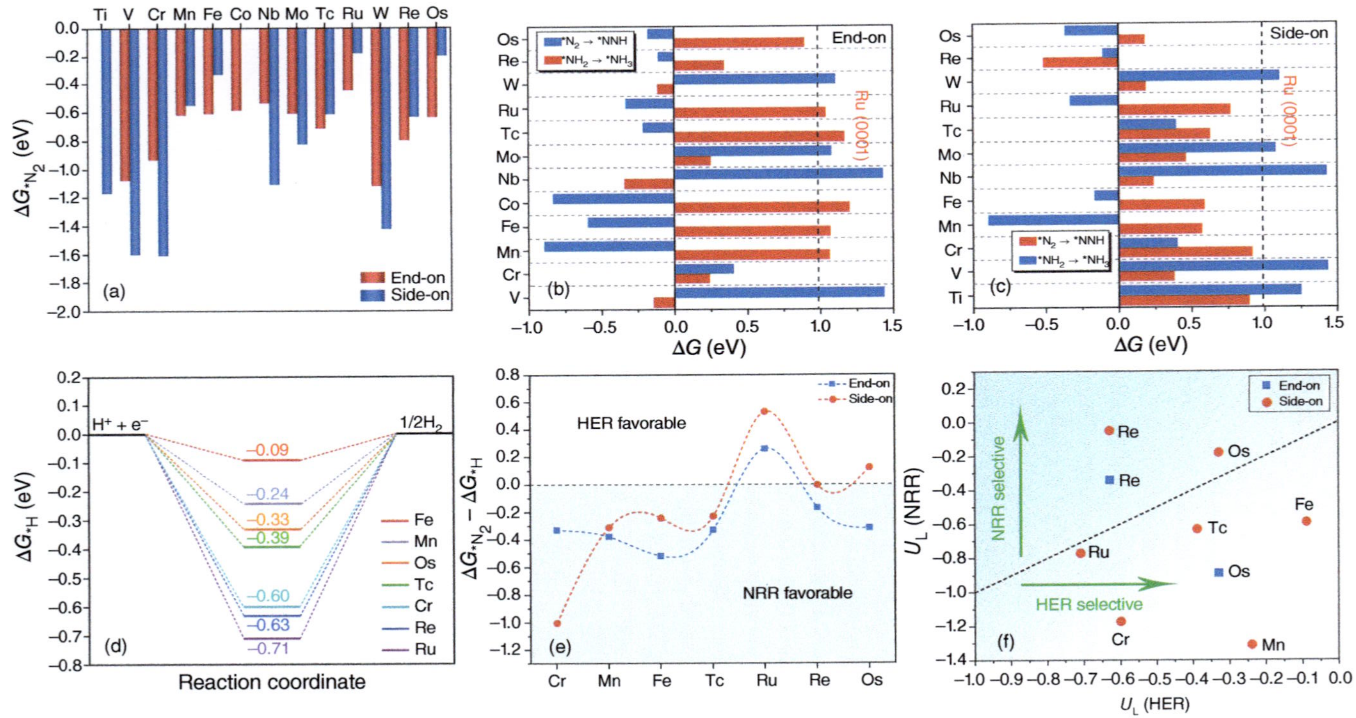

Figure 8.22 (a) Gibbs free energy change of N_2 adsorption on TMB_2 in end-on and side-on patterns. In respect to N_2 adsorption on TiB_2, it transforms spontaneously from end-on to side-on mode, while for CoB_2, N_2 energetically favors the end-on adsorption. Gibbs free energy change in the first ($^*N_2 + H^+ + e^- \rightarrow {}^*N_2H$) and the sixth protonation ($^*NH_2 + H^+ + e^- \rightarrow {}^*NH_3$) in the NRR process for N_2 (b) end-on and (c) side-on adsorption. (d) Gibbs free energy diagram of HER on TMB_2. (e) $\Delta G_{*N_2} - \Delta G_{*H}$ of TMB_2. (f) Limiting potential for NRR (U_L (NRR)) and HER (U_L (HER)). Source: Zhang et al. [73]. Reproduced with permission of The Royal Society of Chemistry.

ReB_2, and OsB_2, respectively. In respect to the N_2 side-on pattern, it can be found that the largest Gibbs free energy changes are 1.18, 1.30, 0.59, 0.63, 0.77, and 0.18 eV, respectively. As for N_2 side-on adsorption on ReB_2, it will be discussed latter.

As a competitive reaction of NRR, HER should be effectively suppressed to achieve high FE toward NRR [175, 176]. As shown in Figure 8.22d, the Gibbs free energy change for HER on TMB_2 indicates excessive hydrogen adsorption with more negative ΔG_{*H}, except for FeB_2 (−0.09 eV). It is interesting to note FeB_2 with low overpotential of 0.09 V, suggesting the promising potential as high-efficiency HER catalyst. In the consideration that hydrogen adsorption may block the active sites, the Gibbs free energy difference between N_2 and hydrogen adsorption ($\Delta G = \Delta G_{*N_2} - \Delta G_{*H}$) is compared in Figure 8.22e. In cases for RuB_2 and OsB_2 (side-on), postive ΔG indicates that hydrogen adsorption is dominant, while negative ΔG for the others illustrate good selectivity for NRR with supressed hydrogen adsorption. In Figure 8.22f, the limiting potentials of HER and NRR are compared, with $U_L(\text{NRR}) > U_L(\text{HER})$ suggesting the selectivity of TMB_2 for NRR with high FE. As the applied electrode potential increases, proton and electron transfer will dramatically facilitate the hydrogen adsorption; however, the free energy of $*N_2$ cannot be significantly influenced due to the absence of the proton and electron transfer in this process. In order to effectively prevent the "H poisoning effect," it is very important to accelerate the first protonization ($*N_2 \rightarrow *N_2H$) to improve the coverage of $*N_2$. On account of this, the Gibbs free energy difference between *H and $*N_2H$ ($\Delta G_{*H} - \Delta G_{*N_2H}$) is calculated to assess the catalytic selectivity, and positive value indicates preferential hydrogenation of $*N_2$ ($*N_2 \rightarrow *N_2H$) and thus good selectivity. In accordance to these screening criteria, ReB_2 can be chosen as a promising candidate electrocatalyst for NRR, with N_2 molecule adopting a side-on adsorption pattern.

In order to have an in-depth understanding on the origin of the catalytic activity of TMB_2, electronic properties are evaluated. It is conclusive that TM atom-d orbitals play a decisive role in the adsorption and activation of N_2, thus the PDOS and the crystal orbital Hamilton population (COHP) of $*N_2$ are calculated to further reveal the d–$2\pi^*$ interaction, see Figure 8.23. In comparison with the free N_2, after adsorption the N_2 $2\pi^*$ orbital moves toward the Fermi level. As for CrB_2, MnB_2, TcB_2, ReB_2, and OsB_2, the antibonding orbital $2\pi^*$ of N_2 is partially filled, indicating weaker N≡N bonding and more efficient N_2 activation. In cases of FeB_2 and RuB_2, the nearly unoccupied antibonding orbital $2\pi^*$ reveals poor N_2 activation. In addition, integrated COHP (ICOHP) as a quantitative descriptor for the activation degree is further discussed. It is of interest to find that the activation degree of N_2 decreases as ICOHP decreases, i.e. the more negative ICOHP, the less activated N_2 molecule [177]. In agreement with above discussion, the more effective N_2 activation on ReB_2 can be correlated with the less nagative ICOHP of −3.02.

In addition, two linear relationships exist, the positive/negative correlation between the excess electrons on $*N_2$ and ICOHP/adsorption energy of $*N_2$, with coefficients of determination (R^2) being 0.78/0.82, Figure 8.24. As the extra electrons on $*N_2$ increase, adsorption strength increases and N–N bonding becomes weak. Thus, the filling of extra electrons can be used as another descriptor to describe N_2 activation.

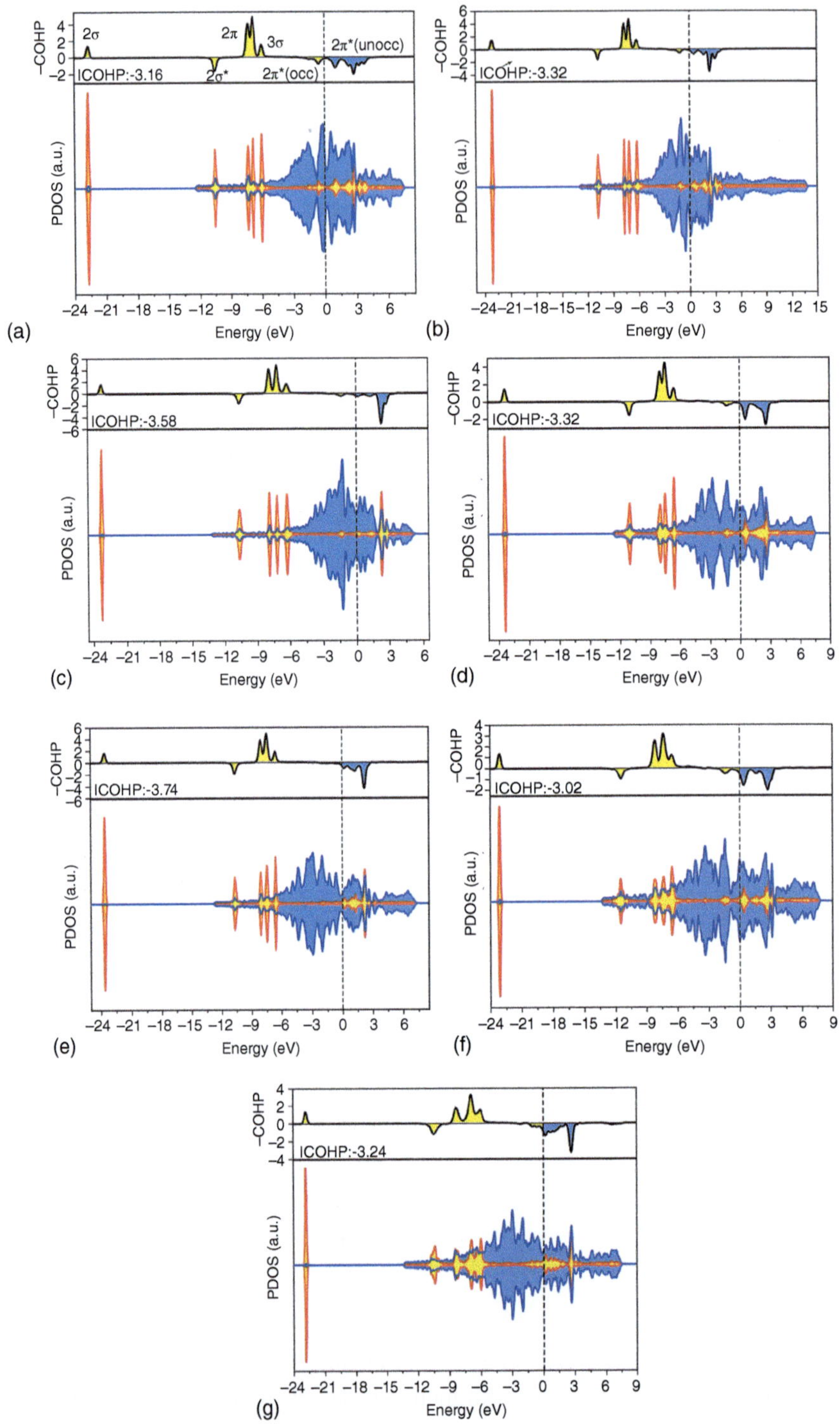

Figure 8.23 Crystal orbital Hamilton population (COHP), integrated COHP (ICOHP), and PDOS for $^*N_2$ on (a) CrB_2, (b) MnB_2, (c) FeB_2, (d) TcB_2, (e) RuB_2, (f) ReB_2, and (g) OsB_2. The vertical dashed line represents the Fermi level. Source: Zhang et al. [73]. Reproduced with permission of The Royal Society of Chemistry.

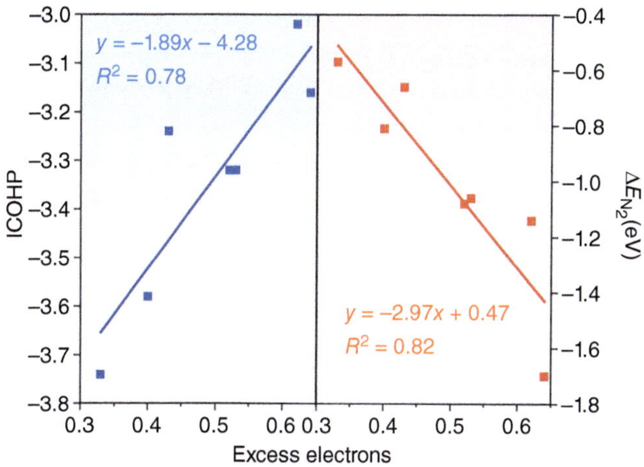

Figure 8.24 Correlation between excess electrons on $^*N_2$ and ICOHP and N_2 adsorption energy. Source: Zhang et al. [73]. Reproduced with permission of The Royal Society of Chemistry.

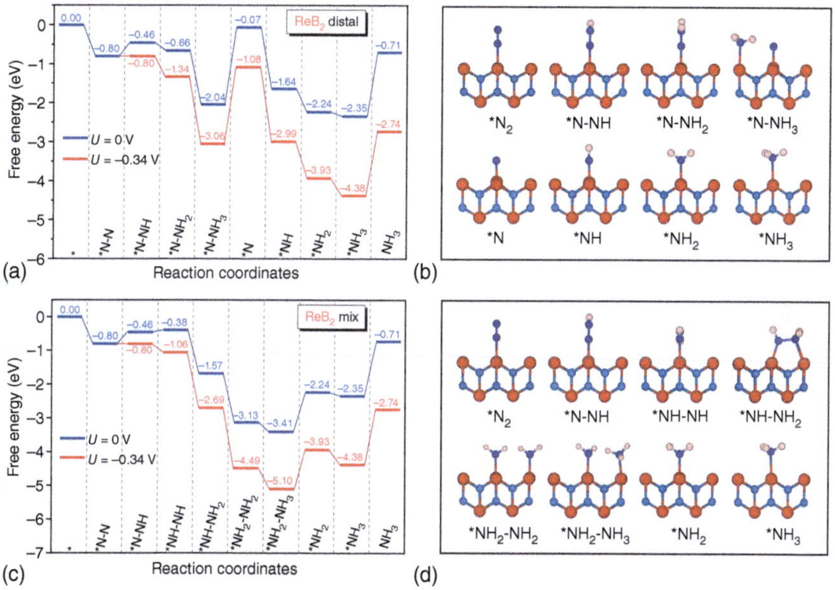

Figure 8.25 Gibbs free energy diagram for NRR on ReB_2 at zero (blue line) and applied potential (red line) along (a) distal and (c) mix mechanisms. The corresponding adsorption configurations are displayed in (b) and (d). Source: Zhang et al. [73]. Reproduced with permission of The Royal Society of Chemistry.

In this section, NRR on ReB_2 following distal and mix routes have been demonstrated, both giving rise to limiting potential of $U_L = -0.34$ V, see Figure 8.25. In Figure 8.26, Gibbs free energy change diagram and relevant intermediates are, respectively, shown for NRR on ReB_2 following the enzymatic pathway. It is stirring

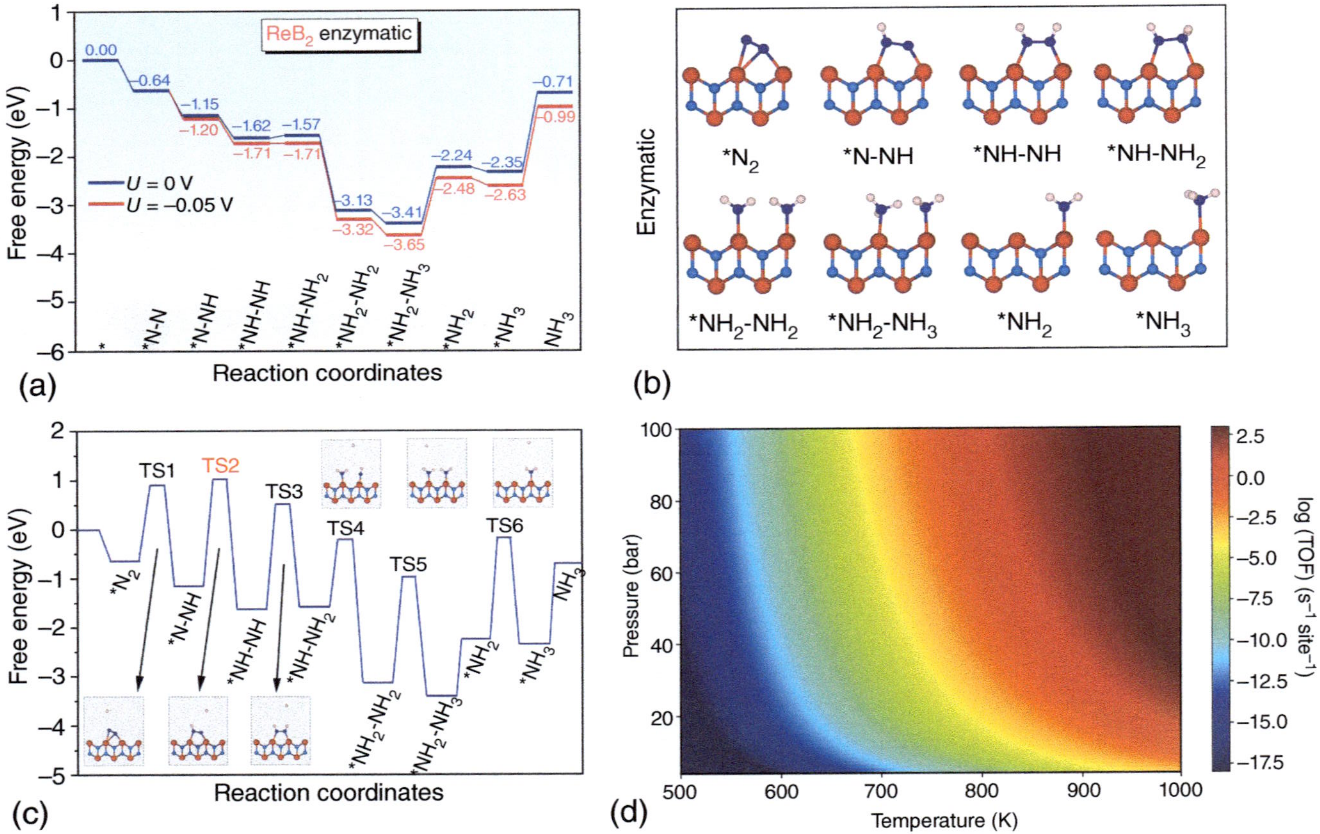

Figure 8.26 (a) Gibbs free energy diagram and (b) intermediates for NRR on ReB_2 along enzymatic route, and (b) corresponding intermediates. (c) Kinetic barriers for N_2 conversion to NH_3 along the enzymatic pathway on ReB_2; insets show the transition states. (d) TOF of NRR for NH_3 synthesis on ReB_2 at various temperature and pressure with N_2:H_2 ratio fixed at 1 : 3. Source: Zhang et al. [73]. Reproduced with permission of The Royal Society of Chemistry.

to find a record-low limiting potential of U_L = −0.05 V, which could be a new theoretical benchmark for NRR. In comparison with TM catalyst Re(111) surface (with a U_L = −0.50 V for NRR) [178], ReB_2 shows significantly improved catalytic activity with U_L = −0.05 V, which can be attributed to the synergistic effect between metal Re and nonmetal B species. In previous works, the bimetallic synergistic effect was intensively discussed [179–182]. It is demonstrated in this work that the metal–nonmetal synergistic effect also works well to significantly promote the NRR catalytic performance in terms of controlling the adsorption configurations and intermediate energies. In order to address the kinetic barrier, CINEB method is employed for every protonation step, with the intermediate images being relaxed until the perpendicular force is smaller than 0.02 eV Å^{-1}. As shown in Figure 8.26c, the maximum kinetic barrier is 2.17 eV for ReB_2 along the enzymatic pathway, which is obviously lower than that of previously reported Cu(111) surface of 2.57 eV [71]. It therefore confirms the thermodynamic and kinetic feasibility for ReB_2 toward N_2 reduction to ammonia, in terms of low limiting potential and moderate kinetic barrier.

In order to evaluate the selectivity, in this case, the FE is calculated to quantitatively describe the catalytic performance:

$$f_{\mathrm{NRR}} = \frac{1}{1 + e^{-\delta G/k_B T}} \times 100\%$$

where k_B is the Boltzmann constant, T = 300 K, and δG is the Gibbs free energy difference between the PDS in HER and NRR. In accordance to this equation, a theoretical FE of 100% can be obtained for ReB_2, indicating the perfect selectivity for NRR.

In consideration of the effects of temperature and pressure on the catalytic performance under realistic conditions, microkinetic modeling is employed to compute the TOF of the NRR on ReB_2. In Figure 8.27, the kinetic barriers and corresponding transition states are shown. It can be found that the RDS occurring on ReB_2 is the hydrogenation of $*N_2H$ to $*N_2H_2$ (TS2), with a maximum kinetic barrier of 0.95 eV. As Figure 8.26d shows, the TOF as the reaction rate of the RDS can be obtained under the QEA [183, 184]. On the basis of the harmonic transition state theory, the rate constant of each elementary step can be calculated by using the kinetic barriers

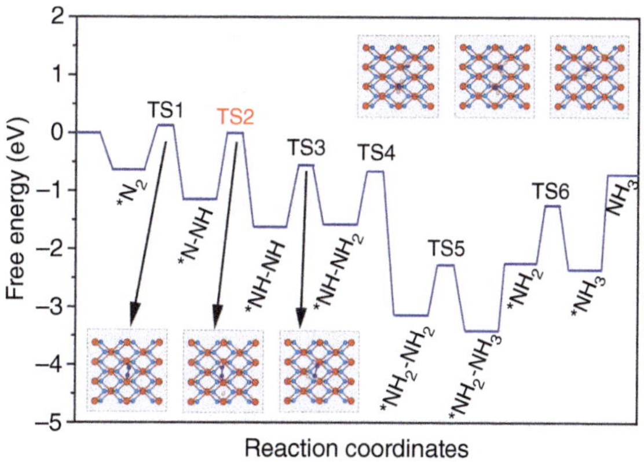

Figure 8.27 Kinetic barriers for N_2 conversion to NH_3 on ReB_2; insets show the transition states. Source: Zhang et al. [73]. Reproduced with permission of The Royal Society of Chemistry.

and thermodynamic Gibbs free reaction energies. As the temperature and pressure increase, the TOF increases significantly, since high pressure/temperature is favorable to N_2 adsorption and kinetic barrier minimization. At 700 K and 100 bar, the TOF is calculated to be $1.24 \times 10^{-2}\ s^{-1}\ site^{-1}$, which is comparable to that of benchmark Fe_3/Al_2O_3 catalysts [172]. In addition, AIMD simulations at 700 K lasting for 5 ps were performed to further examine the thermal stability of ReB_2, and results show that the framework of ReB_2 maintains its pristine structure with negligible distortion and the total energy fluctuates within a small range at an elevated temperature of 700 K, indicative of the thermal stability. It is therefore conclusive that ReB_2 with extremely fast reaction rate could be regarded as the potential NRR catalyst of high efficiency for NH_3 synthesis.

In summary, 13 TM atoms shrouded borophene of new structure with high stability and great experimental accessibility are explored through the first-principles high-throughput screening. It is of interest that the new TMB_2 monolayers with TM atoms exposed on the surface show excellent intrinsic catalytic activity for NRR. In line with the proposed screening strategy, ReB_2 with efficient N_2 activation ($\Delta G_{*N_2} < 0$), record-low energy cost ($U_L = -0.05$ V), and high selectivity (FE= 100%) is picked out as the promising catalyst candidate for NRR. In consideration of the effects of temperature and pressure on the catalytic performance of ReB_2, the maximal TOF calculated at 700 K and 100 bar is $1.24 \times 10^{-2}\ s^{-1}\ site^{-1}$, which is comparable to that of benchmark Fe_3/Al_2O_3 catalysts. It can be highlighted from this work that the intrinsic origin of N_2 activation is related to the $d-2\pi^*$ interaction, explaining the excellent NRR activity of ReB_2 in terms of ICOHP and $2\pi^*$ occupancy. In comparison to defect/doping induced activity in materials, TM borides with exposed metal atom, large specific surface area, and high active site density show advantages as durable and efficient catalysts for specific electrochemical reactions. It is obvious that the results not only identify an efficient NRR electrocatalyst in particular, paving a way for sustainable NH_3 production, but also explain the chemical and physical origin of the activity, advancing the design principle for catalysts for various reactions in general.

References

1 Chen, Y., Ji, S., Chen, C. et al. (2018). *Joule* 2: 1242.
2 Wang, Y., Mao, J., Meng, X. et al. (2019). *Chem. Rev.* 119: 1806.
3 Yang, X.F., Wang, A., Qiao, B. et al. (2013). *Acc. Chem. Res.* 46: 1740–1748.
4 Liu, L. and Corma, A. (2018). *Chem. Rev.* 118: 4981–5079.
5 Liu, Y., Tsunoyama, H., Akita, T. et al. (2011). *ACS Catal.* 1: 2–6.
6 Corma, A., Concepción, P., Boronat, M. et al. (2013). *Nat. Chem.* 5: 775–781.
7 Lang, R., Li, T., Matsumura, D. et al. (2016). *Angew. Chem. Int. Ed.* 55: 16054–16058.
8 Liu, P., Zhao, Y., Qin, R. et al. (2016). *Science* 352: 797–801.
9 Chen, F., Jiang, X., Zhang, L. et al. (2018). *Chin. J. Catal.* 39: 893–898.
10 Qiao, B., Wang, A., Yang, X. et al. (2011). *Nat. Chem.* 3: 634–641.

11 Liu, G., Robertson, A.W., Li, M.M. et al. (2017). *Nat. Chem.* 9: 810–816.
12 Lin, L., Zhou, W., Gao, R. et al. (2017). *Nature* 544: 80–83.
13 Wang, A., Li, J., and Zhang, T. (2018). *Nat. Rev. Chem.* 2: 65–81.
14 Zhang, J., Wu, X., Cheong, W.C. et al. (2018). *Nat. Commun.* 9: 1002.
15 Choi, C.H., Kim, M., Kwon, H.C. et al. (2016). *Nat. Commun.* 7: 10922.
16 Yang, S., Kim, J., Tak, Y.J. et al. (2016). *Angew. Chem. Int. Ed.* 55: 2058–2062.
17 Guo, X., Fang, G., Li, G. et al. (2014). *Science* 344: 616–619.
18 Wang, L., Zhang, W., Wang, S. et al. (2016). *Nat. Commun.* 7: 14036.
19 Wang, Y., Su, H., He, Y. et al. (2020). *Chem. Rev.* 120: 12217–12314.
20 Zhu, Y., Bai, H., and Huang, Y. (2016). *J. Phys. Condens. Matter* 28: 045303.
21 Born, M. and Huang, K. (1954). *Dynamical Theory of Crystal Lattices*. Oxford: Oxford University Press.
22 Lee, C.G., Wei, X.D., Kysar, J.W., and Hone, J. (2008). *Science* 321: 385–388.
23 Gao, Z.B., Dong, X., Li, N.B., and Ren, J. (2017). *Nano Lett.* 17: 772–777.
24 Bertolazzi, S., Brivio, J., and Kis, A. (2011). *ACS Nano* 5: 9703–9709.
25 Peng, R., Ma, Y., He, Z. et al. (2019). *Nano Lett.* 19: 1227–1233.
26 Kou, L., Ma, Y., Tang, C. et al. (2016). *Nano Lett.* 16: 7910–7914.
27 Turner, J.A. (2004). *Science* 305: 972–974.
28 Voiry, D., Salehi, M., Silva, R. et al. (2013). *Nano Lett.* 13: 6222–6227.
29 Zhang, J., Wu, J., Zou, X. et al. (2019). *Mater. Today* 25: 28–34.
30 Zhang, H., Wei, W., Wang, S. et al. (2021). *J. Mater. Chem. A* 9: 4082–4090.
31 Liu, H., Xia, G., Zhang, R. et al. (2017). *RSC Adv.* 7: 3686–3694.
32 Liu, T., Li, P., Yao, N. et al. (2019). *Angew. Chem. Int. Ed.* 58: 4679–4684.
33 Lim, K.R.G., Handoko, A.D., Nemani, S.K. et al. (2020). *ACS Nano* 14: 10834–10864.
34 Tsai, C., Abild-Pedersen, F., and Nørskov, J.K. (2014). *Nano Lett.* 14: 1381–1387.
35 Zhu, Y.-A., Chen, D., Zhou, X.-G., and Yuan, W.-K. (2009). *Catal. Today* 148: 260–267.
36 Nørskov, J.K., Bligaard, T., Logadottir, A. et al. (2005). *J. Electrochem. Soc.* 152: J23–J26.
37 Gao, G., O'Mullane, A.P., and Du, A. (2017). *ACS Catal.* 7: 494–500.
38 Er, D., Ye, H., Frey, N.C. et al. (2018). *Nano Lett.* 18: 3943–3949.
39 Lv, X., Wei, W., Wang, H. et al. (2019). *Appl. Catal. B Environ.* 255: 117743.
40 Norskov, J.K., Rossmeisl, J., Logadottir, A. et al. (2004). *J. Phys. Chem. B* 108: 17886–17892.
41 Lv, X., Wei, W., Wang, H. et al. (2020). *J. Mater. Chem. A* 8: 20047–20053.
42 Huang, X., Wang, J., Bing Tao, H. et al. (2020). *J. Catal.* 389: 461–467.
43 Xue, Z., Zhang, X., Qin, J., and Liu, R. (2019). *J. Mater. Chem. A* 7: 23091–23097.
44 Zhai, X., Li, L., Liu, X. et al. (2020). *Nanoscale* 12: 10035–10043.
45 Wang, S., Wei, W., Lv, X. et al. (2020). *J. Mater. Chem. A* 8: 1378–1385.
46 Ma, Z., Xiao, C., Cui, Z. et al. (2021). *J. Mater. Chem. A* 9: 6945–6954.
47 Honkala, K., Hellman, A., Remediakis, I.N. et al. (2005). *Science* 307: 555–558.
48 Liu, C., Li, Q., Zhang, J. et al. (2021). *J. Mater. Chem. A* 7: 4771–4776.

49 Montoya, J.H., Tsai, C., Vojvodic, A., and Norskov, J.K. (2015). *ChemSusChem* 8: 2180–2186.
50 Wang, C., Zhao, Y.-N., Zhu, C.-Y. et al. (2020). *J. Mater. Chem. A* 8: 23599–23606.
51 Lv, X., Kou, L., and Frauenheim, T. (2021). *ACS Appl. Mater. Interfaces* 13: 14283–14290.
52 Guo, C., Ran, J., Vasileff, A., and Qiao, S.-Z. (2018). *Energy Environ. Sci.* 11: 45–56.
53 Tayyebi, E., Abghoui, Y., and Skulason, E. (2019). *ACS Catal.* 9: 11137–11145.
54 Xu, L., Yang, L.-M., and Ganz, E. (2021). *ACS Appl. Mater. Interfaces* 13: 14091–14101.
55 Han, J., Ji, X., Ren, X. et al. (2018). *J. Mater. Chem. A* 6: 12974–12977.
56 Wang, Z., Gong, F., Zhang, L. et al. (2018). *Adv. Sci.* 6: 1801182.
57 Zhang, R., Guo, H., Yang, L. et al. (2019). *ChemElectroChem* 6: 1014–1018.
58 Zhang, H., Wei, W., Wang, S. et al. (2021). *J. Mater. Chem. A* 9: 4082–4090.
59 Zheng, G., Li, Y., Qian, X. et al. (2021). *ACS Appl. Mater. Interfaces* 13: 16336–16344.
60 Lv, S.-Y., Huang, C.-X., Li, G., and Yang, L.-M. (2021). *ACS Appl. Mater. Interfaces* 13: 29641–29653.
61 Zhao, J. and Chen, Z. (2017). *J. Am. Chem. Soc.* 139: 12480–12487.
62 Nash, J., Yang, X., Anibal, J. et al. (2017). *J. Electrochem. Soc.* 164: F1712–F1716.
63 Qian, J., An, Q., Fortunelli, A. et al. (2018). *J. Am. Chem. Soc.* 140: 6288–6297.
64 Kong, J., Lim, A., Yoon, C. et al. (2017). *ACS Sustain. Chem. Eng.* 5: 10986–10995.
65 Lv, X., Wei, W., Huang, B. et al. (2021). *Nano Lett.* 21: 1871–1878.
66 Zhu, H.-R., Hu, Y.-L., Wei, S.-H., and Hua, D.-Y. (2019). *J. Phys. Chem. C* 123: 4274–4281.
67 Huang, B., Li, N., Ong, W.-J., and Zhou, N. (2019). *J. Mater. Chem. A* 7: 27620–27631.
68 Zhu, D., Zhang, L., Ruther, R.E., and Hamers, R.J. (2013). *Nat. Mater.* 12: 836–841.
69 Anderson, J.S., Rittle, J., and Peters, J.C. (2013). *Nature* 501: 84–87.
70 Yandulov, D.V. and Schrock, R.R. (2003). *Science* 301: 76–78.
71 Chen, Z.W., Lang, X.Y., and Jiang, Q. (2018). *J. Mater. Chem. A* 6: 9623–9628.
72 Liu, C., Li, Q., Zhang, J. et al. (2019). *J. Mater. Chem. A* 7: 4771–4776.
73 Zhang, H., Wang, S., Wang, H. et al. (2021). *Nanoscale* 13: 17331.
74 Foster, S.L., Bakovic, S.I.P., Duda, R.D. et al. (2018). *Nat. Catal.* 1: 490–500.
75 Rosca, V., Duca, M., De Groot, M.T., and Koper, M.T.M. (2009). *Chem. Rev.* 109: 2209–2244.
76 Ling, C., Niu, X., Li, Q. et al. (2018). *J. Am. Chem. Soc.* 140: 14161–14168.
77 Liu, X., Jiao, Y., Zheng, Y. et al. (2019). *J. Am. Chem. Soc.* 141: 9664–9672.
78 Lynggaard, H., Andreasen, A., Stegelmann, C., and Stoltze, P. (2004). *Prog. Surf. Sci.* 77: 71–137.
79 Stoltze, P. (2000). *Prog. Surf. Sci.* 65: 65–150.

80 Liu, J.-C., Ma, X.-L., Li, Y. et al. (2018). *Nat. Commun.* 9: 1610.
81 Lv, X., Wei, W., Li, F. et al. (2019). *Nano Lett.* 19: 6391–6399.
82 Ling, C., Niu, X., Li, Q. et al. (2018). *J. Am. Chem. Soc.* 140: 14161–14168.
83 Zhao, J. and Chen, Z. (2017). *J. Am. Chem. Soc.* 139: 12480–12487.
84 Chen, Z., Zhao, J., Yin, L., and Chen, Z. (2019). *J. Mater. Chem. A* 7: 13284–13292.
85 Yang, X., Shang, C., Zhou, S., and Zhao, J. (2020). *Nanoscale Horiz.* 5: 1106–1115.
86 Yang, T., Song, T.T., Zhou, J. et al. (2020). *Nano Energy* 68: 104304.
87 Liu, C., Li, Q., Zhang, J. et al. (2019). *J. Mater. Chem. A* 7: 4771–4776.
88 Lv, X., Wei, W., Wang, H. et al. (2020). *J. Mater. Chem. A* 8: 20047–20053.
89 Lv, X., Wei, W., Huang, B. et al. (2021). *Nano Lett.* 21: 1871–1878.
90 Li, L., Kong, X., and Peeters, F.M. (2019). *Carbon* 141: 712–718.
91 He, S. and Chen, W. (2015). *Nanoscale* 7: 6957–6990.
92 Chu, S. and Majumdar, A. (2012). *Nature* 488: 294–303.
93 Turner, J.A. (2004). *Science* 305: 972–974.
94 Wu, Q., Ma, Y., Peng, R. et al. (2019). *ACS Appl. Mater. Interfaces* 11: 45818–45824.
95 Lv, X., Wei, W., Wang, H. et al. (2019). *Appl. Catal. B Environ.* 255: 117743.
96 Lv, X., Wei, W., Wang, H. et al. (2020). *Appl. Catal. B Environ.* 264: 118521.
97 Wang, H., Lee, H.-W., Deng, Y. et al. (2015). *Nat. Commun.* 6: 7261.
98 Lee, Y., Suntivich, J., May, K.J. et al. (2012). *J. Phys. Chem. Lett.* 3: 399–404.
99 Seh, Z.W., Fredrickson, K.D., Anasori, B. et al. (2016). *ACS Energy Lett.* 1: 589–594.
100 Jin, H., Wang, J., Su, D. et al. (2015). *J. Am. Chem. Soc.* 137: 2688–2694.
101 Valdes, A., Qu, Z.-W., Kroes, G.-J. et al. (2008). *J. Phys. Chem. C* 112: 9872–9879.
102 Han, A., Zhang, H., Yuan, R. et al. (2017). *ACS Appl. Mater. Interfaces* 9: 2240–2248.
103 Popczun, E.J., McKone, J.R., Read, C.G. et al. (2013). *J. Am. Chem. Soc.* 135: 9267–9270.
104 He, L., Zhou, D., Lin, Y. et al. (2018). *ACS Catal.* 8: 3859–3864.
105 Giovanni, C.D., Reyes-Carmona, A., Coursier, A. et al. (2016). *ACS Catal.* 6: 2626–2631.
106 Liu, J., Wang, J., Zhang, B. et al. (2018). *J. Mater. Chem. A* 6: 2067–2072.
107 Wu, Y., Li, G.-D., Liu, Y. et al. (2016). *Adv. Funct. Mater.* 26: 4839–4847.
108 Liang, Q., Jin, H., Wang, Z. et al. (2019). *Nano Energy* 57: 746–752.
109 Li, Q., Yang, C., Wu, L. et al. (2019). *J. Mater. Chem. A* 7: 5981–5990.
110 Wu, W., Guo, W., and Zeng, X. (2013). *Nanoscale* 5: 9264–9276.
111 Hu, C., Liu, H., Lu, Y. et al. (2019). *Nano Energy* 63: 103874.
112 Ma, D.W., Li, T., Wang, Q. et al. (2015). *Carbon* 95: 756–765.
113 He, T., Matta, S.K., and Du, A. (2019). *Phys. Chem. Chem. Phys.* 21: 1546–1551.
114 He, T., Zhang, L., Kour, G., and Du, A. (2020). *J. CO2 Util.* 37: 272–277.
115 Li, G., Li, Y., Liu, H. et al. (2010). *Chem. Commun.* 46: 3256–3258.
116 Li, Q., Li, Y., Chen, Y. et al. (2018). *Carbon* 136: 248–254.

117 Gao, J., Li, J., Chen, Y. et al. (2018). *Nano Energy* 43: 192–199.
118 Du, Q.-S., Tang, P.-D., Huang, H.-L. et al. (2017). *Sci. Rep.* 7: 40796.
119 Hui, L., Xue, Y., Yu, H. et al. (2019). *J. Am. Chem. Soc.* 141: 10677–10683.
120 Lv, X., Wei, W., Wang, H. et al. (2020). *J. Mater. Chem. A* 8: 20047–20053.
121 Jin, H., Guo, C., Liu, X. et al. (2018). *Chem. Rev.* 118: 6337–6408.
122 Xu, H., Cheng, D., Cao, D., and Zeng, X. (2018). *Nat. Catal.* 1: 339–348.
123 Gao, X., Mei, L., Zhou, Y., and Shen, Z. (2020). *Nanoscale* 12: 7814–7821.
124 Xu, X., Xu, H., and Cheng, D. (2019). *Nanoscale* 11: 20228–20237.
125 Ling, C., Shi, L., Ouyang, Y. et al. (2017). *Nano Lett.* 17: 5133–5139.
126 Kresse, G. and Furthmuller, J. (1996). *Phys. Rev. B* 54: 11169–11186.
127 Kresse, G. and Joubert, D. (1999). *Phys. Rev. B* 59: 1758–1775.
128 Perdew, J.P. and Wang, Y. (1992). *Phys. Rev. B* 45: 13244–13249.
129 Perdew, J.P., Chevary, J.A., Vosko, S.H. et al. (1992). *Phys. Rev. B* 46: 6671–6687.
130 Blochl, P.E. (1994). *Phys. Rev. B* 50: 17953–17979.
131 Perdew, J.P., Burke, K., and Ernzerhof, M. (1996). *Phys. Rev. Lett.* 77: 3865.
132 Liechtenstein, A.I., Anisimov, V.I., and Zaanen, J. (1995). *Phys. Rev. B* 52: R5467–R5470.
133 Martyna, G.J., Klein, M.L., and Tuckerman, M. (1992). *J. Chem. Phys.* 97: 2635–2643.
134 Chen, L., Chen, Z., Wang, Y. et al. (2018). *ACS Catal.* 8: 8107–8114.
135 Ling, C., Ouyang, Y., Shi, L. et al. (2017). *ACS Catal.* 7: 5097–5102.
136 Nørskov, J.K., Bligaard, T., Logadottir, A. et al. (2005). *J. Electrochem. Soc.* 152: J23–J26.
137 Man, I.C., Su, H.-Y., Calle-Valljo, F. et al. (2011). *ChemCatChem* 3: 1159–1165.
138 Song, X., Wang, J., Qi, S. et al. (2019). *J. Phys. Chem. C* 123: 25651–25656.
139 Mohajeri, A. and Dashti, N. (2019). *J. Phys. Chem. C* 123: 30972–30980.
140 Sun, X., Zheng, J., Gao, Y. et al. (2020). *Appl. Surf. Sci.* 526: 146522.
141 Khazaei, M., Wang, J., Estili, M. et al. (2019). *Nanoscale* 11: 11305–11314.
142 Alameda, L.T., Moradifar, P., Metzger, Z.P. et al. (2018). *J. Am. Chem. Soc.* 140: 8833–8840.
143 Zhang, H., Dai, F.-Z., Xiang, H. et al. (2019). *J. Mater. Sci. Technol.* 35: 1593–1600.
144 Zhu, X., Zhou, X., Jing, Y., and Li, Y. (2021). *Nat. Commun.* 12: 4080.
145 Yang, X., Shang, C., Zhou, S., and Zhao, J. (2020). *Nanoscale Horiz.* 5: 1106–1115.
146 Xiao, Y. and Shen, C. (2021). *Small* 17: 2100776.
147 Li, B., Wu, Y., Li, N. et al. (2020). *ACS Appl. Mater. Interfaces* 12: 9261–9267.
148 Yuan, H., Li, Z., and Yang, J. (2019). *J. Phys. Chem. C* 123: 16294–16299.
149 Wang, S., Shi, L., Bai, X. et al. (2020). *ACS Cent. Sci.* 6: 1762–1771.
150 Ling, C., Zhang, Y., Li, Q. et al. (2019). *J. Am. Chem. Soc.* 141: 18264–18270.
151 Liu, C., Li, Q., Wu, C. et al. (2019). *J. Am. Chem. Soc.* 141: 2884–2888.
152 Xiao, Y., Shen, C., and Long, T. (2021). *Chem. Mater.* 33: 4023–4034.
153 Xing, C., Wu, C., Xue, Y. et al. (2020). *Nanoscale Horiz.* 5: 1274–1278.
154 Cai, L., Zhang, N., Qiu, B., and Chai, Y. (2020). *ACS Appl. Mater. Interfaces* 12: 20448–20455.

155 Guo, X., Gu, J., Lin, S. et al. (2020). *J. Am. Chem. Soc.* 142: 5709–5721.

156 Li, L., Martirez, J.M.P., and Carter, E.A. (2020). *ACS Catal.* 10: 12841–12857.

157 Singh, Y., Niazi, A., Vannette, M.D. et al. (2007). *Phys. Rev. B* 76: 214510.

158 Cumberland, R.W., Weinberger, M.B., Gilman, J.J. et al. (2005). *J. Am. Chem. Soc.* 127: 7264–7265.

159 Balan, A.P., Radhakrishnan, S., Woellner, C.F. et al. (2018). *Nat. Nanotechnol.* 13: 602–609.

160 Lv, X., Wei, W., Wang, H. et al. (2020). *J. Mater. Chem. A* 8: 20047–20053.

161 Azofra, L.M., Li, N., MacFarlane, D.R., and Sun, C. (2016). *Energy Environ. Sci.* 9: 2545–2549.

162 Hu, X., Guo, S., Zhang, S. et al. (2019). *J. Mater. Chem. A* 7: 25887–25893.

163 Wang, J., Yu, L., Hu, L. et al. (2018). *Nat. Commun.* 9: 1795.

164 Skulason, E., Bligaard, T., Gudmundsdottir, S. et al. (2012). *Phys. Chem. Chem. Phys.* 14: 1235–1245.

165 Choi, C., Back, S., Kim, N.-Y. et al. (2018). *ACS Catal.* 8: 7517–7525.

166 Logadóttir, Á. and Nørskov, J.K. (2003). *J. Catal.* 220: 273–279.

167 Foster, S.L., Bakovic, S.I.P., Duda, R.D. et al. (2018). *Nat. Catal.* 1: 490–500.

168 Rosca, V., Duca, M., De Groot, M.T., and Koper, M.T.M. (2009). *Chem. Rev.* 109: 2209–2244.

169 Ling, C., Niu, X., Li, Q. et al. (2018). *J. Am. Chem. Soc.* 140: 14161–14168.

170 Liu, X., Jiao, Y., Zheng, Y. et al. (2019). *J. Am. Chem. Soc.* 141: 9664–9672.

171 Li, H., Zhao, Z., Cai, Q. et al. (2020). *J. Mater. Chem. A* 8: 4533–4543.

172 Liu, J.-C., Ma, X.-L., Li, Y. et al. (2018). *Nat. Commun.* 9: 1610.

173 Lv, X., Wei, W., Li, F. et al. (2019). *Nano Lett.* 19: 6391–6399.

174 Guo, X. and Huang, S. (2018). *Electrochim. Acta* 284: 392–399.

175 Cui, X., Tang, C., and Zhang, Q. (2018). *Adv. Energy Mater.* 8: 1800369.

176 Liu, X., Jiao, Y., Zheng, Y., and Qiao, S.-Z. (2020). *ACS Catal.* 10: 1847–1854.

177 Montoya, J.H., Tsai, C., Vojvodic, A., and Norskov, J.K. (2015). *ChemSusChem* 8: 2180–2186.

178 Arachchige, L.J., Xu, Y., Dai, Z. et al. (2021). *J. Mater. Sci.* 77: 244–251.

179 Liu, Y., Deng, P., Wu, R. et al. (2021). *J. Mater. Chem. A* 9: 6694–6709.

180 Yan, M., Dai, Z., Chen, S. et al. (2020). *J. Phys. Chem. C* 124: 13283–13290.

181 Arachchige, L.J., Xu, Y., Dai, Z. et al. (2020). *J. Phys. Chem. C* 124: 15295–15301.

182 Rostamikia, G., Maheshwari, S., and Janik, M.J. (2019). *Catal. Sci. Technol.* 9: 174–181.

183 Lynggaard, H., Andreasen, A., Stegelmann, C., and Stoltze, P. (2004). *Prog. Surf. Sci.* 77: 71–137.

184 Stoltze, P. (2000). *Prog. Surf. Sci.* 65: 65–150.

Index

Calculations and Simulations of Low-Dimensional Materials: Tailoring Properties for Applications, First Edition. Ying Dai, Wei Wei, Yandong Ma, and Chengwang Niu.

f

g

h